Arithmetic, Geometry, Cryptography and Coding Theory

CONTEMPORARY MATHEMATICS

487

Arithmetic, Geometry, Cryptography and Coding Theory

International Conference
November 5–9, 2007
CIRM, Marseilles, France

Gilles Lachaud
Christophe Ritzenthaler
Michael A. Tsfasman
Editors

American Mathematical Society
Providence, Rhode Island

2000 *Mathematics Subject Classification.* Primary 11G10, 11G20, 14G10, 14G15, 14G50, 14Q05, 11M38, 11R42.

Library of Congress Cataloging-in-Publication Data

Arithmetic, geometry, cryptography and coding theory / Gilles Lachaud, Christophe Ritzenthaler, Michael Tsfasman, editors.
    p. cm. — (Contemporary mathematics ; v. 487)
    Includes bibliographical references.
    ISBN 978-0-8218-4716-9 (alk. paper)
    1. Arithmetical algebraic geometry—Congresses. 2. Coding theory—Congresses. 3. Cryptography—Congresses. I. Lachaud, Gilles. II. Ritzenthaler, Christophe, 1976– III. Tsfasman, M.A. (Michael A.), 1954–

QA242.5.A755   2009
510—dc22
                                               2008052063

Contents

Preface

The 11th conference on Arithmetic, Geometry, Cryptography and Coding Theory (AGC^2T 11) was held in Marseilles at the "Centre International de Rencontres Mathématiques" (CIRM), during November 5-9, 2007. This international conference has been a major event in the area of applied arithmetic geometry for more than 20 years and included distinguished guests J.-P. Serre (Fields medal, Abel prize winner), G. Frey, H. Stichtenoth and other leading researchers in the field among its 77 participants.

The meeting was organized by the team "Arithmétique et Théorie de l'Information" (ATI) from the "Institut de Mathématiques de Luminy" (IML). The program consisted of 15 invited talks and 18 communications. Among them, thirteen were selected to form the present proceedings. Twelve are original research articles covering asymptotic properties of global fields, arithmetic properties of curves and higher dimensional varieties, and applications to codes and cryptography. The final article is a special lecture of J.-P. Serre entitled "How to use finite fields for problems concerning infinite fields".

The conference fulfilled its role of bringing together young researchers and specialists. During the conference, we were also happy to celebrate the retirement of our colleague Robert Rolland with a special day of talks.

Finally, we thank the organization commitee of CIRM for their help during the conference and the Calanques for its inspiring atmosphere.

Contemporary Mathematics
Volume **487**, 2009

On the fourth moment of theta functions at their central point

Amadou Diogo BARRY and Stéphane R. LOUBOUTIN

ABSTRACT. Let χ be a Dirichlet character of prime conductor $p \geq 3$. Set $A = (\chi(1) - \chi(-1))/2 \in \{0, 1\}$ and let $\theta(x, \chi) = \sum_{n \geq 1} n^A \chi(n) \exp(-\pi n^2 x/p)$ ($x > 0$) be its associated theta series whose functional equation is used to obtain the analytic continuation and functional equation of the L-series $L(s, \chi) = \sum_{n \geq 1} \chi(n) n^{-s}$. These functional equations depend on some root numbers $W(\chi)$, complex numbers of absolute values equal to one. In particular, if $\theta(1, \chi) \neq 0$, then numerical approximations to $W(\chi) = \theta(1, \chi)/\overline{\theta(1, \chi)}$ can be efficiently computed, which leads to a fast algorithm for computing class numbers and relative class numbers of real or imaginary abelian number fields (see the bibliography). According to numerical computations, it is reasonable to conjecture that $\theta(1, \chi) \neq 0$ for any primitive Dirichlet character χ. One way to prove that this conjecture at least holds true for infinitely many primitive characters is to study the moments $\sum_{\chi} |\theta(1, \chi)|^{2k}$ for $k \in \mathbf{Z}_{\geq 1}$, where χ ranges over all the even or odd primitive Dirichlet characters of conductor p. This paper is devoted to proving a lower bound on these moments for $2k = 4$.

1. Introduction

We restrict ourselves to even characters of prime conductors. Let $p > 3$ be a prime. Let X_p^+ be the set of the $(p-3)/2$ primitive even Dirichlet characters of conductor p. For $\chi \in X_p^+$, set

$$\theta(x, \chi) = \sum_{n \geq 1} \chi(n) e^{-\pi n^2 x/p} \quad (x > 0).$$

The analytic continuation and the functional equation satisfied by the Dirichlet L-series $L(s, \chi) = \sum_{n \geq 1} \chi(n) n^{-s}$, $\Re(s) > 0$, stem from the functional equation satisfied by the associated theta function (see [**Dav**, Chapter 9]):

$$(1) \qquad \theta(x, \chi) = \frac{W(\chi)}{x^{1/2}} \theta(1/x, \bar{\chi}),$$

where $W(\chi) = \tau(\chi)/\sqrt{p}$ (the Artin root number, a complex number of absolute value equal to 1) with $\tau(\chi) = \sum_{k=1}^{p} \chi(k) e^{2\pi i k/p}$ (Gauss sum).

1991 *Mathematics Subject Classification*. 2000 Mathematics Subject Classification. Primary 11R42, 11N37. Secondary 11M06.

Key words and phrases. Dirichlet characters, Artin root numbers, Theta functions, Divisor function.

Define the moments of order $2k$:

$$S_{2k}(p) = \sum_{\chi \in X_p^+} |\theta(1,\chi)|^{2k} \quad (k \in \mathbf{Z}_{\geq 1}).$$

PROPOSITION 1. $S_2(p)$ is asymptotic to $c_2 p^{3/2}$ as $p \to \infty$, where $c_2 = \frac{1}{4\sqrt{2\pi}}$.

PROOF. Recall that

$$\sum_{\chi \in X_p^+} \chi(a)\bar{\chi}(b) = \begin{cases} (p-3)/2 & \text{if } b \equiv \pm a \pmod{p} \text{ and } \gcd(a,p) = \gcd(b,p) = 1, \\ -1 & \text{if } b \not\equiv \pm a \pmod{p} \text{ and } \gcd(a,p) = \gcd(b,p) = 1, \\ 0 & \text{otherwise.} \end{cases}$$

It follows that

$$S_2(p) = \frac{p-1}{2} \sum_{\substack{a,b \\ b \equiv \pm a \pmod{p}}}^{*} e^{-\pi(a^2+b^2)/p} - \left(\sum_a^* e^{-\pi a^2/p}\right)^2$$

(where starred sums indicate sums over indices not divisible by p). The desired result follows. □

PROPOSITION 2. (See [**Lou99**]). There exists $c > 0$ such that $S_4(p) \leq cp^2 \log p$ for $p > 3$.

COROLLARY 3. It holds that that $\theta(1,\chi) \neq 0$ for at least $\gg p/\log p$ of the characters $\chi \in X_p^+$.

PROOF. The Cauchy-Schwarz inequality yields that $\theta(1,\chi) \neq 0$ for at least $S_2(p)^2/S_4(p)$ of the χ's in X_p^+. □

REMARK 4. For such a character χ, we have $W(\chi) = \theta(1,\chi)/\overline{\theta(1,\chi)}$, by (1) (hence numerical approximations to $W(\chi)$ can be efficiently computed, which leads to a fast algorithm for computing class numbers and relative class numbers of real or imaginary abelian number fields (see [**Lou98**], [**Lou02**] and [**Lou07**]).

The aim of this paper is to prove the following new result:

THEOREM 5. There exists $c > 0$ such that $S_4(p) \geq cp^2 \log p$ for $p > 3$.

According to Proposition 2, Corollary 3, Theorem 5 and extended numerical computations, we conjecture that $\theta(1,\chi) \neq 0$ for any primitive Dirichlet character $\chi \neq 1$, and the more precise behavior (see [**Bar**]):

CONJECTURE 6. There exists $c_4 > 0$ such that $S_4(p)$ is asymptotic to $c_4 p^2 \log p$ as $p \to \infty$.

As for the second and fourth moments of the values of Dirichlet L-functions $L(s,\chi)$ at their central point $s = 1/2$, the following asymptotics are known:

$$\sum_{1 \neq \chi \bmod p} |L(1/2,\chi)|^2 \sim p \log p$$

(see [**Rama**, Remark 3]) and

$$\sum_{1 \neq \chi \bmod p} |L(1/2,\chi)|^4 \sim \frac{1}{2\pi^2} p \log^4 p$$

(see [**HB**, Corollary (page 26)]). It is also conjectured that for $k \in \mathbf{Z}_{\geq 1}$ there exists a positive constant $C(k)$ such that

$$\sum_{1 \neq \chi \bmod p} |L(1/2, \chi)|^{2k} \sim C(k)p \log^{k^2} p$$

and it is known that (see [**RS**]))

$$\sum_{1 \neq \chi \bmod p} |L(1/2, \chi)|^{2k} \gg_k p \log^{k^2} p.$$

2. The second moment of the restricted divisor function

Our proof of Theorem 5 is based on a new method: the study of the moment of order 2 of the restricted divisor function (see Proposition 7 below). It is known that (see [**Ten**]):

$$S(x) = \sum_{1 \leq m < x} \sum_{d|m} 1 = x(\log x + 2\gamma - 1) + O(\sqrt{x}).$$

It follows that for $c > 1$ the first moment of the restricted divisor function

$$S_c(x) = \sum_{1 \leq m < x} \sum_{\substack{d|m \\ \frac{1}{c}\sqrt{m} \leq d \leq c\sqrt{m}}} 1 = S(x) - 2\sum_{1 \leq m < x} \sum_{\substack{d|m \\ d < \frac{1}{c}\sqrt{m}}} 1$$

$$= S(x) - 2\sum_{d < \frac{1}{c}\sqrt{x}} \left(\frac{x}{d} - c^2 d + O(1)\right) = 2(\log c)x + O(\sqrt{x})$$

is asymptotic to $2(\log c)x$ as $x \to \infty$. Now,

$$T(n) = \sum_{1 \leq m < x} \left(\sum_{d|m} 1\right)^2$$

is asymptotic to $\frac{1}{\pi^2}x \log^3 x$ as $x \to \infty$. We give an asymptotic for the second moment of the restricted divisor function:

PROPOSITION 7. *Fix $c > 1$. Then,*

$$T_c(x) = \sum_{1 \leq m < x} \left(\sum_{\substack{d|m \\ \frac{1}{c}\sqrt{m} \leq d \leq c\sqrt{m}}} 1\right)^2 = \frac{12 \log^2 c}{\pi^2} x \log x + O(x).$$

It follows that

$$\Lambda(n) = \sum_{1 \leq m < n} \left(\sum_{d|m} e^{-\pi(m/d-d)^2/n}\right)^2 \gg n \log n.$$

REMARK 8. *It holds that $\Lambda(n) \ll n \log n$, by [**Lou99**, Lemme 2].*

PROOF. Let us prove the second assertion. Fix $c > 1$. If $\frac{1}{c}\sqrt{m} \leq d \leq c\sqrt{m}$, then $(m/d - d)^2/n \leq (c - 1/c)^2 m/n \leq (c - 1/c)^2$ for $1 \leq m \leq n$. It follows that

$$\Lambda(n) \geq e^{-2\pi(c-1/c)^2} T_c(n).$$

Let us now prove the first assertion. We have

$$T_c(x) = \sum_{1 \leq m < x} \sum_{\substack{d_1|m \\ \frac{1}{c}\sqrt{m} \leq d_1 \leq c\sqrt{m}}} \sum_{\substack{d_2|m \\ \frac{1}{c}\sqrt{m} \leq d_2 \leq c\sqrt{m}}} 1.$$

Since

$$\sum_{1 \le m < x} \sum_{\substack{d_1 \mid m \\ \frac{1}{c}\sqrt{m} \le d_1 \le c\sqrt{m}}} 1 = S_c(x) = O(x)$$

we may assume that $d_2 \neq d_1$, or even that $d_2 < d_1$. Since

$$\sum_{\substack{\delta \mid D_1 \\ \delta \mid D_2}} \mu(\delta) = \begin{cases} 1 & \text{if } \gcd(D_1, D_2) = 1 \\ 0 & \text{otherwise,} \end{cases}$$

we have

(2) $$T_c(x) = 2U_c(x) + O(x),$$

where

$$\begin{aligned}
U_c(x) &= \sum_{1 \le m < x} \sum_{\substack{d_1 \mid m \\ \frac{1}{c}\sqrt{m} \le d_1 \le c\sqrt{m}}} \sum_{\substack{d_2 \mid m \\ \frac{1}{c}\sqrt{m} \le d_2 < d_1}} 1 \\
&= \sum_{d < c\sqrt{x}} \sum_{D_1 < \frac{c\sqrt{x}}{d}} \sum_{\substack{D_2 < D_1 \\ \gcd(D_1,D_2)=1}} \sum_{\substack{m = kdD_1D_2 < x \\ d^2 D_i^2/c^2 \le m \le c^2 d^2 D_i^2, \ i \in \{1,2\}}} 1 \\
&= \sum_{d < c\sqrt{x}} \sum_{\delta < \frac{c\sqrt{x}}{d}} \mu(\delta) \sum_{D_1 < \frac{c\sqrt{x}}{d\delta}} \sum_{D_2 < D_1} \sum_{\substack{m = kd\delta^2 D_1 D_2 < x \\ d^2\delta^2 D_i^2/c^2 \le m \le c^2 d^2 \delta^2 D_i^2, \ i \in \{1,2\}}} 1
\end{aligned}$$

(write $d_i = dD_i$ with $\gcd(D_1, D_2) = 1$ and $m = kdD_1D_2$, then change D_i into δD_i and m into $kd\delta^2 D_1 D_2$).
Hence,

(3) $$U_c(x) = \sum_{d < c\sqrt{x}} \sum_{\delta < \frac{c\sqrt{x}}{d}} \mu(\delta) M_c(d, \delta, x),$$

where

$$M_c(d, \delta, x) = \sum_{D_1 < \frac{c\sqrt{x}}{d\delta}} \sum_{D_2 < D_1} N_c(d, \delta, D_1, D_2, x),$$

with $N_c(d, \delta, D_1, D_2, x)$ being the number of positive integers k such that

(4) $$k < \frac{x}{d\delta^2 D_1 D_2}$$

and $\frac{1}{c}\sqrt{kd\delta^2 D_1 D_2} \le d\delta D_i \le c\sqrt{kd\delta^2 D_1 D_2}$ for $i \in \{1, 2\}$, i.e. such that

(5) $$\frac{dD_1}{c^2 D_2} \le k \le \frac{c^2 dD_2}{D_1},$$

which implies $D_1/c^2 \le D_2$.
To begin with, $U_c(x)$ is not too large. Indeed, by (5), we have $d/c^2 \le k \le c^2 d$ and $N_c(d, \delta, D_1, D_2, x) \le (c^2 - 1/c^2 + 1)d$. Therefore,

$$M_c(d, \delta, x) \le (c^2 - 1/c^2 + 1)d \sum_{D_1 < \frac{c\sqrt{x}}{d\delta}} \sum_{D_2 < \frac{c\sqrt{x}}{d\delta}} 1 \le (c^4 + c^2 - 1)\frac{x}{d\delta^2}$$

and $U_c(x) \le \sum_{d < c\sqrt{x}} \sum_{\delta < c\sqrt{x}/d} M_c(d, \delta, x) \ll x \log x$. Thus, by (2), we have

$$T_c(x) \ll x \log x.$$

To prove our Proposition, we use (2), (3) and the following Lemma 9.   □

LEMMA 9. *It holds that* $M_c(d, \delta, x) = (2 \log^2 c)\frac{x}{d\delta^2} + O(\frac{\sqrt{x}}{\delta}) + O(\frac{x}{d^2\delta^2})$.

PROOF. In the range $D_1 < \frac{c\sqrt{x}}{d\delta}$ we have

$$\frac{dD_1}{c^2 D_2} < \frac{x}{d\delta^2 D_1 D_2}.$$

Hence, (4) and (5) are equivalent to

(6)
$$\frac{dD_1}{c^2 D_2} \leq k \leq \begin{cases} \frac{c^2 dD_2}{D_1} & \text{if } D_2 \leq \frac{\sqrt{x}}{cd\delta} \\ \frac{x}{d\delta^2 D_1 D_2} & \text{otherwise.} \end{cases}$$

In setting

$$X = \frac{\sqrt{x}}{cd\delta},$$

we obtain that $N_c(d, \delta, D_1, D_2, x)$ is the number of k's such that

(7)
$$\frac{dD_1}{c^2 D_2} \leq k \leq \begin{cases} \frac{c^2 dD_2}{D_1} & \text{if } D_2 \leq X \\ \frac{c^2 dX^2}{D_1 D_2} & \text{otherwise.} \end{cases}$$

Therefore, we have:

$$M_c(d, \delta, x) = \sum_{D_1 < c^2 X} \sum_{D_1/c^2 \leq D_2 < D_1} N_c(d, \delta, D_1, D_2, x)$$

$$= \sum_{D_1 < X} \sum_{D_1/c^2 \leq D_2 < D_1} \left(\frac{c^2 dD_2}{D_1} - \frac{dD_1}{c^2 D_2} + O(1) \right)$$

$$+ \sum_{X \leq D_1 < c^2 X} \left\{ \sum_{D_1/c^2 \leq D_2 \leq X} \left(\frac{c^2 dD_2}{D_1} - \frac{dD_1}{c^2 D_2} + O(1) \right) \right.$$

$$\left. + \sum_{X \leq D_2 < D_1} \left(\frac{c^2 dX^2}{D_1 D_2} - \frac{dD_1}{c^2 D_2} + O(1) \right) \right\}$$

$$= c^2 d \sum_{D_1 < X} \sum_{D_1/c^2 \leq D_2 < D_1} \frac{D_2}{D_1} + c^2 d \sum_{X \leq D_1 < c^2 X} \sum_{D_1/c^2 \leq D_2 \leq X} \frac{D_2}{D_1}$$

$$- \frac{d}{c^2} \sum_{D_1 < c^2 X} \sum_{D_1/c^2 \leq D_2 < D_1} \frac{D_1}{D_2}$$

$$+ c^2 dX^2 \sum_{X \leq D_1 < c^2 X} \sum_{X \leq D_2 < D_1} \frac{1}{D_1 D_2}$$

$$+ O(X^2)$$

$$= \frac{c^2 d}{2} \sum_{D_1 < X} \left(1 - \frac{1}{c^4} \right)(D_1 + O(1)) + \frac{c^2 d}{2} \sum_{X \leq D_1 < c^2 X} \left(\frac{X^2}{D_1} - \frac{1}{c^4}D_1 + O(1) \right)$$

$$- \frac{d}{c^2} \sum_{D_1 < c^2 X} \left(D_1 \log(c^2) + O(1) \right)$$

$$+ c^2 dX^2 \sum_{X \leq D_1 < c^2 X} \left(\frac{\log(D_1/X)}{D_1} + O\left(\frac{1}{D_1 X}\right) \right)$$

$$+ O(X^2)$$

and

$$
\begin{aligned}
M_c(d,\delta,x) &= \frac{c^2 d}{4}\Big(1-\frac{1}{c^4}\Big)\big(X^2+O(X)\big)\\
&\quad +\frac{c^2 d}{2}\Big(\big(X^2\log(c^2)+O(X)\big)-\frac{1}{2c^4}\big((c^4-1)X^2+O(X)\big)+O(X)\Big)\\
&\quad -\frac{d\log(c^2)}{2c^2}\big(c^4 X^2+O(X)\big)\\
&\quad +c^2 dX^2\Big(\int_X^{c^2 X}\frac{\log(u/X)}{u}du+O(\tfrac{1}{X})\Big)\\
&\quad +O(X^2)\\
&= 2c^2 d(\log^2 c)X^2+O(dX)+O(X^2),
\end{aligned}
$$

which proves the Lemma. $\square$

3. Proof of Theorem 5

Set

$$
0<\lambda_n(m)=\sum_{d\mid m}e^{-\pi(m/d-d)^2/n}.
$$

Then,

$$
\begin{aligned}
S_4(p) &= \sum_{\chi\in X_p^+}\sum_{a,b,c,d}\chi(ab)\bar\chi(cd)e^{-\pi(a^2+b^2+c^2+d^2)/p}\\
&= -\sideset{}{^*}\sum_{a,b,c,d}e^{-\pi(a^2+b^2+c^2+d^2)/p}+\frac{p-1}{2}\sideset{}{^*}\sum_{\substack{a,b,c,d\\ cd\equiv\pm ab\ (\mathrm{mod}\ p)}}e^{-\pi(a^2+b^2+c^2+d^2)/p}\\
&\geq -\Big(\sideset{}{^*}\sum_{a}e^{-\pi a^2/p}\Big)^4+\frac{p-1}{2}\sum_{j=1}^{p-1}\Big(\sum_{\substack{a,b\\ ab\equiv j\ (\mathrm{mod}\ p)}}e^{-\pi(a^2+b^2)/p}\Big)^2\\
&\geq -\Big(\int_0^\infty e^{-\pi u^2/p}du\Big)^4+\frac{p-1}{2}\sum_{j=1}^{p-1}\Big(\sum_{\substack{m\geq 1\\ m\equiv j\ (\mathrm{mod}\ p)}}\sum_{a\mid m}e^{-\pi(a^2+m^2/a^2)/p}\Big)^2\\
&= -p^2/16+\frac{p-1}{2}\sum_{j=1}^{p-1}\Big(\sum_{\substack{m\geq 1\\ m\equiv j\ (\mathrm{mod}\ p)}}\lambda_p(m)e^{-2\pi m/p}\Big)^2\\
&\geq -p^2/16+\frac{p-1}{2}\sum_{j=1}^{p-1}\lambda_p(j)^2 e^{-4\pi j/p}\\
&\geq -p^2/16+\frac{p-1}{2}e^{-4\pi}\sum_{j=1}^{p-1}\lambda_p(j)^2 = -p^2/16+\frac{p-1}{2}e^{-4\pi}\Lambda(p).
\end{aligned}
$$

By Proposition 7, it follows that

$$
S_4(p)\gg p^2\log p.
$$

References

[Bar] A. D. Barry. Moments of theta functions at their central point. *PhD Thesis*, ongoing work.

[Dav] H. Davenport. *Multiplicative Number Theory.* Springer-Verlag, Grad. Texts Math. **74**, Third Edition, 2000.

[HB] D. R. Heath-Brown. *The fourth power mean of Dirichlet's L-function.* Analysis **1** (1981), 25–32.

[Lou98] S. Louboutin. *Computation of relative class numbers of imaginary abelian number fields.* Experimental Math. **7** (1998), 293–303.

[Lou99] S. Louboutin. *Sur le calcul numérique des constantes des équations fonctionnelles des fonctions L associées aux caractères impairs.* C. R. Acad. Sci. Paris Sér. I Math. **329** (1999), 347–350.

[Lou02] S. Louboutin. *Efficient computation of class numbers of real abelian number fields.* Algorithmic Number Theory (Sydney, 2002), Lectures Notes in Computer Science **2369** (2002), 134–147.

[Lou07] S. Louboutin. *Efficient computation of root numbers and class numbers of parametrized families of real abelian number fields.* Math. Comp. **76** (2007), 455–473.

[Rama] K. Ramachandra. *Some remarks on a Theorem of Montgomery and Vaughan.* J. Number Theory **11** (1979), 465–471.

[RS] Z. Rudnick and K. Soundararajan. *Lower bounds for moments of L-functions.* Proc. Natl. Acad. Sci. USA **102** (2005), 6837–6838.

[Ten] G. Tenenbaum. *Introduction à la théorie analytique et probabiliste des nombres.* Cours Spécialisés. Société Mathématique de France, Paris, 1995.

INSTITUT DE MATHÉMATIQUES DE LUMINY, UMR 6206. 163, AVENUE DE LUMINY. CASE 907. 13288 MARSEILLE CEDEX 9, FRANCE

E-mail address: `barry@iml.univ-mrs.fr, loubouti@iml.univ-mrs.fr`

Contemporary Mathematics
Volume **487**, 2009

On the Construction of Galois Towers

Alp Bassa and Peter Beelen

ABSTRACT. In this paper we study an asymptotically optimal tame tower over the field with p^2 elements introduced by Garcia-Stichtenoth. This tower is related with a modular tower, for which explicit equations were given by Elkies. We use this relation to investigate its Galois closure. Along the way, we obtain information about the structure of the Galois closure of $X_0(p^n)$ over $X_0(p^r)$, for integers $1 < r < n$ and prime p and the Galois closure of other modular towers $(X_0(p^n))_n$.

1. Introduction

Using Goppa's construction of codes from curves over finite fields, Tsfasman–Vladut–Zink [13] constructed sequences of codes of increasing length with limit parameters above the Gilbert–Varshamov bound and hence better than those of all previously known such sequences. Their construction is mainly based on the existence of curves over a finite field of high genus with many rational points. This enhanced the interest in towers of curves over finite fields. Subsequently, other applications of such towers in coding theory and cryptography were discovered, for instance for the construction of hash functions, low discrepancy sequences etc.

A natural idea is to search for such sequences of curves, with some additional structure, which would reflect itself in some additional structure of the objects constructed from them. Stichtenoth [12] constructed for example sequences of self-dual and transitive codes attaining the Tsfasman–Vladut–Zink bound over finite fields with square cardinality. This was done by using a tower of function fields $E_0 \subseteq E_1 \subseteq E_2 \subseteq \ldots$, where all extensions E_n/E_0 are Galois.

Motivated by this, we study Galois closures of the modular towers $(X_0(p^n))_n$. In particular, we investigate the Galois closure of a tower $\mathcal{M}$ over $\mathbb{F}_{p^2}$ introduced by Garcia–Stichtenoth [3], which is recursively defined by

$$Y^2 = \frac{X^2 + 1}{2X}.$$

This tower corresponds to the modular tower $(X_0(2^n))_n$, for which explicit equation were given by Elkies [2]. Using this interpretation of $\mathcal{M}$ as a modular tower, we find the exact degrees of extensions in the Galois closure of it and study the Galois

---

1991 *Mathematics Subject Classification.* Primary:14H05, Secondary:11R32.
Key words and phrases. Function field, modular curve, Galois tower.

groups that appear. We show that the function fields of the Galois closure can be obtained as a compositum of three different embeddings of the function fields in the tower $\mathcal{M}$.

For more definitions and further details about (explicit) towers of algebraic function fields, we refer to [**5**].

2. Groups of Galois closure

In this section the field of definition is always assumed to be $\mathbb{C}$, the field of complex numbers. Let p be a prime number and $n > 1$ an integer. The following group is standard in the theory of modular curves:

$$\Gamma_0(p^n) := \left\{ \left(\begin{array}{cc} a & b \\ c & d \end{array} \right) \in SL(2, \mathbb{Z}) : c \equiv 0 \pmod{p^n} \right\}.$$

Associated to this group is the modular curve $X_0(p^n)$ which has been studied extensively in the literature, cf. [**7**, **8**].

Let $0 < r < n$ be integers. The Galois closure of $X_0(p^n)$ over $X_0(p^r)$ has Galois group $\Gamma_0(p^r)/\Delta_r(p^n)$ with

$$\Delta_r(p^n) := \bigcap_{\sigma \in \Gamma_0(p^r)} \sigma \Gamma_0(p^n) \sigma^{-1}.$$

The group $\Delta_r(p^n)$ is the largest normal subgroup of $\Gamma_0(p^r)$ contained in $\Gamma_0(p^n)$, since if $H \trianglelefteq \Gamma_0(p^r)$ and $H \subset \Gamma_0(p^n)$, then $H \subset \bigcap_{\sigma \in \Gamma_0(p^r)} \sigma \Gamma_0(p^n) \sigma^{-1} = \Delta_r(p^n)$. The maximality of $\Delta_r(p^n)$ with respect to the above property will be used later.

The goal of this section is to compute the order of the groups $\Gamma_0(p^r)/\Delta_r(p^n)$ and to obtain information about its group structure. We start by describing the group $\Delta_r(p^n)$ in more detail.

PROPOSITION 2.1.

$$\Delta_r(p^n) = \left\{ \left(\begin{array}{cc} a & b \\ c & d \end{array} \right) \in \Gamma_0(p^n) : p^{n-r} | a - d - bp^r \text{ and } p^{n-r} | 2bp^r \right\}$$

PROOF. We denote by H the group on the right-hand side of above equality. For an element

$$h = \left(\begin{array}{cc} a & b \\ c & d \end{array} \right)$$

of $SL(2, \mathbb{Z})$ to be in H it needs to satisfy three things:
1) $p^n | c$,
2) $p^{n-r} | a - d - bp^r$ and
3) $p^{n-r} | 2bp^r$.

Clearly $H \subset \Gamma_0(p^n)$, so to prove the proposition it is enough to show that $H \trianglelefteq \Gamma_0(p^r)$ and that $\Delta_r(p^n) \subset H$, since then $\Delta_r(p^n) \supset H$ follows from the maximality of $\Delta_r(p^n)$.

First we prove that $H \trianglelefteq \Gamma_0(p^r)$. Conjugating an element $h \in H$ with a matrix

$$m = \left(\begin{array}{cc} \alpha & \beta \\ \gamma p^r & \delta \end{array} \right),$$

from $\Gamma_0(p^r)$ we find that

$$mhm^{-1} = \left(\begin{array}{cc} -p^r \gamma(\alpha b + \beta d) + (\alpha a + \beta c)\delta & \alpha^2 b + \alpha\beta(d - a) - \beta^2 c \\ p^r \gamma(a - d)\delta - bp^{2r}\gamma^2 + c\delta^2 & p^r \gamma(\alpha b - \beta a) + (\alpha d - \beta c)\delta \end{array} \right).$$

We need to check that this an element of H. First we show that it is an element of $\Gamma_0(p^n)$. We have

$$p^r(a-d)\gamma\delta - bp^{2r}\gamma^2 + c\delta^2 \equiv bp^{2r}\gamma(\delta-\gamma) \equiv 0 \pmod{p^n}.$$

The first equality follows from properties 1) and 2) of h listed above. The last equality follows directly from property 3) of h if $p \neq 2$. If $p = 2$, then it only implies that 2^{n-1} divides bp^{2r}, but δ has to be odd if $p = 2$, implying that in this case 2 divides $\gamma(\delta-\gamma)$.

Using again that $a \equiv d + bp^r \pmod{p^{n-r}}$ and $c \equiv 0 \pmod{p^{n-r}}$, we see that the second condition for mhm^{-1} to be in H is equivalent to the statement

$$bp^r(p^r\beta(\alpha+\gamma) - \alpha(\alpha+2\gamma-\delta)) \equiv 0 \pmod{p^{n-r}}.$$

From property 3) of h we see that this is satisfied if $p \neq 2$, while if $p = 2$, then $2^{n-r-1}|bp^r$ and $2|\alpha(\alpha-\delta)$, since δ is odd if $p = 2$.

It remains to check that the third condition is satisfied, but this can easily be seen to hold as well. We conclude that $mhm^{-1} \in H$ and hence that $H \trianglelefteq \Gamma_0(p^r)$.

Now we wish to prove that $\Delta_r(p^n) \subset H$. In order to do this we introduce the element

$$A := \begin{pmatrix} 1 & 0 \\ p^r & 1 \end{pmatrix}.$$

Then for any $h \in \Gamma_0(p^n)$ we have that

$$AhA^{-1} = \begin{pmatrix} a-bp^r & b \\ c+p^r(a-d-bp^r) & d+bp^r \end{pmatrix},$$

which implies that if $AhA^{-1} \in \Gamma_0(p^n)$, then $p^{n-r}|a-d-bp^r$. Similarly, if $A^{-1}hA \in \Gamma_0(p^n)$, then $p^{n-r}|a-d+bp^r$. Therefore, if $h \in \Gamma_0(p^n)\cap A\Gamma_0(p^n)A^{-1}\cap A^{-1}\Gamma_0(p^n)A$, then $p^n|c$, $p^{n-r}|a-d-bp^r$ and $p^{n-r}|a-d+bp^r$, which is equivalent to conditions 1),2) and 3) above. In other words:

$$\Delta_r(p^n) \subset \left(\Gamma_0(p^n) \cap A\Gamma_0(p^n)A^{-1} \cap A^{-1}\Gamma_0(p^n)A\right) \subset H,$$

which concludes the proof. □

COROLLARY 2.2. *We have*

$$\Delta_r(p^n) = \Gamma_0(p^n) \cap A\Gamma_0(p^n)A^{-1} \cap A^{-1}\Gamma_0(p^n)A,$$

with

$$A := \begin{pmatrix} 1 & 0 \\ p^r & 1 \end{pmatrix}.$$

PROOF. In the above proposition we saw that

$$\Delta_r(p^n) \subset \left(\Gamma_0(p^n) \cap A\Gamma_0(p^n)A^{-1} \cap A^{-1}\Gamma_0(p^n)A\right) \subset H,$$

but we have also seen that $H \subset \Delta_r(p^n)$. □

The group $\Delta_r(p^n)$ has some further properties we wish to ascertain. For a group G, we denote by $[G,G]$ its commutator subgroup.

LEMMA 2.3. *Suppose that $n > r > 0$. We have*

$$[\Delta_r(p^n), \Delta_r(p^n)] \subset \Delta_r(p^{n+1})$$

and

$$g \in \Delta_r(p^n) \Rightarrow g^p \in \Delta_r(p^{n+1}).$$

As a consequence, the group $\Delta_r(p^n)/\Delta_r(p^{n+1})$ is an elementary abelian p-group.

PROOF. A direct calculation shows that $[\Delta_r(p^n), \Delta_r(p^n)] \subset \Gamma_0(p^{n+1})$. Also, since $\Delta_r(p^n) \trianglelefteq \Gamma_0(p^r)$, we find that for any $\sigma \in \Gamma_0(p^r)$ we have

$$\sigma[\Delta_r(p^n), \Delta_r(p^n)]\sigma^{-1} = [\sigma\Delta_r(p^n)\sigma^{-1}, \sigma\Delta_r(p^n)\sigma^{-1}] = [\Delta_r(p^n), \Delta_r(p^n)].$$

This implies that

$$[\Delta_r(p^n), \Delta_r(p^n)] = \bigcap_{\sigma \in \Gamma_0(p^r)} \sigma[\Delta_r(p^n), \Delta_r(p^n)]\sigma^{-1} \subset \Delta_r(p^{n+1}).$$

To prove the second item, we use that

$$\begin{pmatrix} a & b \\ cp^n & d \end{pmatrix}^k = \begin{pmatrix} a^k + \mathcal{O}(p^n) & b\frac{a^k-d^k}{a-d} + \mathcal{O}(p^n) \\ cp^n\frac{a^k-d^k}{a-d} + \mathcal{O}(p^{2n}) & d^k + \mathcal{O}(p^n) \end{pmatrix},$$

where $\mathcal{O}(p^m)$ denotes some number divisible by p^m. This can be showed directly using induction on k. If $k = p$, then $cp^n(a^p - d^p)/(a-d) \equiv cp^n(a-d)^{p-1} \bmod p^{n+1}$. Since $g \in \Delta_r(p^n)$ we have that $p^{n-r}|a - d - bp^r$, implying that $p|a - d$. Hence $g^p \in \Gamma_0(p^{n+1})$. By definition of $\Delta_r(p^n)$, we have that for any $\sigma \in \Gamma_0(p^r)$, the element $\sigma^{-1}g\sigma$ is in $\Gamma_0(p^n)$, implying that $(\sigma^{-1}g\sigma)^p = \sigma^{-1}g^p\sigma \in \Gamma_0(p^{n+1})$. This implies that $g^p \in \bigcap_{\sigma \in \Gamma_0(p^r)} \sigma\Gamma_0(p^{n+1})\sigma^{-1} = \Delta_r(p^{n+1})$.

The final statement of the lemma follows directly from the first two statements.
□

3. The order of the group $\Delta_r(p^n)/\Gamma(p^n)$

The following congruence group is well-known:

$$\Gamma(p^n) := \left\{ \begin{pmatrix} a & b \\ c & d \end{pmatrix} \in SL(2, \mathbb{Z}) : \begin{pmatrix} a & b \\ c & d \end{pmatrix} \equiv \begin{pmatrix} 1 & 0 \\ 0 & 1 \end{pmatrix} \pmod{p^n} \right\}.$$

It is the kernel of the reduction modulo p^n map: $\varphi : SL(2, \mathbb{Z}) \to SL(2, \mathbb{Z}/p^n\mathbb{Z})$ and one can show that this map is surjective ([9, section 1.6]). Also it is well known [9] that

$$(3.1) \qquad\qquad \#SL(2, \mathbb{Z}/p^n\mathbb{Z}) = p^{3n} - p^{3n-2}.$$

Note that by Proposition 2.1 the group $\Gamma(p^n)$ is a (normal) subgroup of $\Delta_r(p^n)$. The goal of this section is to compute the cardinality of the group $\Delta_r(p^n)/\Gamma(p^n)$. We will start by giving several lemmas.

LEMMA 3.1. We have that

$$\Delta_r(p^n)/\Gamma(p^n) \cong \left\{ \begin{pmatrix} a & b \\ 0 & d \end{pmatrix} \in SL(2, \mathbb{Z}/p^n\mathbb{Z}) : p^{n-r}|a - d - bp^r \text{ and } p^{n-r}|2bp^r \right\}.$$

PROOF. This follows directly from Proposition 2.1 using the reduction modulo p^n map φ.
□

LEMMA 3.2. Let $n > r > 0$ be integers and suppose that p is an odd prime. Then we have that

$$\#\Delta_r(p^n)/\Gamma(p^n) = \begin{cases} 2p^{r+n} & \text{if } n \leq 2r, \\ 2p^{3r} & \text{else.} \end{cases}$$

PROOF. Using Lemma 3.1 and the assumption that p is odd, it is enough to count the number of triples $(a, b, d) \in (\mathbb{Z}/p^n\mathbb{Z})^3$ satisfying $p^n|ad-1$, $p^{n-r}|a-d$ and $p^{n-r}|bp^r$.

We claim that the number of $(a, d) \in (\mathbb{Z}/p^n\mathbb{Z})^2$ satisfying $p^n|ad-1$ and $p^{n-r}|a-d$ equals $2p^r$. From the conditions, it is clear that $p^{n-r}|a^2 - 1$, which implies that $a \equiv \pm 1 \pmod{p^{n-r}}$. This leaves exactly $2p^r$ possibilities for a. Given any a satisfying the last congruence, there exists exactly one $d \in \mathbb{Z}/p^n\mathbb{Z}$ such that $p^n|ad-1$ and by reducing modulo p^{n-r} we see that $d \equiv \pm 1 \equiv a$. This means that $p^{n-r}|a - d$ is satisfied for this d as well.

We claim that the number of $b \in \mathbb{Z}/p^n\mathbb{Z}$ such that $p^{n-r}|bp^r$ is equal to p^n if $n \le 2r$ and equal to p^{2r} if $n > 2r$. Indeed, if $n \le 2r$, the condition $p^{n-r}|bp^r$ is always satisfied, so that all b's in $\mathbb{Z}/p^n\mathbb{Z}$ are possible. If $n > 2r$, then the condition simplifies to $p^{n-2r}|b$, meaning that all p^{2r} multiples of p^{n-2r} in $\mathbb{Z}/p^n\mathbb{Z}$ are solutions.

Multiplying the number of possibilities for (a, d) with that for b, the lemma follows. $\square$

LEMMA 3.3. *Let $n > r > 0$ be integers. Then we have that*

$$\#\Delta_r(2^n)/\Gamma(2^n) = \begin{cases} 2^{2r+1} & \text{if } n - r = 1, \\ 2^4 & \text{if } n = 3 \text{ and } r = 1, \\ 2^{2r+3} & \text{if } n - r = 2 \text{ and } r > 1, \\ 2^5 & \text{if } n = 4 \text{ and } r = 1, \\ 2^8 & \text{if } n = 5 \text{ and } r = 2, \\ 2^{2r+5} & \text{if } n - r = 3 \text{ and } r > 2, \\ 2^{n+r+2} & \text{if } n - r > 3 \text{ and } n \le 2r, \\ 2^{3r+3} & \text{if } n - r > 3 \text{ and } n > 2r. \end{cases}$$

PROOF. Using Lemma 3.1 it is enough to count the number of triples $(a, b, d) \in (\mathbb{Z}/2^n\mathbb{Z})^3$ satisfying
1) $2^n|ad-1$,
2) $2^{n-r}|a - d - b2^r$ and
3) $2^{n-r-1}|b2^r$.
Since $2^{n-r-1}|b2^r$, we see that $b2^r \equiv 0 \bmod 2^{n-r}$ or $b2^r \equiv 2^{n-r-1} \bmod 2^{n-r}$. Combining with 2), we see that $d \equiv a \bmod 2^{n-r}$ or $d \equiv a + 2^{n-r-1} \bmod 2^{n-r}$. Substituting in 1) gives that $a^2 \equiv 1 \bmod 2^{n-r}$ or $a^2 \equiv 1 + a2^{n-r-1} \bmod 2^{n-r}$. Since from 1), we can deduce that a is odd, the latter congruence simplifies to $a^2 \equiv 1 + 2^{n-r-1} \bmod 2^{n-r}$. We now distinguish several cases.

Case 1, $n - r = 1$. In this case all solutions are characterized by choosing $a \in \mathbb{Z}/2^n\mathbb{Z}$ to be odd, d its multiplicative inverse modulo 2^n and arbitrary $b \in \mathbb{Z}/2^n\mathbb{Z}$. Thus there are $2^{2n-1} = 2^{2r+1}$ possibilities.

Case 2, $n - r = 2$. We have seen that $a^2 \equiv 1 \bmod 4$ or $a^2 \equiv 3 \bmod 4$. The latter is not possible, so we deduce that $a^2 \equiv 1 \bmod 4$, which implies that $a \equiv d \bmod 4$ and $b2^r \equiv 0 \bmod 4$. All in all we get that we can choose $a \equiv \pm 1 \bmod 4$, $b2^r \equiv 0 \bmod 4$ and $d \equiv a^{-1} \bmod 2^n$. For $r = 1$ this gives 16 possibilities for (a, b, d) and for $r > 1$ exactly 2^{2r+3}.

Case 3, $n - r = 3$. First we get that $a^2 \equiv 1 \bmod 8$ or $a^2 \equiv 5 \bmod 8$, but the latter is again not possible, since $8|a^2 - 1$ for any odd number a. This means that $b2^r \equiv 0 \bmod 8$. Moreover, the condition that $a^2 \equiv 1 \bmod 8$ implies that $a \equiv \pm 1$ or $\pm 3 \bmod 8$. Counting similarly as above, we find that there are 32 possibilities for (a,b,d) if $r = 1$, 256 if $r = 2$ and 2^{2r+5} if $r > 2$.

Case 4, $n - r > 3$. First we assume that $b2^r \equiv 0 \bmod 2^{n-r}$, which means that there are 2^{2r} possibilities for b if $n > 2r$ and 2^n otherwise. Then we found that $a^2 \equiv 1 \bmod 2^{n-r}$, which implies that $a \equiv \pm 1$ or $\pm 1 + 2^{n-r-1} \bmod 2^{n-r}$, leaving $4 \cdot 2^r$ possibilities for a. Now we can choose d to be the multiplicative inverse of a modulo 2^n and a direct computation shows that $a \equiv d \bmod 2^{n-r}$. All in all we find 2^{n+r+2} possibilities if $n \leq 2r$ and 2^{3r+2} if $n > 2r$, still assuming that $b2^r \equiv 0 \bmod 2^{n-r}$. Now assume that $b2^r \equiv 2^{n-r-1} \bmod 2^{n-r}$. This can only occur if $r \leq n - r - 1$, or equivalently if $n > 2r$ and then the number of possibilities for b is 2^{2r}. We saw that $a^2 \equiv 1 + 2^{n-r-1} \bmod 2^{n-r}$, implying that $a \equiv \pm 1 + 2^{n-r-2}$ or $\pm 1 - 2^{n-r-2} \bmod 2^{n-r}$. As before we choose d to be the inverse of a, but now we find that $d \equiv a + 2^{n-r-1} \bmod 2^{n-r}$, so that condition 2) is satisfied. Condition 3) is satisfied automatically. We find 2^{3r+2} possibilities if $n > 2r$, but none if $n \leq 2r$. In total for case 4, we find 2^{n+r+2} possibilities for (a, b, d) if $n \leq 2r$ and 2^{3r+3} otherwise. $\square$

4. Degrees and structure of Galois closure

Given $n > r > 0$ and a prime p, we will now determine the degree of the Galois closure of $X_0(p^n)$ over $X_0(p^r)$. We quote the following well-known facts [**9**, section 1.6]: Let m be a positive integer. The degree of the covering $X(p^m) \to X(1)$ equals $p^{3m-2}(p^2 - 1)/2$, unless $p = 2$ and $m = 1$ in which case it equals 6. The degree of the extension $X_0(p^m) \to X(1)$ equals $(p+1)p^{m-1}$. As a consequence we see that the degree of $X(p^{m+1}) \to X(p^m)$ equals p^3 unless $p = 2$ and $m = 1$, in which case it equals 4. Also the degree of $X_0(p^{m+1}) \to X_0(p^m)$ equals p. This together with the previous results enables us to compute all degrees in the tower obtained by taking the Galois closure of $X_0(p^n)$ over $X_0(p^r)$ for running n and fixed r.

LEMMA 4.1. *Let $n > r > 0$ be integers, p an odd prime and let $\tilde{X}_0^r(p^n)$ denote the Galois closure of $X_0(p^n)$ over $X_0(p^r)$. Then*

$$\deg(\tilde{X}_0^r(p^n) \to \tilde{X}_0^r(p^{n-1})) = \begin{cases} p(p-1)/2 & \text{if } n = r+1, \\ p^2 & \text{if } r+1 < n \leq 2r, \\ p^3 & \text{if } n > 2r. \end{cases}$$

For $n > r+1$, the covering $\tilde{X}_0^r(p^n) \to \tilde{X}_0^r(p^{n-1})$ is elementary abelian.

PROOF. From Lemma 3.2 we can calculate all degrees of the coverings $X(p^n) \to \tilde{X}_0^r(p^n)$. Indeed, since $-I \in \Delta_r(p^n)$ and $-I \notin \Gamma(p^n)$, the only thing we need to do is divide $\#\Delta_r(p^n)/\Gamma(p^n)$ by 2. Further, since $\deg(X_0(p^r) \to X(1)) = (p+1)p^{r-1}$ and $\deg(X(p^r) \to X(1)) = (p^2 - 1)p^{3r-2}/2$, we find that $\deg(X(p^r) \to X_0(p^r)) = (p-1)p^{2r-1}/2$. All in all we now know all degrees of the coverings $X(p^m) \to \tilde{X}_0^r(p^m)$ for $m \geq r$. Combined with the fact that $\deg(X(p^{m+1}) \to X(p^m)) = p^3$, the first part of the lemma follows. The second part follows directly from Lemma 2.3. $\square$

LEMMA 4.2. *Let $n > r > 0$ be integers and let $\tilde{X}_0^r(2^n)$ denote the Galois closure of $X_0(2^n)$ over $X_0(2^r)$. Then*

$$\deg(\tilde{X}_0^r(2^n) \to \tilde{X}_0^r(2^{n-1})) = \begin{cases} 2 & \text{if } n = r+1, \\ 2 & \text{if } n = r+2 \text{ and } r > 1, \\ 2 & \text{if } n = r+3 \text{ and } r > 2, \\ 4 & \text{if } n = r+2 \text{ and } r = 1, \\ 4 & \text{if } n = r+3 \text{ and } r = 1,2, \\ 4 & \text{if } n = r+4 \text{ and } r = 1,2, \\ 4 & \text{if } r+4 \le n \le 2r+1, \\ 8 & \text{if } n > 2r+3 \text{ and } r = 1, \\ 8 & \text{if } n > 2r+2 \text{ and } r = 2, \\ 8 & \text{if } n > 2r+1 \text{ and } r > 2. \end{cases}$$

PROOF. The proof is similar to that of the previous lemma, but now we use Lemma 3.3. ☐

LEMMA 4.3. *Let $n > r > 0$ be integers and p a prime. The extension $X_0(p^n) \to X_0(p^r)$ is Galois if and only if*

(1) *$p = 2$ and $n - r = 1$,*
(2) *$p = 2$, $r > 1$ and $n - r = 2$,*
(3) *$p = 2$, $r > 2$ and $n - r = 3$,*
(4) *$p = 3$ and $n - r = 1$.*

In all of these cases the Galois group is cyclic.

PROOF. Since $\deg(X_0(p^n) \to X_0(p^r)) = p^{n-r}$, we can use Lemmas 4.1 and 4.2 to check when this degree is the same as $\deg(\tilde{X}_0^r(p^n) \to X_0(p^r))$. Assuming the covering $X_0(p^n) \to X_0(p^r)$ is Galois of order p^{n-r}, we also see that its Galois group is $\Gamma_0(p^r)/\Delta_r(p^n)$. However, the element $A \bmod \Delta_r(p^n)$, with A as in Corollary 2.2, has order p^{n-r}. ☐

5. Reduction mod ℓ.

Let p be a prime. Until now, we have assumed that all the modular curves we considered were defined over the field $\mathbb{C}$. However, it is well known that the curves $X_0(p^n)$ have a model defined over $\mathbb{Q}$ [6]. Denote by ζ_{p^n} a primitive p^n-th root of unity. The curve $X(p^n)$ has a model defined over $\mathbb{Q}(\zeta_{p^n})$ and the covering $X(p^n) \to X(1)$ is still Galois and has the same degree as when working over $\mathbb{C}$. Since the Galois closure of $X_0(p^n)$ over $X_0(p^r)$ is contained in $X(p^n)$, it also has a model defined over $\mathbb{Q}(\zeta_{p^n})$ and all degrees computed before are still correct when working over this field. These models have good reduction modulo a prime ℓ if $\ell \neq p$. The Galois covering $X(p^n) \to X_0(p^r)$ is not necessarily Galois after this reduction, but will be so when we consider it over a field containing a p^n-th root of unity. After having done so, the Galois group will be the same as before reducing and in particular its degree is the same. All group theoretic arguments used before are then still valid for the reductions, as long as the field of definition contains a p^n-th root of unity.

The following Lemmas will be useful:

LEMMA 5.1. *Let F be a function field over a perfect field K and let $f(T) \in F[T]$ be a separable irreducible polynomial over F. Let $\alpha \in \Omega$ be a root of $f(T)$ in some fixed algebraically closed field $\Omega \supset F$. Let K' be a separable algebraic* *extension of K. Suppose that there exists a place P of F, which splits completely in the extension $F(\alpha)/F$. Then the polynomial $f(T)$ is irreducible in $FK'[T]$ and $\mathcal{G}(f, F) \cong \mathcal{G}(f, FK')$ where $\mathcal{G}(f, F)$ and $\mathcal{G}(f, FK')$ denote the Galois group of f over F and FK', respectively.*

PROOF. Since there exists a place P of F splitting completely in the extension $F(\alpha)/F$, the field K is algebraically closed in $F(\alpha)$. So the polynomial $f(T)$ is irreducible in $FK'[T]$ (cf. [**11**, Proposition III.6.6]). Denote by Z (respectively Z') the splitting field of $f(T)$ over F (respectively FK'). Let $\alpha_1, \ldots, \alpha_n$ be all conjugates of α over F. We have $Z = F(\alpha_1, \ldots, \alpha_n)$. Since $f(T)$ is irreducible in $FK'[T]$, the conjugates of α over FK' are also given by $\alpha_1, \ldots, \alpha_n$, and hence $Z' = FK'(\alpha_1, \ldots, \alpha_n) = ZK'$ and therefore

$$\mathcal{G}(f, FK') \cong \mathcal{G}(Z'/FK') \cong \mathcal{G}(Z/Z \cap FK').$$

Since the place P of F splits completely in the extension $F(\alpha)/F$, it will also split in the Galois closure Z/F. So the field K is algebraically closed in Z and hence $Z \cap FK' = F$. We obtain

$$\mathcal{G}(f, FK') \cong \mathcal{G}(Z/F) \cong \mathcal{G}(f, F).$$

$\square$

By use of the primitive element theorem, we immediately get the following

LEMMA 5.2. *Let F be a function field over a perfect field K and let E be a finite separable extension of F. Let K' be a separable algebraic extension of K. Suppose that there exists a place P of F splitting completely in the extension E/F. Consider the constant field extensions FK' and EK' of F and E, respectively. Denote by $\mathcal{GC}(E/F)$ respectively $\mathcal{GC}(EK'/FK')$ the Galois closure of the extension E/F respectively EK'/FK'. Then*

$$\mathcal{GC}(EK'/FK') = \mathcal{GC}(E/F)K';$$

i.e., taking the Galois closure of such an extension commutes with extending the field of constants. Moreover the Galois groups of $\mathcal{GC}(EK'/FK')/FK'$ and $\mathcal{GC}(E/F)/F$ are isomorphic.

Denote by F_n the function field of the curve $X_0(p^n)$ reduced modulo a prime ℓ. Note that its constant field is $\mathbb{F}_\ell$. We would like to use Lemma 5.2 in order to gain information on $\mathcal{GC}(F_n/F_r)$. In order to do so, we need that the extension F_n/F_r contains a completely splitting place, but this is not true in general. It is well known however, that if we extend the constant field to $\mathbb{F}_{\ell^2}$, the tower $F_r \subset F_{r+1} \subset \cdots$ is asymptotically optimal. So if the constant field is $\mathbb{F}_{\ell^2}$, we can expect completely splitting places. The following lemma confirms this for a large class of cases.

LEMMA 5.3. *Suppose that ℓ and p are two primes such that $\ell \geq 13$ and $\ell \neq p$, and let $0 < r < n$ be two integers. The extension $F_n\mathbb{F}_{\ell^2}/F_r\mathbb{F}_{\ell^2}$ contains a completely splitting place.*

PROOF. Let $F_{-1}\mathbb{F}_{\ell^2}$ denote the function field arising by reducing the modular curve $X(1)$ modulo ℓ and then extending the constant field. The reason the function fields $F_n\mathbb{F}_{\ell^2}$ have many rational places is that the supersingular j-invariants in $F_{-1}\mathbb{F}_{\ell^2}$ are $\mathbb{F}_{\ell^2}$-rational and all places in $F_n\mathbb{F}_{\ell^2}$ lying above any of these j-invariants different from 0 and 1728 are $\mathbb{F}_{\ell^2}$-rational as well (see Lemma 5.3 in [**1**]). On the other hand, it is well known that the only branching in F_n/F_{-1} occurs at $j = 0$, $j = 1728$ and $j = \infty$.

In order to prove the lemma it is therefore enough to show that there exists a supersingular j-invariant different from 0 and 1728. Such a j-invariant always exists if $\ell \geq 13$ (see [10], section V.4]). $\qquad\square$

6. Galois closure of a tame tower

Explicit equations for some of the towers considered above (and also for some other modular towers) were given by Elkies (see [2]). In particular let $p = 2$ and consider the tower

$$\ldots \to \ldots \to X_0(2^6) \to X_0(2^5) \to X_0(2^4).$$

Let ℓ be a prime such that $\ell \geq 13$. For $i \geq 0$ let M_i be the function field of the curve $X_0(2^{i+4})$, with the field of constants extended to $\mathbb{F}_{\ell^2}$. Hence we have a corresponding tower of function fields

$$\mathcal{M} = (M_0, M_1, M_2, \ldots)$$

over $\mathbb{F}_{\ell^2}$. This tower can be recursively defined as follows (see [2] and [3, Remark 5.9]): $M_0 = \mathbb{F}_{\ell^2}(x_0)$ is the rational function field and $M_n = M_{n-1}(x_n)$ where

$$x_n^2 = \frac{x_{n-1}^2 + 1}{2x_{n-1}}$$

for $n \geq 1$. This tower was studied in detail in [3]. It is an asymptotically optimal tower over $\mathbb{F}_{\ell^2}$. Following Section 5, we can use the group theoretical arguments considered before to obtain detailed information about the Galois closure of the tower $\mathcal{M}$. For $i \geq 0$ let G_i be the Galois closure of M_i over M_0 and consider the sequence of function fields $\mathcal{G} = (G_0, G_1, \ldots)$ called the Galois closure of the tower $\mathcal{M}$ over M_0. Let

$$\mathfrak{M} = \bigcup_{j=0}^{\infty} M_j \quad \text{and} \quad \mathfrak{G} = \bigcup_{j=0}^{\infty} G_j,$$

and let Ω be a fixed algebraically closed field containing $\mathfrak{M}$.

THEOREM 6.1. (1) $\mathcal{G}$ is a tower over $\mathbb{F}_{\ell^2}$.

(2) The tower $\mathcal{G}$ is optimal; i.e., for the limit $\lambda(\mathcal{G})$ of $\mathcal{G}$ we have

$$\lambda(\mathcal{G}) = \lambda(\mathcal{M}) = \ell - 1.$$

(3) There exist two embeddings σ, τ of $\mathfrak{M}$ into Ω over M_0, such that

$$G_i = M_i \cdot \sigma(M_i) \cdot \tau(M_i), \text{ for } i \geq 0,$$

and

$$\mathfrak{G} = \mathfrak{M} \cdot \sigma(\mathfrak{M}) \cdot \tau(\mathfrak{M}).$$

(4) The extension M_3/M_0 (and more generally the extension M_{i+3}/M_i for $i \geq 0$) is a cyclic Galois extension of degree 8. In particular we have $G_3 = M_3$.

(5) We have

$$[G_i : G_{i-1}] = \begin{cases} 2 & \text{if } 1 \leq i \leq 3, \\ 4 & \text{if } 3 < i \leq 5, \\ 8 & \text{if } 5 < i. \end{cases}$$

(6) The G_i/G_{i-1} is an elementary abelian 2-extension for $i \geq 1$.

PROOF.   (1) Since there is a place of M_0 splitting completely in the tower $\mathcal{M}$, the result follows from [**4**, Prop. 2.1]

(2) See [**4**, Rem. 2.4].

(3) This follows directly from Corollary 2.2 by noting that the matrix

$$A = \begin{pmatrix} 1 & 0 \\ 2^4 & 1 \end{pmatrix}$$

is the same at every step.

(4) This follows from Lemma 4.3, with $p = 2, r = 4$.

(5) This is just a special case of Lemma 4.2.

(6) The Galois groups of the extension G_i/G_{i-1} is given by $\Delta_r(p^i)/\Delta_r(p^{i-1})$. So it follows from Lemma 2.3 that this extension is an elementary abelian 2-extension.

$\square$

Next we will give some alternative generators and equations for the tower considered above. Let $\mathcal{F} = (F_0, F_1, F_2, \ldots)$ be a tower of function fields over a field K, which is recursively defined by the polynomial $f(X, Y) \in K[X, Y]$; i.e., $F_0 = K(x_0)$ is the rational function field, and for every $n \geq 1$, we have $F_n = F_{n-1}(x_n)$ with $f(x_{n-1}, x_n) = 0$.

So in particular we have

$$F_n = F_0(x_1, x_2, \ldots, x_n)$$

for all $n \geq 1$. It turns out that for most of the interesting towers we in fact have

$$F_n = F_0(x_n).$$

The following Lemma gives an easy criterion for this to be the case.

LEMMA 6.2. *Let $\mathcal{F} = (F_0, F_1, F_2, \ldots)$ be a tower of function fields over a field K recursively defined by $f(X, Y) \in K[X, Y]$. Suppose that there exists a place P of F_0 such that the place P is totally ramified in the extension F_n/F_0 and the unique place Q of F_n lying above P is unramified over $K(x_n)$. Then $F_n = F_0(x_n)$.*

PROOF. We have $F_0 \subseteq F_0(x_n) \subseteq F_n$. Since the place Q of F_n is unramified over $K(x_n)$, it will be unramified over $F_0(x_n)$. But since Q is totally ramified in the extension F_n/F_0, we have $F_0(x_n) = F_n$. $\square$

Let ℓ be an odd prime and consider the tower $\mathcal{M} = (M_0, M_1, \ldots)$ over $\mathbb{F}_{\ell^2}$ above, which is recursively defined by

$$Y^2 = \frac{X^2 + 1}{2X}.$$

From the ramification in the tower $\mathcal{M}$ and Lemma 6.2 it follows that $M_n = M_0(x_n)$.

By Theorem 6.1 for $n \geq 0$ the extension M_{n+3}/M_n is a cyclic Galois extension of degree 8. It is hence natural to consider the tower $\mathcal{M}' = (M_0', M_1', M_2' \ldots)$ with $M_n' = M_{3n}$. This is in fact just an alternative way of defining the tower $\mathcal{M}$, where a step in this new description corresponds to 3 steps in the tower $\mathcal{M}$. For $n \geq 0$ let $x_n' = x_{3n}$. Clearly for $n \geq 1$ we then have $M_n' = M_0'(x_n') = M_{n-1}'(x_n')$. It can be verified that the minimal polynomial of x_n' over M_{n-1}' is given by

$$T^8 - T^4 - \frac{(x_{n-1}' - 1)^8}{128 x_{n-1}'(x_{n-1}'^2 + 1)(x_{n-1}' + 1)^4} \in M_{n-1}'[T].$$

Letting $f(T) = T^8 - T^4$ we note that

$$\frac{(x'_{n-1} - 1)^8}{128 x'_{n-1} (x'^2_{n-1} + 1)(x'_{n-1} + 1)^4} = \frac{1}{16 f\left(\frac{x'_{n-1}+1}{x'_{n-1}-1}\right)}.$$

It hence follows that the tower $\mathcal{M}' = (M'_0, M'_1, M'_2, \ldots)$ can be recursively defined by

$$f(Y) = \frac{1}{16 f\left(\frac{X+1}{X-1}\right)},$$

where $f(T) = T^8 - T^4$.

As mentioned before, the steps M'_n/M'_{n-1} are cyclic Galois extensions of degree 8. This can be seen explicitly. Define

$$\alpha_n = \frac{2 x'_n}{x_{3n-2} + 1}.$$

Then $\alpha_n \in M'_n$ and it can be shown by an explicit calculation that

(6.1) $$\alpha_n^8 = \frac{32 x'^3_{n-1}}{(x'^2_{n-1} + 1)(x'_{n-1} + 1)^4}.$$

Denote by P the place of M'_0 that is the pole of function x'_0, by Q a place of M'_{n-1} lying above P in the extension M'_{n-1}/M'_0 and by R the place $Q \cap \mathbb{F}_{\ell^2}(x'_{n-1})$. From the ramification structure of the tower $\mathcal{M}$, it follows that R is the pole of the function x'_{n-1} and $e(Q|R) = 1$, see [3]. From this and equation (6.1) it then follows immediately that $M'_n = M'_{n-1}(\alpha_n)$.

We conclude by analyzing a property of the splitting locus of the tower $\mathcal{M}'$. For an odd prime number ℓ we define the Deuring polynomial

$$H(X) = \sum_{i=0}^{(\ell-1)/2} \left(\frac{\frac{\ell-1}{2}}{i}\right)^2 \cdot X^i \in \mathbb{F}_\ell[X].$$

The splitting locus of the tower $\mathcal{M}$ (and hence of $\mathcal{M}'$) is given by all $\alpha \in \bar{\mathbb{F}}_\ell$ such that $H(\alpha^4) = 0$, see [3]. By the recursive definition of the tower this means that if $H(\alpha^4) = 0$ and $\beta \in \bar{\mathbb{F}}_\ell$ satisfies $\beta^2 = (\alpha^2 + 1)/2\alpha$, then $H(\beta^4) = 0$. Considering the above description of the tower $\mathcal{M}'$, we then get the following property for the Deuring polynomial $H(X)$: let $\alpha \in \bar{\mathbb{F}}_\ell$ such that $H(\alpha^4) = 0$ and let $\beta \in \bar{\mathbb{F}}_\ell$ such that $f(\beta) = 1/(16 \cdot f(\frac{\alpha+1}{\alpha-1}))$. Then $H(\beta^4) = 0$.

References

[1] P. Beelen and I.I. Bouw, *Asymptotically good towers and differential equations*, Compositio Math. 141, pp. 1405–1424, 2005.

[2] N.D. Elkies, *Explicit modular towers*, in Proc. 35th Ann. Allerton Conf. on Communication, Control and Computing, Urbana, IL, 1997, pp. 23–32.

[3] A. Garcia, H. Stichtenoth and H. Rück, *On tame towers over finite fields*, J. Reine Angew. Math. 557, pp. 53–80, 2003.

[4] A. Garcia and H. Stichtenoth, *On the Galois closure of towers*, in Recent trends in coding theory and its applications, pp. 83–92, AMS/IP Stud. Adv. Math., 41, Amer. Math. Soc., 2007.

[5] A. Garcia and H. Stichtenoth (eds.), *Topics in geometry, coding theory and cryptography*, Algebr. Appl. 6, Springer-Verlag, 2007.

[6] J. Igusa, *Kroneckerian model of fields of elliptic modular functions*, Amer. J. Math. 81,, pp. 561–577, 1959.

[7] W. Maak, *Elliptische Modulfunktionen*, unter Benutzung einer Vorlesung von E. Hecke aus dem Jahre 1935, lecture notes, Göttingen Univ. 1955/56.

[8] B. Schoeneberg, *Elliptic Modular Functions*, Springer Verlag, 1974.

[9] G. Shimura, *Introduction to the arithmetic theory of automorphic functions*, Iwanami Shoten Publishers and Princeton University Press, 1971.

[10] J.H. Silverman, *The arithmetic of elliptic curves*, Springer Verlag, 1986.

[11] H. Stichtenoth, *Algebraic function fields and codes*, Springer Verlag, 1993.

[12] H. Stichtenoth, *Transitive and self-dual codes attaining the Tsfasman–Vladut–Zink bound*, IEEE Trans. Inform. Theory 52, no. 5, pp. 2218–2224, 2006.

[13] M.A. Tsfasman, S.G. Vladut and T. Zink, *Modular curves, Shimura curves and Goppa codes, better than the Varshamov–Gilbert bound*, Math. Nachr. 109, pp. 21–28, 1982.

ECOLE POLYTECHNIQUE FÉDÉRALE DE LAUSANNE, EPFL-SFB-IMB-CSAG, STATION 8, 1015, LAUSANNE, SWITZERLAND
E-mail address: `alp.bassa@epfl.ch`

DTU-MATHEMATICS, TECHNICAL UNIVERSITY OF DENMARK, MATEMATIKTORVET, BUILDING 303S, DK-2800, LYNGBY, DENMARK
E-mail address: `p.beelen@mat.dtu.dk`

Contemporary Mathematics
Volume **487**, 2009

Codes defined by forms of degree 2 on quadric varieties in $\mathbb{P}^4(\mathbb{F}_q)$

Frédéric A. B. Edoukou

ABSTRACT. We study the functional codes of second order defined by G. Lachaud on $\mathcal{X} \subset \mathbb{P}^4(\mathbb{F}_q)$ a quadric of $\text{rank}(\mathcal{X})=3,4,5$. We give some bounds for the number of points of quadratic sections of $\mathcal{X}$, which are the best possible and show that codes defined on non-degenerate quadrics are better than those defined on degenerate quadrics. We also show the geometric structure of the minimum weight codewords and estimate the second weight of these codes. We also prove by using the theorem of Ax on the zeros of polynomials over a finite field that all the weights of the codewords of the codes $C_2(\mathcal{X})$ defined in any quadric in $\mathbb{P}^n(\mathbb{F}_q)$ are divisible by q. The paper ends with a conjecture on the number of points of two quadrics in $\mathbb{P}^n(\mathbb{F}_q)$ with no common hyperplane.

1. Introduction

We denote by $\mathbb{F}_q$ the field with q elements. Let $V = A^{n+1}(\mathbb{F}_q)$ be the affine space of dimension $n+1$ over $\mathbb{F}_q$ and $\mathbb{P}^n(\mathbb{F}_q) = \Pi_n$ the corresponding projective space. Then

$$\pi_n = \#\mathbb{P}^n(\mathbb{F}_q) = q^n + q^{n-1} + \ldots + 1.$$

We use the term forms of degree h to describe homogeneous polynomials f of degree h, and $Z(f)$ denotes the zeros of f in the projective space $\mathbb{P}^n(\mathbb{F}_q)$. Let $\mathcal{F}_h(V)$ be the vector space of forms of degree h in $V = \mathbb{A}^{n+1}(\mathbb{F}_q)$, $X \subset \mathbb{P}^n(\mathbb{F}_q)$ a variety and $|X|$ the number of rational points of X over $\mathbb{F}_q$. Let W_i be the set of points with homogeneous coordinates $(x_0 : \ldots : x_n) \in \mathbb{P}^n(\mathbb{F}_q)$ such that $x_j = 0$ for $j < i$ and $x_i \neq 0$. The family $\{W_i\}_{0 \leq i \leq n}$ is a partition of $\mathbb{P}^n(\mathbb{F}_q)$. The code $C_h(X)$ is the image of the linear map $c : \mathcal{F}_h(V) \longrightarrow \mathbb{F}_q^{|X|}$, defined by $c(f) = (c_x(f))_{x \in X}$, where $c_x(f) = f(x_0, \ldots, x_n)/x_i{}^h$ with $x = (x_0 : \ldots : x_n) \in W_i$. The length of $C_h(X)$ is equal to $|X|$. The dimension of $C_h(X)$ is equal to $\dim \mathcal{F}_h(V) - \dim \ker c$, where

$$(1.1) \qquad \dim \mathcal{F}_h(V) = \binom{n+h}{h}.$$

The minimum distance of $C_h(X)$ is equal to the minimum over all non null polynomials f of $|X| - |X \cap Z(f)|$.

2000 *Mathematics Subject Classification.* Primary 05B25, 11T71, 14J29.

Key words and phrases. finite fields, functional codes, projective index, quadric varieties, regulus, weight.

We study the codes $C_2(\mathcal{X})$ defined on the quadric $\mathcal{X}$ in $\mathbb{P}^4(\mathbb{F}_q)$. We will consider only the cases where rank$(\mathcal{X})$= 3, 4, 5, since otherwise, the computation of the mininum distance of the codes $C_2(\mathcal{X})$ is reduced to the study of plane or hyperplane sections of quadrics which is well known.

The case where $\mathcal{X}$ is a non-degenerate quadric in $\mathbb{P}^4(\mathbb{F}_q)$ (i.e. rank$(\mathcal{X}) = 5$, $\mathcal{X}$ is a parabolic quadric), and $\mathcal{Q}$ another quadric (distinct from $\lambda.\mathcal{X}$, $\lambda \in \mathbb{F}_q^*$) a bound for the minimum distance is given by D. B. Leep and L. M. Schueller [11, p.172]:

$$|\mathcal{X} \cap \mathcal{Q}| \leq 3q^2 + q + 1.$$

It is not optimal. When $\mathcal{X}$ is a degenerate quadric of rank 3 or 4, the hypothesis of D. B. Leep and L. M. Schueller is too much restrictive and we cannot find the minimal distance. In this case, however, we have a bound given by G. Lachaud [10, proposition 2.3]:

$$|\mathcal{X} \cap \mathcal{Q}| \leq 4(q^2 + q + 1)$$

which is not the best possible.

In this paper we find some optimal bounds except the case where $\mathcal{X}$ is a degenerate quadric with rank$(\mathcal{X}) = 4$ and projective index g$(\mathcal{X})$=1.

The paper is organized as follows. First of all we recall some generalities on quadrics. Secondly by using the projective classification of quadrics in [9, p.4] and their geometric structure, we find some interesting bounds on the number of rational points on the intersection of two quadrics. Thus, we show that for $\mathcal{X}$ a non-degenerate quadric:

$$|\mathcal{X} \cap \mathcal{Q}| \leq 2q^2 + 3q + 1$$

is the best possible. We also show that for $\mathcal{X}$ degenerate and rank $(\mathcal{X}) = 3$, we get

$$|\mathcal{X} \cap \mathcal{Q}| \leq 4q^2 + q + 1,$$

and this bound is optimal. Identically, for rank$(\mathcal{X}) = 4$ and g$(\mathcal{X})$=2, we have the optimal bound:

$$|\mathcal{X} \cap \mathcal{Q}| \leq 4q^2 + 1.$$

For $\mathcal{X}$ of rank$(\mathcal{X}) = 4$ and g$(\mathcal{X})$=1, we get the bound

$$|\mathcal{X} \cap \mathcal{Q}| \leq 3q^2 + q + 1.$$

Next, we find the exact parameters of the codes $C_2(\mathcal{X})$, the geometric structure of the minimum weight codewords and show that the performances of the codes $C_2(\mathcal{X})$ defined on the non-degenerate quadrics are better than the ones defined on the degenerate quadrics.

We also prove by using the theorem of Ax on the zeros of polynomials over a finite field that all the weights of the codewords of the codes $C_2(\mathcal{X})$ defined in any quadric in $\mathbb{P}^n(\mathbb{F}_q)$ are divisible by q. The paper ends with a conjecture which generalizes the result on the number of points of two quadrics with no common hyperplane in $\mathbb{P}^n(\mathbb{F}_q)$.

2. Generalities

Let $\mathcal{Q}$ be a quadric in $\mathbb{P}^n(\mathbb{F}_q)$(i.e. $\mathcal{Q} = Z(F)$ where F is a form of degree 2). The rank of $\mathcal{Q}$, denoted r$(\mathcal{Q})$, is the smallest number of indeterminates appearing in F under any change of coordinate system. The quadric $\mathcal{Q}$ is said to be degenerate if r$(\mathcal{Q}) < n + 1$; otherwise it is non-degenerate. For $\mathcal{Q}$ a degenerate quadric and

$r(\mathcal{Q})$=r, $\mathcal{Q}$ is a cone $\Pi_{n-r}\mathcal{Q}_{r-1}$ with vertex Π_{n-r} (the set of the singular points of $\mathcal{Q}$) and base $\mathcal{Q}_{r-1}$ in a subspace Π_{r-1} skew to Π_{n-r}.

DEFINITION 2.1. *For $\mathcal{Q} = \Pi_{n-r}\mathcal{Q}_{r-1}$ a degenerate quadric with $r(\mathcal{Q}) = r$, $\mathcal{Q}_{r-1}$ is called the non-degenerate quadric associated to $\mathcal{Q}$. The degenerate quadric $\mathcal{Q}$ will be said to be of hyperbolic type (resp. elliptic, parabolic) if its associated non-degenerate quadric is of that type.*

DEFINITION 2.2. *[11, p.158] Let $\mathcal{Q}_1$ and $\mathcal{Q}_2$ be two quadrics. The order $w(\mathcal{Q}_1, \mathcal{Q}_2)$ of the pair $(\mathcal{Q}_1, \mathcal{Q}_2)$, is the minimum number of variables, after invertible linear change of variables, necessary to write $\mathcal{Q}_1$ and $\mathcal{Q}_2$. When $w(\mathcal{Q}_1, \mathcal{Q}_2) = t + 1$, we assume that $\mathcal{Q}_1$ and $\mathcal{Q}_2$ are defined with the first $t + 1$ indeterminates $x_0, \ldots, x_t$ and we define $\mathbb{E}_t(\mathbb{F}_q) = \{x \in \mathbb{P}^n(\mathbb{F}_q) \,|x_{t+1} = \ldots = x_n = 0\}$.*

DEFINITION 2.3. *For any projective algebraic variety $\mathcal{V}$, the maximum dimension $g(\mathcal{V})$ of linear subspaces lying on $\mathcal{V}$, is called the projective index of $\mathcal{V}$. The largest dimensional spaces contained in $\mathcal{V}$ are called the generators of $\mathcal{V}$.*

3. Intersection of two quadrics

In this section, we will estimate the number of points in the intersection of two quadrics $\mathcal{X}$ and $\mathcal{Q}$ in $\mathbb{P}^4(\mathbb{F}_q)$. Some bounds in $\mathbb{P}^4(\mathbb{F}_q)$ have been given by Y. Aubry [1], G. Lachaud [10], D. B. Leep and L. M. Schueller [11]. These bounds are not the best possible even in the case of $\mathbb{P}^4(\mathbb{F}_q)$. Here $\mathcal{X}$ denotes a quadric of rank$(\mathcal{X}) = 3, 4, 5$ in $\mathbb{P}^4(\mathbb{F}_q)$.
From the result of D. B. Leep and L. M. Schueller [11, p.172] , we deduce the following result.

COROLLARY 3.1. *Let $\mathcal{Q}_1$ and $\mathcal{Q}_2$ be two quadrics in $\mathbb{P}^4(\mathbb{F}_q)$ with $w(\mathcal{Q}_1, \mathcal{Q}_2) = 5$. Then,*

$$|\mathcal{Q}_1 \cap \mathcal{Q}_2| \leq 3q^2 + q + 1.$$

In [11, p.172] the hypothesis on the order of the two quadrics $\mathcal{Q}_1$ and $\mathcal{Q}_2$ is too restrictive and does not work in the general case. Let us recall another property, free from this condition.

LEMMA 3.2. *[6, pp.70-71] Let $\mathcal{Q}_1$ and $\mathcal{Q}_2$ be two quadrics in $\mathbb{P}^n(\mathbb{F}_q)$ and l an integer such that $1 \leq l \leq n-1$. Suppose that $w(\mathcal{Q}_1, \mathcal{Q}_2) = n-l+1$ (i.e. there exists a linear transformation such that $\mathcal{Q}_1$ and $\mathcal{Q}_2$ are defined with the indeterminates $x_0, x_1, ..., x_{n-l}$) and $|\mathcal{Q}_1 \cap \mathcal{Q}_2 \cap \mathbb{E}_{n-l}(\mathbb{F}_q)| \leq m$. Then*

$$|\mathcal{Q}_1 \cap \mathcal{Q}_2| \leq mq^l + \pi_{l-1}.$$

This bound is optimal as soon as m is optimal in $\mathbb{E}_{n-l}(\mathbb{F}_q)$.

LEMMA 3.3. *[6, p.71] For $\mathcal{Q}_1$ and $\mathcal{Q}_2$ two quadrics in $\mathbb{P}^n(\mathbb{F}_q)$, we get $w(\mathcal{Q}_1, \mathcal{Q}_2) \geq \sup\{r(\mathcal{Q}_1), r(\mathcal{Q}_2)\}$.*

Let us recall the classification of quadrics in $\mathbb{P}^4(\mathbb{F}_q)$ according to J. W. P. Hirschfeld [9, p.4]. From the work of Ray-Chaudhuri [13, pp.132-136] we deduce that a non-degenerate quadric $\mathcal{P}_4$ in $\mathbb{P}^4(\mathbb{F}_q)$ contains exactly $\alpha_q = \pi_3$ lines and there are exactly $q + 1$ lines contained in $\mathcal{P}_4$ through a given point in $\mathcal{P}_4$.
A regulus, defined in the work of Hirschfeld [8, p.4], is the set of transversals of three skew lines in $\mathbb{P}^3(\mathbb{F}_q)$; it consists of $q+1$ skew lines. Thus, a hyperbolic quadric $\mathcal{Q}$ in $\mathbb{P}^3(\mathbb{F}_q)$ is a pair of complementary reguli (each one of the two reguli generates

r($\mathcal{Q}$)	Description	$|\mathcal{Q}|$	g($\mathcal{Q}$)
1	repeated hyperplane $\Pi_3\mathcal{P}_0$	$q^3 + q^2 + q + 1$	3
2	pair of distinct hyperplanes $\Pi_2\mathcal{H}_1$	$2q^3 + q^2 + q + 1$	3
2	plane $\Pi_2\mathcal{E}_1$	$q^2 + q + 1$	2
3	cone $\Pi_1\mathcal{P}_2$	$(q + 1)(q^2 + 1)$	2
4	cone $\Pi_0\mathcal{H}_3(\mathcal{R}, \mathcal{R}')$	$q(q + 1)^2 + 1$	2
4	cone $\Pi_0\mathcal{E}_3$	$q(q^2 + 1) + 1$	1
5	parabolic quadric $\mathcal{P}_4$	$(q + 1)(q^2 + 1)$	1

TABLE 1. Quadrics in $\mathbb{P}^4(\mathbb{F}_q)$

the whole hyperbolic quadric). It is denoted by $\mathcal{Q} = \mathcal{H}_3(\mathcal{R}, \mathcal{R}')$ where $\mathcal{R}$ and $\mathcal{R}'$ are the two reguli.

3.1. Section of $\mathcal{X}$ by the quadrics $\mathcal{Q}$ containing planes as generators: g($\mathcal{Q}$)=2. Here we study the section of the quadric $\mathcal{X}$ by a quadric $\mathcal{Q}$ with g($\mathcal{Q}$)=2. We have three cases: $\mathcal{Q} = \Pi_2\mathcal{E}_1$, $\mathcal{Q} = \Pi_1\mathcal{P}_2$ and $\mathcal{Q} = \Pi_0\mathcal{H}_3(\mathcal{R}, \mathcal{R}')$. Here $\mathcal{Q} = \Pi_1\mathcal{P}_2$ and $\mathcal{Q} = \Pi_0\mathcal{H}_3(\mathcal{R}, \mathcal{R}')$ are respectively a set of $q+1$ planes through a common line or a common point.

3.1.1. *$\mathcal{Q}$ is a quadric with r($\mathcal{Q}$)=2: $\mathcal{Q} = \Pi_2\mathcal{E}_1$.* Here $\mathcal{Q}$ consists of one plane of $\mathbb{P}^4(\mathbb{F}_q)$. If this plane is not contained in $\mathcal{X}$, then $\mathcal{Q} \cap \mathcal{X}$ is a plane quadric. In the book of J. W. P. Hirschfeld [7, p.156], we have the following classsification of plane quadrics $\mathcal{Q}'$.

r($\mathcal{Q}'$)	Description	$|\mathcal{Q}'|$	g($\mathcal{Q}'$)
1	repeated line $\Pi_1\mathcal{P}_0$	$q + 1$	1
2	pair of lines $\Pi_0\mathcal{H}_1$	$2q + 1$	1
2	point $\Pi_0\mathcal{E}_1$	1	0
3	parabolic $\mathcal{P}_2$	$q + 1$	0

TABLE 2. Plane quadrics in $\mathbb{P}^2(\mathbb{F}_q)$

Therefore we deduce that if $\mathcal{Q}$ is a plane of $\mathbb{P}^4(\mathbb{F}_q)$:
– If it is contained in $\mathcal{X}$, we get $|\mathcal{X} \cap \mathcal{Q}| = q^2 + q + 1$.
– Otherwise, we get from the table above that $|\mathcal{X} \cap \mathcal{Q}| \leq 2q + 1$.

3.1.2. *$\mathcal{Q}$ is a quadric with r($\mathcal{Q}$)=3,4: $\mathcal{Q} = \Pi_1\mathcal{P}_2$ and $\mathcal{Q} = \Pi_0\mathcal{H}_3$.* For $\mathcal{X}$ a non-degenerate quadric, there is no plane in $\mathcal{X}$. Thus each of the $q+1$ planes of $\mathcal{Q}$ intersects $\mathcal{X}$ in at most $2q+1$ points. Therefore, we deduce that $|\mathcal{X} \cap \mathcal{Q}| \leq 2q^2 + 3q + 1$.

For $\mathcal{X}$ degenerate of r($\mathcal{X}$) = 3, 4, let us consider the order $w(\mathcal{X}, \mathcal{Q})$ of the pair $\{\mathcal{X}, \mathcal{Q}\}$. We have three possibilities $w(\mathcal{X}, \mathcal{Q}) = 3, 4, 5$; the case $w(\mathcal{X}, \mathcal{Q}) = 5$ is solved by Corollary 3.1.

    (i) If $w(\mathcal{X}, \mathcal{Q}) = 4$, then there exists a linear transformation such that $\mathcal{X}$ and $\mathcal{Q}$ are defined with the indeterminates x_0, x_1, x_2, x_3.
        Suppose rank($\mathcal{X}$)=3. From [5, IV-D-3,4], we get $|\mathcal{X} \cap \mathcal{Q} \cap \mathbb{E}_3(\mathbb{F}_q)| = 4q + 1$ or $|\mathcal{X} \cap \mathcal{Q} \cap \mathbb{E}_3(\mathbb{F}_q)| \leq 3q$. From Lemma 3.2 we deduce that either $|\mathcal{X} \cap \mathcal{Q}| = 4q^2 + q + 1$ in the case where $\mathcal{X}$ and $\mathcal{Q}$ have exactly four common

planes through a line or $|\mathcal{X} \cap \mathcal{Q}| \leq 3q^2 + 1$ otherwise.

If rank$(\mathcal{X})=4$ and $g(\mathcal{X}) = 1$, from [5, IV-D-2] we get $|\mathcal{X} \cap \mathcal{Q} \cap \mathbb{E}_3(\mathbb{F}_q)| \leq 2(q+1)$. Therefore Lemma 3.2 gives $|\mathcal{X} \cap \mathcal{Q}| \leq 2q^2 + 2q + 1$.

If rank$(\mathcal{X})=4$ and $g(\mathcal{X}) = 2$, [5, IV-D-3 and IV-E-1] gives $|\mathcal{X} \cap \mathcal{Q} \cap \mathbb{E}_3(\mathbb{F}_q)| = 4q$ or $|\mathcal{X} \cap \mathcal{Q} \cap \mathbb{E}_3(\mathbb{F}_q)| \leq 3q + 1$. In fact, in the case $|\mathcal{X} \cap \mathcal{Q} \cap \mathbb{E}_3(\mathbb{F}_q)| = 4q$, $\mathcal{X} \cap \mathcal{Q} \cap \mathbb{E}_3(\mathbb{F}_q)$ is exactly a set of four lines with two lines in each regulus of $\mathcal{H}_3$. Therefore, from Lemma 3.2 and the geometric structure of $\mathcal{X}$ and $\mathcal{Q}$, we get $|\mathcal{X} \cap \mathcal{Q}| = 4q^2 + 1$ when $\mathcal{X}$ and $\mathcal{Q}$ have the same vertex Π_0 and contain four common planes $\Pi_2^{(1)}$, $\Pi_2^{(2)}$, $\Pi_2^{(3)}$ and $\Pi_2^{(4)}$ with the following configuration:

($\Pi_2^{(1)}$ and $\Pi_2^{(2)}$ meet at the point Π_0, $\Pi_2^{(3)}$ and $\Pi_2^{(4)}$ also meet at the point Π_0; $\Pi_2^{(1)}$ meets $\Pi_2^{(3)}$ and $\Pi_2^{(4)}$ respectively at two distinct lines $\mathcal{D}_{1,3}$ and $\mathcal{D}_{1,4}$; $\Pi_2^{(2)}$ meets $\Pi_2^{(3)}$ and $\Pi_2^{(4)}$ respectively at two distinct lines $\mathcal{D}_{2,3}$ and $\mathcal{D}_{2,4}$; $\mathcal{D}_{1,3}$, $\mathcal{D}_{1,4}$, $\mathcal{D}_{2,3}$ and $\mathcal{D}_{2,4}$ passes through Π_0). For example, for the two quadrics defined by $f_{\mathcal{X}} = x_0 x_1 + x_2 x_3$ and $f_{\mathcal{Q}} = x_3 x_0 + x_1 x_2$ we have $|\mathcal{X} \cap \mathcal{Q}| = 4q^2 + 1$. Otherwise, we get $|\mathcal{X} \cap \mathcal{Q}| \leq 3q^2 + q + 1$.

(ii) If $w(\mathcal{X}, \mathcal{Q}) = 3$, from Lemma 3.3, we get r$(\mathcal{X})$=r$(\mathcal{Q})$=3. Here $\mathcal{Q} \cap \mathbb{E}_2(\mathbb{F}_q)$ and $\mathcal{X} \cap \mathbb{E}_2(\mathbb{F}_q)$ are two irreducible curves (conics). Therefore from the theorem of Bézout (the fact that two plane conics have exactly four common points or less than three points) and Lemma 3.2, we deduce that $|\mathcal{X} \cap \mathcal{Q}| = 4q^2 + q + 1$, or otherwise $|\mathcal{X} \cap \mathcal{Q}| \leq 3q^2 + q + 1$.

3.2. Section of $\mathcal{X}$ by the quadrics $\mathcal{Q}$ containing hyperplanes: g $(\mathcal{Q})$=3.
This paragraph deals with the section of $\mathcal{X}$ by $\mathcal{Q}$ in the case g$(\mathcal{Q})$=3 i.e. $\mathcal{Q}$ contains a hyperplane. We have two cases: r$(\mathcal{Q})$=1, or r$(\mathcal{Q})$=2 with $\mathcal{Q}$ a pair of hyperplanes.

3.2.1. $\mathcal{Q}$ is a quadric with r$(\mathcal{Q})$=1: $\mathcal{Q} = \Pi_3 P_0$. Here $\mathcal{Q}$ is a repeated hyperplane H. From [15, p.264] a degenerate quadric of rank r of $\mathbb{P}^n(\mathbb{F}_q)$ meets a hyperplane in a quadric of $\mathbb{P}^{n-1}(\mathbb{F}_q)$ of rank r, $r-1$, or $r-2$. From [12, pp.299-300] a non-degenerate quadric of $\mathbb{P}^n(\mathbb{F}_q)$ meets a hyperplane in either a non-degenerate quadric or in a degenerate quadric of rank $n-1$ in $\mathbb{P}^{n-1}(\mathbb{F}_q)$; and from [16, pp.191-192] this degenerate quadric is of the same type as the non-degenerate (in the sense of the definition 2.1). Let $H \subset \mathbb{P}^4(\mathbb{F}_q)$ be a hyperplane. If $\mathcal{X}$ is a non-degenerate quadric, from [12, pp.299-300] , we get

$$(\star) \quad |\mathcal{X} \cap H| = \begin{cases} (q+1)^2 \quad \text{or} \quad q^2 + 1 & \text{if H is not tangent to } \mathcal{X}, \\ q^2 + q + 1 & \text{if H is tangent to } \mathcal{X}. \end{cases}$$

(3.1) If $\mathcal{X}$ is a degenerate quadric, from [15, p.264] we get $|\mathcal{X} \cap H| \leq 2q^2 + q + 1$.

3.2.2. $\mathcal{Q}$ is a quadric with r$(\mathcal{Q})$=2 and $\mathcal{Q}$ is a pair of hyperplanes. Let H_1 and H_2 be the two distinct hyperplanes generating $\mathcal{Q}$ (i.e. $\mathcal{Q} = H_1 \cup H_2$), $\mathcal{P} = H_1 \cap H_2$ be the plane of the intersection of H_1 and H_2. Let $\hat{\mathcal{X}}_1 = H_1 \cap \mathcal{X}$, and $\hat{\mathcal{X}}_2 = H_2 \cap \mathcal{X}$, we have:

(3.2) $$|\mathcal{Q} \cap \mathcal{X}| = |H_1 \cap \mathcal{X}| + |H_2 \cap \mathcal{X}| - |\mathcal{P} \cap \mathcal{X}|.$$

(3.3) $$\mathcal{P} \cap \mathcal{X} = \mathcal{P} \cap \hat{\mathcal{X}}_1 = \mathcal{P} \cap \hat{\mathcal{X}}_2.$$

3.2.2.1 If $\mathcal{X}$ is non-degenerate (i.e. parabolic).

(i) In the case where each hyperplane is tangent to $\mathcal{X}$, we know that $\hat{\mathcal{X}}_1$, and $\hat{\mathcal{X}}_2$ are quadric cones of rank 3. We get that $|\hat{\mathcal{X}}_1| = |\hat{\mathcal{X}}_2| = q^2 + q + 1$. From [15, p.264] and table 2, we deduce that $|\mathcal{P} \cap \hat{\mathcal{X}}_1| \geq 1$. Therefore, from relation 3.2, we deduce that $|\mathcal{X} \cap \mathcal{Q}| \leq 2q^2 + 2q + 1$.

(ii) In the case where one hyperplane H_2 is tangent to $\mathcal{X}$, and the second hyperplane H_1 is non-tangent to $\mathcal{X}$, $\hat{\mathcal{X}}_1$ is a non-degenerate quadric surface in $\mathbb{P}^3(\mathbb{F}_q)$: elliptic or hyperbolic.

–If $\hat{\mathcal{X}}_1$ is an elliptic quadric, from [12, pp.299-300] [16, pp.191-192] and table 2, we get that $\mathcal{P} \cap \hat{\mathcal{X}}_1$ is either a plane conic (parabolic), or a single point according to $\mathcal{P}$ being non-tangent or tangent to $\hat{\mathcal{X}}_1$. Therefore, from relations (3.3), (3.2) and table 2, we deduce that $|\mathcal{X} \cap \mathcal{Q}| \leq 2q^2 + q + 1$.

–If $\hat{\mathcal{X}}_1$ is an hyperbolic quadric, from [12, pp.299-300] [16, pp.191-192] and table 2, we get that $\mathcal{P} \cap \hat{\mathcal{X}}_1$ is either a plane conic (parabolic), or a pair of two distinct lines according to $\mathcal{P}$ being non-tangent or tangent to $\hat{\mathcal{X}}_1$. Therefore, from (3.3) and (3.2), we deduce that $|\mathcal{X} \cap \mathcal{Q}| \leq 2q^2 + q + 1$.

(iii) In the case where each hyperplane is non-tangent to $\mathcal{X}$, $\hat{\mathcal{X}}_1$, and $\hat{\mathcal{X}}_2$ are non-singular quadric surfaces: elliptic or hyperbolic. They can be of the same type or of different types.

–If one of the two quadrics is elliptic, from [12, pp.299-300] and table 2, we deduce that $|\mathcal{P} \cap \hat{\mathcal{X}}_1| \geq 1$. Therefore from (3.2) and ($\star$) we get that $|\mathcal{X} \cap \mathcal{Q}| \leq 2q^2 + 2q + 1$.

–If the two quadrics are hyperbolic, from [12, pp.299-300] [16, pp.191-192] and table 2, we deduce that $|\mathcal{P} \cap \hat{\mathcal{X}}_1| \geq q + 1$. Therefore, from (3.3) and (3.2), we get $|\mathcal{X} \cap \mathcal{Q}| \leq 2q^2 + 3q + 1$. And this upper bound is reached when $\mathcal{P}$ is non-tangent to $\hat{\mathcal{X}}_1$ and $\hat{\mathcal{X}}_2$. For example, this upper bound is reached for the two quadrics defined by $f_{\mathcal{X}} = x_0 x_1 + x_2 x_3 + x_4^2$ and $f_{\mathcal{Q}} = (x_0 + x_1)(x_2 + x_3)$ with $\mathrm{car}(\mathbb{F}_q) \neq 2$

3.2.2.2 If $\mathcal{X}$ is degenerate (i.e. rank $(\mathcal{X})$=3,4). .

Here we have $w(\mathcal{X}, \mathcal{Q}) = 3, 4, 5$; the case $w(\mathcal{X}, \mathcal{Q}) = 5$ is solved by Corollary 3.1.

(i) If $w(\mathcal{X}, \mathcal{Q}) = 4$. Suppose rank$(\mathcal{X})$=3. From [5, IV-C] we get $|\mathcal{X} \cap \mathcal{Q} \cap \mathbb{E}_3(\mathbb{F}_q)| = 4q + 1$ or $|\mathcal{X} \cap \mathcal{Q} \cap \mathbb{E}_3(\mathbb{F}_q)| \leq 3q + 1$. From Lemma 3.2 we deduce that either $|\mathcal{X} \cap \mathcal{Q}| = 4q^2 + q + 1$ in the case where $\mathcal{Q}$ is union of two hyperplanes (non-tangent) each through a pair of planes of $\mathcal{X}$ and the plane of intersection of the two hyperplanes intersecting $\mathcal{X}$ in a line or $|\mathcal{X} \cap \mathcal{Q}| \leq 3q^2 + q + 1$ otherwise.

If rank $(\mathcal{X})$=4 and $g(\mathcal{X}) = 1$ from [5, IV-B] we get $|\mathcal{X} \cap \mathcal{Q} \cap \mathbb{E}_3(\mathbb{F}_q)| \leq 2(q + 1)$. Therefore Lemma 3.2 gives $|\mathcal{X} \cap \mathcal{Q}| \leq 2q^2 + 2q + 1$.

If rank $(\mathcal{X})$=4 and $g(\mathcal{X}) = 2$, the result of [5, IV-B] gives $|\mathcal{X} \cap \mathcal{Q} \cap \mathbb{E}_3(\mathbb{F}_q)| = 4q$ or $|\mathcal{X} \cap \mathcal{Q} \cap \mathbb{E}_3(\mathbb{F}_q)| \leq 3q + 1$. Therefore from Lemma 3.2, we get either $|\mathcal{X} \cap \mathcal{Q}| = 4q^2 + 1$ when each hyperplane is tangent to $\mathcal{X}$ with the plane of intersection meeting $\mathcal{X}$ at two lines or $|\mathcal{X} \cap \mathcal{Q}| \leq 3q^2 + q + 1$ otherwise.

(ii) If $w(\mathcal{X}, \mathcal{Q}) = 3$, from Lemma 3.3, we get r$(\mathcal{X})$=3. Here, the quadric $\mathcal{Q}$ describes a pair of lines in $\mathbb{E}_2(\mathbb{F}_q)$. The number of points in the intersection of two secant lines with a conic (non-singular plane quadric) is exactly four or less than three. From table 2 and Lemma 3.2 we get that either $|\mathcal{X} \cap \mathcal{Q}| = 4q^2 + q + 1$ or $|\mathcal{X} \cap \mathcal{Q}| \leq 3q^2 + q + 1$.

3.3. Section of $\mathcal{X}$ by the quadrics $\mathcal{Q}$ containing lines as generators: $g(\mathcal{Q})=1$. In this section we estimate the number of points in the intersection of $\mathcal{X}$ and a quadric $\mathcal{Q}$ with $g(\mathcal{Q}) = 1$. In this case $\mathcal{Q}$ is the degenerate quadric $\Pi_0\mathcal{E}_3$ or the non-degenerate quadric $\mathcal{P}_4$.

3.3.1. $\mathcal{Q}$ *is a quadric with* $r(\mathcal{Q})=4$: $\mathcal{Q} = \Pi_0\mathcal{E}_3$.

   (i) For $\mathcal{X}$ a non-degenerate quadric, we get two possibilities:

   – If there is a line of $\mathcal{X} \cap \mathcal{Q}$ through the vertex Π_0 of the cone $\mathcal{Q} = \Pi_0\mathcal{E}_3$, it is obvious that this vertex is a point of $\mathcal{X}$. Therefore there are at most $q + 1$ lines of the cone $\mathcal{Q}$ through Π_0 contained in $\mathcal{X}$; the other lines of $\mathcal{Q}$ meet $\mathcal{X}$ in at most two points each. Thus, we get $|\mathcal{X} \cap \mathcal{Q}| \leq 2q^2 + 2q + 1$.

   – If there is no line of $\mathcal{X}\cap\mathcal{Q}$ through the vertex of the cone $\mathcal{Q} = \Pi_0\mathcal{E}_3$, each line of $\mathcal{Q}$ intersecting $\mathcal{X}$ in at most two points, we deduce that $|\mathcal{X} \cap \mathcal{Q}| \leq 2(q^2 + 1)$.

   (ii) For $\mathcal{X}$ a degenerate quadric (i.e. $r(\mathcal{X}) = 3, 4$), from Lemma 3.3, we have $w(\mathcal{X}, \mathcal{Q}) = 4, 5$. The case $w(\mathcal{X}, \mathcal{Q}) = 5$ follows from Corollary 3.1. Let us consider now that $w(\mathcal{X}, \mathcal{Q}) = 4$.

   If $r(\mathcal{X})=3,4$ and $g(\mathcal{X})=2$, or $r(\mathcal{X})=4$ and $g(\mathcal{X})=1$, from [5, IV-D-2, IV-E-2] we get $|\mathcal{X} \cap \mathcal{Q} \cap \mathbb{E}_3(\mathbb{F}_q)| \leq 2(q + 1)$. Thus, we deduce from Lemma 3.2 that $|\mathcal{X} \cap \mathcal{Q}| \leq 2q^2 + 2q + 1$.

REMARK 3.4. *Let $\mathcal{Q}$ be a non-degenerate quadric (i.e. $\mathcal{Q} = \mathcal{P}_4$) and $r(\mathcal{X})=3,4$. The cases $r(\mathcal{X})=3,4$ and $g(\mathcal{X})=2$, correspond to the first part of 3.1.2. with $\mathcal{Q}$ at the place of $\mathcal{X}$. In the same way $r(\mathcal{X})=4$ and $g(\mathcal{X})=1$, corresponds to 3.3.1.(i).*

3.3.2. *Intersection of two non-degenerate quadrics.* Here we study the number of points in the intersection of two non-degenerate quadrics $\mathcal{X}$ and $\mathcal{Q}$.

PROPOSITION 3.5. *Let $\mathcal{X}$ and $\mathcal{Q}$ be two non-degenerate quadrics in $\mathbb{P}^4(\mathbb{F}_q)$. If there is no line in $\mathcal{X} \cap \mathcal{Q}$, then $|\mathcal{X} \cap \mathcal{Q}| \leq 2(q^2 + 1)$.*

Proof We use the fact that α_q is the number of lines of $\mathcal{X}$, each one of them meeting $\mathcal{Q}$ in at most two points. Moreover there pass exacly $q + 1$ lines through each point contained in $\mathcal{X}$.

Now we will study the section of two non-degenerate quadrics containing a common line. Let us consider a line $\mathcal{D}$ contained in $\mathcal{X} \cap \mathcal{Q}$ and $\mathcal{P}$ a plane through $\mathcal{D}$. By the principle of duality in the projective space [14, pp.49-51], or [7, p.33, Theorem 3.1 p.85], we deduce that there are exactly $q+1$ hyperplanes H_i through $\mathcal{P}$. These $q + 1$ hyperplanes H_i generate $\mathbb{P}^4(\mathbb{F}_q)$.

For $i = 1, ..., q + 1$, we denote $\hat{\mathcal{X}}_i = H_i \cap \mathcal{X}$ and $\hat{\mathcal{Q}}_i = H_i \cap \mathcal{Q}$. Thus, we get:

$$(3.4) \qquad |\mathcal{X} \cap \mathcal{Q}| \leq |\hat{\mathcal{X}}_1 \cap \hat{\mathcal{Q}}_1| + \sum_{i=2}^{q+1} |(\hat{\mathcal{X}}_i \cap \hat{\mathcal{Q}}_i) - \mathcal{D}|.$$

From [12, pp.299-300] we deduce that $\hat{\mathcal{X}}_i$ and $\hat{\mathcal{Q}}_i$ are quadrics of rank 3 or 4 in $\mathbb{P}^3(\mathbb{F}_q)$. Since they contain the line $\mathcal{D}$, they can not be elliptic. They are either hyperbolic or cone quadrics (of rank 3). Thus one has to study the three types of intersection in $\mathbb{P}^3(\mathbb{F}_q)$ given by the table 3.

LEMMA 3.6. *Let $\mathcal{X}$ and $\mathcal{Q}$ be two non-degenerate quadrics in $\mathbb{P}^n(\mathbb{F}_q)$, and $\mathcal{K}$ a linear space of codimension 2. Then there exist at most two hyperplanes H_i, $i = 1, 2$ through $\mathcal{K}$ such that $\hat{\mathcal{X}}_i = \hat{\mathcal{Q}}_i$.*

Types	$\hat{\mathcal{X}}_i \cap \hat{\mathcal{Q}}_i$
1	(hyperbolic quadric) $\cap$ (cone quadric)
2	(cone quadric) $\cap$ (cone quadric)
3	(hyperbolic quadric) $\cap$ (hyperbolic quadric)

TABLE 3. Intersection of $\hat{\mathcal{X}}_i \cap \hat{\mathcal{Q}}_i$ in $\mathbb{P}^3(\mathbb{F}_q)$

Proof Let

$$\mathcal{Q} = \sum_{0 \le i \le j \le n} a_{ij} x_i x_j \text{ and } \mathcal{X} = \sum_{0 \le i \le j \le n} a'_{ij} x_i x_j.$$

Without loss of generality, we can choose a system of coordinates, such that $H_1 = \{x_n = 0\}$, $H_2 = \{x_{n-1} = 0\}$. For $H_1 \cap \mathcal{X} = H_1 \cap \mathcal{Q}$, we get that $a_{ij} = a'_{ij}$ except maybe for (i, n) with $i = 0, ..., n$. For $H_2 \cap \mathcal{X} = H_2 \cap \mathcal{Q}$, we get that $a_{ij} = a'_{ij}$ except maybe for $(i, n-1)$ with $i = 0, ..., n-1$ and $(n-1, n)$. Therefore, we deduce that $a_{ij} = a'_{ij}$ except for $(i, j) = (n-1, n)$; and there exists $(\alpha, \beta) \in \mathbb{F}_q^2$ $(\alpha \ne \beta)$ such that:

$$\begin{cases} \mathcal{Q} = \mathcal{Q}_0(x_0, x_1, ..., x_{n-1}, x_n) + \alpha x_{n-1} x_n \\ \mathcal{X} = \mathcal{Q}_0(x_0, x_1, ..., x_{n-1}, x_n) + \beta x_{n-1} x_n. \end{cases}$$

Let us suppose that there is a third hyperplane H_3 through $\mathcal{K}$ such that $H_3 \cap \mathcal{X} = H_3 \cap \mathcal{Q}$. We can suppose that $H_3 = \{a x_{n-1} + b x_n = 0\}$ and $b \ne 0$, therefore $x_n = -\frac{a}{b} x_{n-1}$. From the above system and the fact that $H_3 \cap \mathcal{X} = H_3 \cap \mathcal{Q}$ we deduce that $-\alpha \frac{a}{b} x_{n-1}^2 = -\beta \frac{a}{b} x_{n-1}^2$. Since $\alpha \ne \beta$, we deduce that $a = 0$, which leads to $H_3 = H_2$.

PROPOSITION 3.7. *Let $\mathcal{X}$ and $\mathcal{Q}$ be two non-degenerate quadrics in $\mathbb{P}^4(\mathbb{F}_q)$ containing a line $\mathcal{D}$. There is a plane $\mathcal{P}$ containing $\mathcal{D}$ such that $\hat{\mathcal{X}}_i \ne \hat{\mathcal{Q}}_i$ for $i = 1, ..., q + 1$.*

Proof Just consider a plane which meets the two quadrics in different point sets (such a plane exists since the two quadrics do not coincide).

REMARK 3.8. *As suggested by professor Hendrick Van Maldeghem, the results of [5,§IV-D-4 and IV-E-1] have been revised and completed. In fact if two cone quadrics (of rank 3) in $\mathbb{P}^3(\mathbb{F}_q)$ intersect in at least two lines, then they have the same vertex (which is not in [5,§IV-D-4). Hence it suffices to look at the size of the intersection of two conics. This can be 2, 3 or 4. Consequently, the cones meet in exactly $2q + 1$, $3q + 1$ and $4q + 1$. Secondly two hyperbolic quadrics in $\mathbb{P}^3(\mathbb{F}_q)$ can also intersect in exactly three lines (two lines in a regulus and the third in the complementary regulus).*

From the results of [5, §IV], Remark 3.8 and table 3 above, we deduce the following table 4.

Let us explain the table 4. For type 1, the intersection of a hyperbolic quadric and a cone quadric, contains at most two lines; therefore $3q$ and $2q + 1$ are respectively the maximum number of this intersection when containing (exactly or) at most two lines or exactly one line. From the above remark, types 2 and 3 are

Types	4 lines	3 lines	2 lines	1 line
1			$3q$	$2q+1$
2	4q+1	3q+1	$2q+1$	$2q+1$
3	$4q$	3q+1	$3q+1$	$2(q+1)$

TABLE 4. Number of points and lines in $\hat{\mathcal{X}}_i \cap \hat{\mathcal{Q}}_i$

described as before.

Now an estimation on the number of points in the intersection of the two non-degenerate quadrics $\mathcal{X}$ and $\mathcal{Q}$ containing a common line is reduced to the two following simple cases.

If $\mathcal{X} \cap \mathcal{Q}$ contains exactly one common line: Each $\hat{\mathcal{X}}_i \cap \hat{\mathcal{Q}}_i$ contains exactly one line, and from the table 4, we get for $i = 1, ..., q+1$ $|\hat{\mathcal{X}}_i \cap \hat{\mathcal{Q}}_i| \leq 2(q+1)$. Therefore from relation (3.4), we deduce that $|\mathcal{X} \cap \mathcal{Q}| \leq q^2 + 3q + 2$.

If $\mathcal{X} \cap \mathcal{Q}$ contains at least two common lines: We distinguish the two following cases:

1. In the case where $\mathcal{X} \cap \mathcal{Q}$ contains only skew lines: from table 4, we get for $i = 1, ..., q+1$ that $|\hat{\mathcal{X}}_i \cap \hat{\mathcal{Q}}_i| \leq 2q+1$ for types 1 and 2. Indeed, if $\hat{\mathcal{X}}_i \cap \hat{\mathcal{Q}}_i$ contains more than one line, from the fact that one of the two quadrics $\hat{\mathcal{X}}_i$ or $\hat{\mathcal{Q}}_i$ is a cone quadric, they are secant. For type 3, if $\hat{\mathcal{X}}_i \cap \hat{\mathcal{Q}}_i$ contains exactly four common lines, then it contains necessarily two secant lines. This is a contradiction. From Remark 3.8, $\hat{\mathcal{X}}_i \cap \hat{\mathcal{Q}}_i$ can only contain at most two lines. Thus, from table 4, $3q + 1$ is an upper bound for the number of points in $\hat{\mathcal{X}}_i \cap \hat{\mathcal{Q}}_i$. The lines of $\hat{\mathcal{X}}_i \cap \hat{\mathcal{Q}}_i$ are skew; so they belong to the same regulus. From [5, IV-E-1] we get $|\hat{\mathcal{X}}_i \cap \hat{\mathcal{Q}}_i| \leq 2(q+1)$. Finally, from relation (3.4) we deduce that $|\mathcal{X} \cap \mathcal{Q}| \leq q^2 + 3q + 2$.

2. In the case where there exist secant lines in $\mathcal{X} \cap \mathcal{Q}$, let $\{\mathcal{D}_1, \mathcal{D}_2\}$ denote a pair of secant lines and $\mathcal{P}$ be the plane generated by this pair of secants. Let

$$\mathcal{Q} = \sum_{0 \leq i \leq j \leq 4} a_{ij} x_i x_j \text{ and } \mathcal{X} = \sum_{0 \leq i \leq j \leq 4} a'_{ij} x_i x_j.$$

(i) If there are exactly two hyperplanes H_i $i = 1, 2$ such that $\hat{\mathcal{X}}_i = \hat{\mathcal{Q}}_i$, then we can choose a system of coordinates, such that $H_1 = \{x_4 = 0\}$ and $H_2 = \{x_3 = 0\}$. From the proof of Lemma 3.6, there exists $(\alpha, \beta) \in \mathbb{F}_q^2$ $(\alpha \neq \beta)$ such that:

$$\begin{cases} \mathcal{Q} = \mathcal{Q}_0(x_0, x_1, x_2, x_3, x_4) + \alpha x_3 x_4 \\ \mathcal{X} = \mathcal{Q}_0(x_0, x_1, x_2, x_3, x_4) + \beta x_3 x_4. \end{cases}$$

We have $\mathcal{X} \cap \mathcal{Q} = (H_1 \cap \mathcal{X}) \cup (H_2 \cap \mathcal{X})$ and from the relation (3.2) we get that $|\mathcal{X} \cap \mathcal{Q}| \leq 2q^2 + 2q + 1$.

(ii) If there is exactly one hyperplane H_1 such that $\hat{\mathcal{X}}_1 = \hat{\mathcal{Q}}_1$, then by the same reasoning as above, we can choose two distinct linear forms $h_1(x_0, x_1, x_2, x_3, x_4)$ and $h_2(x_0, x_1, x_2, x_3, x_4)$ such that:

$$\begin{cases} \mathcal{Q} = \mathcal{X}_0(x_0, x_1, x_2, x_3) + x_4 h_1(x_1, x_2, x_3, x_4) \\ \mathcal{X} = \mathcal{X}_0(x_0, x_1, x_2, x_3) + x_4 h_2(x_1, x_2, x_3, x_4). \end{cases}$$

We have $\mathcal{X} \cap \mathcal{Q} = (H_1 \cap \mathcal{X}) \cup (H_2 \cap \mathcal{X} \cap \mathcal{Q})$ where H_2 is the hyperplane defined by the linear form $h_1 - h_2$. We also have $|H_2 \cap \mathcal{X} \cap \mathcal{Q}| \leq 4q + 1$ from table 4 and $|H_1 \cap \mathcal{X}| \leq q^2 + 2q + 1$. Therefore we get that $|\mathcal{X} \cap \mathcal{Q}| \leq q^2 + 6q + 2$.

(iii) If for $i = 1, ..., q + 1$ we get $\hat{\mathcal{X}}_i \neq \hat{\mathcal{Q}}_i$, one has two possibilities: If the $q + 1$ hyperplanes H_i are all non-tangent to $\mathcal{X}$, $\hat{\mathcal{X}}_i$ are non-degenerate quadrics and only types 1 and 3 of the table 3 can appear. From table 4 and Remark 3.8, we deduce that $|\hat{\mathcal{X}}_i \cap \hat{\mathcal{Q}}_i| \leq 4q$. If there exists a hyperplane tangent to $\mathcal{X}$, it is unique. Indeed the two secant lines $(\mathcal{D}_1)$ and $(\mathcal{D}_2)$ intersect in a point P. The plane $\mathcal{P} = < \mathcal{D}_1, \mathcal{D}_2 >$ defined by these two secant lines through P only lies in the tangent hyperplane to $\mathcal{X}$ in P. Let H_1 be this tangent hyperplane, $\hat{\mathcal{X}}_1$ is a cone quadric and $\hat{\mathcal{X}}_1 \cap \hat{\mathcal{Q}}_1$ is of type 1 or 2. Therefore we get $|\hat{\mathcal{X}}_1 \cap \hat{\mathcal{Q}}_1| \leq 4q + 1$. Finally by applying relation (3.4) where $(\mathcal{D}_1 \cup \mathcal{D}_2)$ taking the place of $\mathcal{D}$, we deduce that $|\mathcal{X} \cap \mathcal{Q}| \leq 2q^2 + 3q + 1$.

4. The structure of the code $C_2(\mathcal{X})$ defined on the quadric $\mathcal{X}$

When $\mathcal{X} = Z(F) \subset \mathbb{P}^n(\mathbb{F}_q)$ is a quadric, the map $c : \mathcal{F}_2 \longrightarrow \mathbb{F}_q^{|\mathcal{X}|}$ is not injective and we have: $\dim C_2(\mathcal{X}) = \frac{n(n+3)}{2}$. From the results of section 3, we deduce the following results.

PROPOSITION 4.1. *Let $\mathcal{Q}$ be a quadric in $\mathbb{P}^4(\mathbb{F}_q)$ and $\mathcal{X}$ another quadric in $\mathbb{P}^4(\mathbb{F}_q)$.*
–If $\mathcal{X}$ is non-degenerate, then

$$|\mathcal{X} \cap \mathcal{Q}| \leq 2q^2 + 3q + 1$$

and this bound is the best possible.
–If $\mathcal{X}$ is degenerate with $r(\mathcal{X}) = 3$, then

$$|\mathcal{X} \cap \mathcal{Q}| = 4q^2 + q + 1 \quad \text{or} \quad |\mathcal{X} \cap \mathcal{Q}| \leq 3q^2 + q + 1.$$

–If $\mathcal{X}$ is degenerate with $r(\mathcal{X})=4$ and $g(\mathcal{X}) = 2$, then

$$|\mathcal{X} \cap \mathcal{Q}| = 4q^2 + 1 \quad \text{or} \quad |\mathcal{X} \cap \mathcal{Q}| \leq 3q^2 + q + 1.$$

THEOREM 4.2. *The code $C_2(\mathcal{X})$ defined on the parabolic quadric $\mathcal{X}$ is a $[n, k, d]_q$-code where: $n = (q + 1)(q^2 + 1)$, $k = 14$, $d = q^3 - q^2 - 2q$.*
The minimum weight codewords of the code $C_2(\mathcal{X})$ correspond to:
–either quadrics which are union of two (non-tangent) hyperplanes each intersecting $\mathcal{X}$ at a hyperbolic quadric such that their plane of intersection intersects $\mathcal{X}$ in a plane conic.
–or quadrics with $r(\mathcal{Q}) = 3$ (i.e. $q + 1$ planes through a line) and each plane containing exactly two lines of $\mathcal{X}$.
–or non-degenerate (i.e. parabolic) quadrics containing two secant lines of $\mathcal{X}$ defining a plane contained in the $q + 1$ hyperplanes which has one tangent hyperplane, the q other non-tangent to $\mathcal{X}$, and with maximal hyperplane section.

THEOREM 4.3. *The code $C_2(\mathcal{X})$ defined on the degenerate quadric $\mathcal{X}$ in $\mathbb{P}^4(\mathbb{F}_q)$ with with $r(\mathcal{X})=3$ is a $[n, k, d]_q$-code where: $n = q^3 + q^2 + q + 1$, $k = 14$, $d = q^3 - 3q^2$.*
The minimum weight codewords of the code $C_2(\mathcal{X})$ correspond to:
–quadrics which are union of two hyperplanes (non-tangent) each through a pair of

planes and the plane of intersection of the two hyperplanes intersecting $\mathcal{X}$ in a line.
–quadrics with $r(\mathcal{Q}) = 3$ and a plane containing exactly four lines of $\mathcal{X}$.
The code $C_2(\mathcal{X})$ defined on the degenerate quadric $\mathcal{X}$ in $\mathbb{P}^4(\mathbb{F}_q)$ with $r(\mathcal{X})=3$ and $g(\mathcal{X}) = 2$ is a $[n, k, d]_q$-code where: $n = q^3 + 2q^2 + q + 1$, $k = 14$, $d = q^3 - 2q^2 + q$.
The minimum weight codewords of the code $C_2(\mathcal{X})$ correspond to:
–quadrics which are union of two tangent hyperplanes to $\mathcal{X}$ and the plane of inter-section of the two hyperplanes meeting $\mathcal{X}$ at two secant lines.
–quadrics with $r(\mathcal{Q}) = 4$ and $g(\mathcal{Q}) = 2$ containing exactly four planes $\Pi_2^{(1)}$, $\Pi_2^{(2)}$, $\Pi_2^{(3)}$ and $\Pi_2^{(4)}$ of $\mathcal{X}$ with the following configuration: ($\Pi_2^{(1)}$ and $\Pi_2^{(2)}$ meet at the point Π_0, and $\Pi_2^{(3)}$ and $\Pi_2^{(4)}$ meet at Π_0; $\Pi_2^{(1)}$ meets $\Pi_2^{(3)}$ and $\Pi_2^{(4)}$ respectively at two distinct lines $\mathcal{D}_{1,3}$ and $\mathcal{D}_{1,4}$; $\Pi_2^{(2)}$ meets $\Pi_2^{(3)}$ and $\Pi_2^{(4)}$ respectively at two distinct lines $\mathcal{D}_{2,3}$ and $\mathcal{D}_{2,4}$; $\mathcal{D}_{1,3}$, $\mathcal{D}_{1,4}$, $\mathcal{D}_{2,3}$ and $\mathcal{D}_{2,4}$ pass through Π_0).

PROPOSITION 4.4. *Let $\mathcal{Q}$ be a quadric in $\mathbb{P}^4(\mathbb{F}_q)$ and $\mathcal{X}$ a degenerate quadric in $\mathbb{P}^4(\mathbb{F}_q)$ of rank $(\mathcal{X}) = 4$ with $g(\mathcal{X}) = 1$. We get :*

$$|\mathcal{X} \cap \mathcal{Q}| \leq 3q^2 + q + 1.$$

The code $C_2(\mathcal{X})$ defined on $\mathcal{X}$ is a $[n, k, d]_q$-code where $n = q^3 + q + 1$, $k = 14$, $d \geq q^3 - 3q^2$.

THEOREM 4.5. *Let $\mathcal{X}$ be a quadric in $\mathbb{P}^n(\mathbb{F}_q)$ where $n \geq 4$. All the weights w_i of the code $C_2(\mathcal{X})$ defined on $\mathcal{X}$ are divisible by q.*

Proof: Let F and f be two forms of degree 2 in $n + 1$ indeterminates with $n \geq 4$ and N the number of common zeros of F and f in $\mathbb{F}_q^{n+1}$. By the theorem of Ax [2, p.260] N is divisible by q since $n + 1 > 2 + 2$. On the other hand, F and f are homogeneous polynomials, therefore $N - 1$ is divisible by $q - 1$. Let $\mathcal{X}$ and $\mathcal{Q}$ be the projective quadrics associated to F and f, one has $|\mathcal{X} \cap \mathcal{Q}| = \frac{N-1}{q-1}$. Let $M = \frac{N-1}{q-1}$, one has $N \equiv (1 - M)$ (mod. q). Since N is divisible by q, we deduce that $M \equiv 1$ (mod. q). By the theorem of Ax [2, p.260] again, we get that the number of zeros of the polynomials F in $\mathbb{F}_q^{n+1}$ is divisible by q, so that $|\mathcal{X}| \equiv 1$ (mod. q). The weight of a codeword associated to the quadric $\mathcal{Q}$ is equal to:

$$w = |\mathcal{X}| - |\mathcal{X} \cap \mathcal{Q}| = |\mathcal{X}| - M \equiv 0 (\text{mod.} q).$$

In $\mathbb{P}^3(\mathbb{F}_q)$, the above theorem fails. For instance let us consider the code $C_2(\mathcal{X})$ constructed on a hyperbolic quadric. From [5, IV-E, p.862] two hyperbolic quadrics meeting in four lines have exactly $4q$ points. Therefore there is a codeword of weight equal to $q^2 - 2q + 1$ which is not divisible by q.

REMARK 4.6. *From the study of the parameters of the codes $C_2(\mathcal{X})$, we assert that the performances of those defined on non-degenerate quadrics are better than the ones defined on degenerate quadrics.*

Conjecture: Let $\mathcal{Q}_1$ and $\mathcal{Q}_2$ be two quadrics in $\mathbb{P}^n(\mathbb{F}_q)$ with no common hyperplane. Then

$$|\mathcal{Q}_1 \cap \mathcal{Q}_2| \leq 4q^{n-2} + \pi_{n-3}.$$

And this bound is the best possible.

REMARK 4.7. *The preceding conjecture is true for $n=2,3$, and 4.*

With the convention that $\pi_{-1} = 0$, in the case $n = 2$, this is the theorem of Bezout. For $n = 3$ see [5, V-VI]. For $n = 4$, see Theorem 4.1.

The results obtained on the intersection of quadrics in the projective spaces $\mathbb{P}^3(\mathbb{F}_q)$ [5] and $\mathbb{P}^4(\mathbb{F}_q)$ are generalized to projective spaces of any dimension $n > 4$ (odd and even) in another paper which should be submitted for publication.

Acknowledgment The author would like to thank Prof. F. Rodier and Prof. H. Van Maldeghem for their precious remarks and comments on a preliminary version of this paper.

References

[1] Y. Aubry, Reed-Muller codes associated to projective algebraic varieties. In "Algebraic Geomertry and Coding Theory ". (Luminy, France, June 17-21, 1991). Lecture Notes in Math. 1518, Springer-Verlag, Berlin, (1992), 4-17.

[2]J. Ax, Zeroes of polynomials over finite fields. Amer. J. Math., 86 (1964), 255-261.

[3] M. Boguslavsky, On the number of solutions of polynomial systems. Finite Fields and Their Applications 3 (1997), 287-299.

[4] I. M. Chakravarti, Some properties and applications of Hermitian varieties in finite projective space PG(N,q^2) in the construction of strongly regular graphs (two-class association schemes) and block designs, Journal of Comb. Theory, Series B, 11(3) (1971), 268-283.

[5] F. A. B. Edoukou, Codes defined by forms of degree 2 on quadric surfaces. IEEE. Trans. Inform. Theory, Vol. 54, N0. 2 (2008), 860-864.

[6] F. A. B. Edoukou, Codes correcteurs d'erreurs construits à partir des variétés algébriques, Ph. D. Thesis, Université de la Méditerranée (Aix-Marseille II), Marseille, France 2007.

[7] J. W. P. Hirschfeld, Projective Geometries Over Finite Fields (Second Edition) Clarendon Press. Oxford 1998.

[8] J. W. P. Hirschfeld, Finite projective spaces of three dimensions, Clarendon press. Oxford 1985.

[9] J. W. P. Hirschfeld, General Galois Geometies, Clarendon press. Oxford 1991.

[10] G. Lachaud, Number of points of plane sections and linear codes defined on algebraic varieties; in " Arithmetic, Geometry, and Coding Theory ". (Luminy, France, 1993), Walter de Gruyter, Berlin-New York, (1996), 77-104.

[11] D. B. Leep and L. M. Schueller, Zeros of a pair of quadric forms defined over finite field. Finite Fields and Their Applications 5 (1999), 157-176.

[12] E. J. F. Primrose, Quadrics in finite geometries, Proc. Camb. Phil. Soc., 47 (1951), 299-304.

[13] D. K. Ray-Chaudhuri, Some results on quadrics in finite projective geometrie based on Galois fields, Canadian J. Math. Vol. 14 (1962), 129-138.

[14] P. Samuel, Géométrie projective, Presses Universitaires de France, 1986.

[15] H. P. F. Swinnerton-Dyer, Rational zeros of two quadratics forms, Acta Arith. 9 (1964), 261-270.

[16] J. Wolfmann, Codes projectifs à deux ou trois poids associés aux hyperquadriques d'une géométrie finie, Discrete Mathematics 13 (1975), 185-211.

CNRS, INSTITUT DE MATHÉMATIQUES DE LUMINY, LUMINY CASE 907 - 13288 MARSEILLE CEDEX 9 - FRANCE
E-mail address: edoukou@iml.univ-mrs.fr

Contemporary Mathematics
Volume **487**, 2009

Curves of genus 2 with elliptic differentials and associated Hurwitz spaces

Gerhard Frey and Ernst Kani

1. Motivation

In the whole paper K is a field of finite type over its prime field K_0 of characteristic $p \geq 0$ and not equal to 2. As typical cases we can take K as a number field or as a function field of one variable over a finite field. With K_s we denote its separable closure. With n we denote an *odd* number larger than 5 and prime to p.

We begin by stating three motivations.

1.1. Rational Fundamental Groups. Let C be a smooth projective geometrically irreducible curve over K with function field $F(C)$.

We are interested in unramified Galois extension $U(C)$ of $F(C)$ with the additional property that K is algebraically closed in $U(C)$.

The finite quotients of the Galois group $G(U(C)/F(C))$ correspond to covers

$$ D \xrightarrow{f} C $$

where D is a smooth projective absolutely irreducible curve defined over K and f is Galois (and hence étale).

Base Points. Take $P \in C(K)$ and consider a connected Galois cover $D \xrightarrow{f} C$ in which P splits completely. Then D is absolutely irreducible. Moreover, normalizations of fibre products of such covers (for P fixed) satisfy the same conditions, and so there is a maximal unramified extension of $F(C)$ in which P is totally split, or equivalently, an absolutely irreducible étale (pro-)cover of C over K in which the fibre over P consists of K-rational points, which is Galois with group $\Pi_K(C, P)$, the K-rational fundamental group of C *with base point P*.

1991 *Mathematics Subject Classification.* Primary 14H30, 14G32; Secondary 11G99, 14G05.
Key words and phrases. Galois Representations, Fundamental Groups, Hurwitz Spaces.
In this paper we shall discuss results presented at the Conference on Arithmetic, Geometry, Cryptography and Coding Theory (AGCT 11). We would like to thank the organizers for the opportunity to participate in this very interesting and inspiring conference.
The second author gratefully acknowledges receipt of funding from the Natural Sciences and Engineering Research Council of Canada (NSERC).

1.1.1. *Questions*. Let g be a non-negative integer.

Q1: Are there finitely generated fields K, curves C of genus g defined over K, prime numbers ℓ and points $P \in C(K)$ such that $\Pi_K(C, P)$ has an ℓ-adic subrepresentation with infinite image?

Q2: Are there finitely generated fields K, curves C of genus g defined over K and points $P \in C(K)$ such that $\Pi_K(C, P)$ is infinite?

Q3: Are there finitely generated fields K and curves C of genus g defined over K such that there is a finite extension K_1/K and an infinite tower of projective absolutely irreducible Galois covers of C defined over K_1?

Q4: For given $n \in \mathbf{N}$ and fixed K and g are there curves C of genus g defined over K such that there are projective absolutely irreducible Galois covers of C of degree $\geq n$?

1.1.2. *Known Results*.

(1) For $g = 0$ there are no unramified extensions of C and so the answer to all questions is negative.
(2) The same result is true for $g = 1$, as was explained in [**FKV**], §2. However, in this case the negative answer to **Q4** (which implies negative answers to all others) relies on the deep result of Merel on the uniform bounds of torsion points of elliptic curves over fixed number fields.
(3) In [**FK1**] it is shown that there are examples for positive answers for Question **Q1**. In all these examples K has positive characteristic and the genus of C is larger than 2.[1]
(4) If $\operatorname{char}(K) \equiv 3 \bmod 4$ and K contains a fourth root of unity then for all $g \geq 3$ there is an explicitly given subvariety of positive dimension in the moduli space of curves of curves g and curves C such that **Q1** is true for curves with moduli point in this set ([**FK1**]).
(5) For any field K containing the fourth roots of unity and for all $g \geq 3$ there is an explicitly given subvariety of positive dimension in the moduli space of curves of curves g and curves C such that **Q2** is true for curves with moduli point in this set ([**FKV**]).
(6) There are examples (see [**I**] and [**Ki**]) for curves of genus 2 over finite fields for which **Q1** is true.

We want to ask in this paper whether there is, as in the case of curves of higher genus, a subspace of positive dimension in the moduli space of curves of genus 2 for which some of the questions can be answered positively. As we will see we can get partial answers by restricting us to curves of genus 2 with elliptic differentials, i.e. with Jacobians isogenous to a product of elliptic curves. In particular, **Q4** has a positive answer (Proposition 6.10) for any K, and **Q3** is true in positive characteristic (Corollary 7.15), in both cases for non-constant curves.

---

[1] We remark that if we replace C by a number field the Conjecture of Fontaine-Mazur implies the negative answer to **Q1**.

1.2. Isomorphic Galois Representations. One of the most efficient tools for studying the absolute Galois group G_K of K is the investigation of representations induced by the action on torsion groups of abelian varieties, or more generally, on cohomology groups attached to varieties. In fact the results mentioned in 1.1.2 are obtained in this way.

Conversely, it is natural to ask how much information about these representations is needed to characterize the isogeny class of abelian varieties. By celebrated results of Tate (for finite fields), Faltings (for number fields) and Zarhin (for fields of positive characteristic) we know that ℓ-adic representations attached to Tate modules as well as infinitely many representations on torsion points of different prime orders are enough. For a discussion and some generalizations we refer to [**FJ**].

This question is closely related to arithmetic properties of abelian varieties. A discussion in a rather general frame can be found in [**Fr3**].

In this paper we restrict ourselves to the case that we look at pairs of elliptic curves E and E' defined over K.

The representations of G_K induced on the points $E[n]$ respectively $E'[n]$ of order dividing n of E respectively E' are denoted by $\rho_{E/K,n}$ respectively $\rho_{E'/K,n}$.

CONJECTURE 1 (Darmon[**Da**]).
There is a number $n_0(K)$ such that for all elliptic curves E, E' over K and all $n \geq n_0(K)$ we get:

If $\rho_{E/K,n} \cong \rho_{E'/K,n}$ then E is isogenous to E'.

A variant is

CONJECTURE 2 (Darmon/Kani).
There is a number n_0 such that for every $n \geq n_0$ and every number field K there are, up to simultaneous twists, only finitely many pairs (E, E') of elliptic curves over K which are not isogenous yet whose mod n Galois representations are isomorphic: $\rho_{E/K,n} \cong \rho_{E'/K,n}$. Moreover, for prime numbers n the bound n_0 can be chosen to be 23.

Much weaker is

CONJECTURE 3 (Frey).
We fix an elliptic curve E_0/K. Let $S(E_0)$ denote the set of primes ℓ for which there exists a K-rational cyclic isogeny of degree ℓ of E_0.

There is a number $n_0(E, K)$ such that for all elliptic curves E over K and all $n \geq n_0(E, K)$ and prime to elements in $S(E_0)$ we get:

E is isogenous to E_0 iff $\rho_{E/K,n} \cong \rho_{E_0/K,n}$.

There is a close relation of these conjectures with other conjectures in Diophantine geometry. We shall give some Diophantine motivation for Conjecture 2 in Subsection 6.4. There it is shown that the conjecture predicts properties of rational points on explicitly given surfaces of general type, and Lang's conjecture together with a conjectural modular description of curves of genus ≤ 1 (cf. Conjecture 4) on these surfaces would imply Conjecture 2.

Conjecture 3 is true if the height conjecture for elliptic curves holds; cf. [**Fr2**]. In particular, it is true in the case that K is a function field over a finite field.

For $K = \mathbb{Q}$ the height conjecture is equivalent with the ABC-conjecture and with the degree conjecture for modular parameterizations of elliptic curves ([**Fr2**]).

Moreover, by switching from n to $2n$, we get an equivalence of Conjecture 3 for *even* numbers with the asymptotic Fermat conjecture.

1.3. A "Special" Hurwitz Space. Our third motivation comes from group theory. By Hurwitz spaces one parameterizes covers of a base curve with given monodromy group and given ramification type. Over $\mathbb{C}$ as ground field and (mainly) $\mathbb{P}^1$ as base curve this is classical theory due to Hurwitz[**Hu**]. By the work of Fulton, Fried–Völklein and many others, the theory of Hurwitz spaces is extended to a powerful tool in Arithmetic Geometry.

We want to study covers
$$\varphi : \mathbb{P}^1_K \to \mathbb{P}^1_K$$
of degree n which are primitive (i.e. have no proper intermediate subcovers) and which are ramified at 5 points with a certain ramification structure (as described below). In order to classify such covers, we shall use the following terminology.

DEFINITION 1.1. Let $f_i : C_i \to C$ be two covers of the curve C.

(a) f_1 is *isomorphic* to f_2 ($f_1 \cong f_2$) if there is an isomorphism $\alpha : C_1 \to C_2$ with $f_1 = f_2 \circ \alpha$.

(b) f_1 is *equivalent* to f_2 ($f_1 \sim f_2$) if there is an isomorphism $\alpha_1 : C_1 \to C_2$ and an automorphism $\alpha_2 : C \to C$ with $\alpha_2 \circ f_1 = f_2 \circ \alpha_1$.

Thus, the set of equivalence classes of covers $\varphi : \mathbb{P}^1_K \to \mathbb{P}^1_K$ as above has two points which can move "freely" and hence the associated Hurwitz space is a surface.

Let us now assume in addition that the (different) points $P_1, \ldots, P_5 \in \mathbb{P}^1(K_s)$ ramify with ramification order at most 2. We want to make sure that the monodromy group of φ is S_n and so we assume that P_5 has as ramification cycle a transposition. Together with the condition that φ is primitive this implies by group theory ([**Wi**] or [**Hup**]) that the Galois closure of φ is a regular cover with Galois group S_n; cf. [**Fr1**] or Corollary 3.7 below for more detail. Note that this condition clearly implies that P_5 has exactly one ramified extension in the cover.

For $1 \le i \le 5$ let r_i denote the number of *unramified* extensions of P_i after base change to K_s. By assumption $r_5 = n - 2$. The Hurwitz genus formula for φ yields:
$$-2 = -2n + 1 + \sum_{1 \le i \le 4} \frac{(n - r_i)}{2} = 1 - \sum_{1 \le i \le 4} \frac{r_i}{2}.$$
Thus $\sum r_i = 6$, and hence, since n is odd, r_i is odd and so there are three points, say P_1, P_2, P_3, with $r_i = 1$ and one point P_4 with $r_4 = 3$. Since this numbering convention will be in force throughout the paper, we formulate it here for easy reference

CONVENTION 1.2. The points $P_1, \ldots, P_4$ have been numbered in such a way that $r_1 = r_2 = r_3 = 1$ and $r_4 = 3$. Thus, there is a unique unramified extension above P_i for $i = 1, 2, 3$, and there are precisely 3 unramified extensions above P_4.

Since $\varphi : \mathbb{P}^1_K \to \mathbb{P}^1_K$ is defined over K, the discriminant divisor $\operatorname{disc}(\varphi) = P_1 + \ldots + P_5$ is K-rational. Since $n - 2 > 3$ the points P_5 and P_4 have to be K-rational and P_1, P_2, P_3 are permuted under the Galois action.

Summarizing, we have:

PROPOSITION 1.3. *Let φ be a primitive K-cover of degree n of the projective line by itself ramified in the following way.*

(*): φ is ramified in five points $P_1, \ldots, P_5$ in $\mathbb{P}^1(K_s)$ with ramification orders ≤ 2 and the ramification cycle of P_5 in the monodromy group is a transposition.

Then the Galois closure $\overline{\varphi} : \overline{C} \to \mathbb{P}^1_K$ of φ has Galois group S_n and is regular over K. The ramification structure induces five involutions $\sigma_i \in S_n$ which generate S_n with $\prod \sigma_i = 1$, where σ_5 is a transposition, σ_4 is a product of $\frac{n-3}{2}$ transpositions and each of σ_i, $i = 1, 2, 3$ is a product of $\frac{n-1}{2}$ transpositions (after a suitable numeration).

Thus, the ramification structure of φ (or, more precisely, of $\overline{\varphi}$) is described by the tuple

$$\mathbf{C} = (cl(\sigma_1), \ldots, cl(\sigma_5)) = ((2)^{\frac{n-1}{2}}, (2)^{\frac{n-1}{2}}, (2)^{\frac{n-1}{2}}, (2)^{\frac{n-3}{2}}, (2)),$$

where $cl(\sigma_i)$ denotes the conjugacy class of σ_i in S_n (which is determined by its cycle decomposition), and the symbol $(2)^k$ means that cycle decomposition of σ_i consists of k transpositions.

Conversely we have the following result from Galois theory.

PROPOSITION 1.4. *Suppose* $\overline{\varphi} : \overline{C} \to \mathbb{P}^1$ *is Galois cover with Galois group* $G \simeq S_n$ *which has ramification structure given by* $\mathbf{C}$ *as above. Then* $\overline{\varphi}$ *is the Galois closure of a (primitive) cover* $\varphi : \mathbb{P}^1 = H_x \backslash \overline{C} \to \mathbb{P}^1$ *of degree* n *which satisfies condition* (*). *Here* $H_x \simeq S_{n-1}$ *is the stabilizer subgroup of an element* $x \in \{1, \ldots, n\}$.

We can express the above in terms of the associated *Hurwitz spaces* as follows. Let $P_n^*(K)$ denote the set of *isomorphism classes* of primitive covers $\varphi : \mathbb{P}^1_K \to \mathbb{P}^1_K$ of degree n which satisfy (*), and let $H^{ab}(S_n, \mathbf{C})(K)$ denote the set of isomorphism classes of (regular) Galois covers $\overline{\varphi} : \overline{C} \to \mathbb{P}^1_K$ with group S_n and ramification structure $\mathbf{C}$ as above. Then the above Propositions 1.3 and 1.4 show that the map $\varphi \mapsto \overline{\varphi}$ induces a bijection $P_n^*(K) \xrightarrow{\sim} H^{ab}(S_n, \mathbf{C})(K)$. Now since S_n has no outer automorphisms (as $n \neq 6$), it follows that $H^{ab}(S_n, \mathbf{C})(K)$ can be identified with the set of K-rational points of the Hurwitz space $H^{in}(S_n, \mathbf{C})$ as defined in [**V1**], ch. 10.

Since S_n has trivial centre, the Hurwitz space $H^{in}(S_n, \mathbf{C})$ *finely* represents the associated Hurwitz functor $\mathcal{H}_{\mathbb{P}^1}(S_n, \mathbf{C})$ which classifies covers φ of type $\mathbf{C}$; cf. Wewers[**We**], p. 66, 70. However, this moduli space (which has dimension 5) is not quite the space we want because we want to study equivalence classes rather than isomorphism classes of covers. Thus, we want to consider instead the quotient space

$$H_n^* := H^{in}(S_n, \mathbf{C})/\text{Aut}(\mathbb{P}^1)$$

(which has dimension 2). It can be shown (cf. Proposition 4.8 below) that H_n^* *coarsely represents* the functor $\mathcal{H}_n^*$ which classifies equivalence classes of such covers.

REMARK 1.5. (a) Since H_n^* is only a coarse moduli space for the functor $\mathcal{H}_n^*$ (which is the quotient functor $\mathcal{H}_{\mathbb{P}^1}(S_n, \mathbf{C})/\text{Aut}(\mathbb{P}^1)$), it is more difficult to characterize the K-rational points of H_n^*. It is clear, however, that if $\varphi : \mathbb{P}^1_K \to \mathbb{P}^1_K$ is a K-cover of type (*), then φ defines a K-rational point on $H^{in}(S_n, \mathbf{C})$ and hence also on H_n^*. We thus have a natural map $\mathcal{H}_n^*(K) \to H_n^*(K)$ (which is a bijection if K is algebraically closed).

(b) Since $H^{in}(S_n, \mathbf{C})$ is finite and étale over the smooth affine space $U_5 \subset \mathbb{P}^5$ (cf. [**We**], p. 7), it follows that it is also smooth and affine. Thus, H_n^* is normal and affine. This argument, however, does not guarantee that $H^{in}(S_n, \mathbf{C})$ and H_n^* are connected; this will be established later by another method (cf. Theorem 4.9).

(c) The system $(\sigma_1, \cdots \sigma_5)$ is not rigid (in the sense of [**V1**], p. 38). Of course it is not difficult to see this by group theory but it is more difficult to compute the number of non-conjugate isomorphism classes. We shall get an answer by using geometrical arguments in Theorem 3.11.

We shall see in the next sections that points on H_n^* carry much more arithmetical information. In fact, we shall associate to such points covers of elliptic curves by curves of genus 2, and so one is naturally led to a *Hurwitz moduli functor* (in the spirit of Fulton [**Fu**]). A detailed discussion is given in [**Ka5**].

By applying this we can approach some of the problems mentioned in Remark 1.5 and see that there is a close connection between Subsection 1.2 and Subsection 1.3.

Acknowledgement. The authors would like to thank Helmut Völklein very much for numerous inspiring discussions and his great help to clarify the interplay between group theory and geometry. In addition, we would like to thank the referee for his careful reading of the manuscript and for pointing out numerous typos and other inaccuracies.

2. Genus 2 Covers of Elliptic Curves

2.1. Normalized Covers. Let C be a curve of genus 2 over K with Weierstraß points $\mathcal{W}_C = \{W_1, \ldots, W_6\}$ and *Weierstraß divisor* $W_C = W_1 + \ldots + W_6$. (Note that the divisor W_C is K-rational, whereas the individual points W_i need not be.)

We assume that C has an elliptic differential over K. This means that there is an elliptic curve E over K and a non-constant K-morphism

$$f : C \to E.$$

In addition we assume that f has odd[2] degree n.

DEFINITION 2.1. A cover

$$f : C \to E$$

(of the above type) is called *normalized* ([**Ka5**]) if

    (a) f is *minimal*, i.e. there is no non-trivial isogeny $\eta : E' \to E$ and a cover $f' : C \to E'$ with $f = \eta \circ f'$.

    (b) The norm (direct image) of W_C has the form

(2.1) $$f_* W_C = 3 \cdot 0_E + P_1' + P_2' + P_3'.$$

where P_i' are the points of (exact) order 2 of E.

Note that every minimal cover $f_0 : C \to E$ is equivalent to a normalized cover; more precisely, we have (cf. [**Ka5**], Proposition 2.2):

---

[2]In fact, the case that n is even is also important (see Sections 3 and 6.4), and is analyzed in [**Ka5**] and [**Di**].

LEMMA 2.2. *Assume that $f_0 : C \to E$ is a minimal cover of an elliptic curve E by a curve of genus 2 of odd degree n defined over K. Then there is a unique translation $\tau : E \to E$ such that $f = \tau \circ f_0$ is normalized.*

For our purposes it is important to note that a normalized cover also satisfies the condition

$$(2.2) \qquad f \circ \omega_C = [-1]_E \circ f,$$

where ω_C denotes the hyperelliptic involution of C; cf. [**Ka5**], Corollary 2.3. It follows that f factors over the hyperelliptic cover $\pi' : C \to C/\langle \omega \rangle = \mathbb{P}^1$.

By the Riemann-Hurwitz genus formula we see that the different of f has to be a divisor of degree 2.

2.2. Associated $\mathbb{P}^1$-Covers. We return to covers $\varphi : \mathbb{P}^1 \to \mathbb{P}^1$ of type (*) and discuss their relation to curves of genus 2 with elliptic differentials. Indeed, it is this connection which motivated the consideration of this specific group theoretical situation (*); cf. [**Fr1**].

To explain this connection, note first that the points $P_1, \cdots, P_4$ determine an elliptic curve E/K which is unique up to a quadratic twist. This curve is constructed as follows. Choose a coordinate x of $\mathbb{P}^1_K$ such that $(x)_\infty = P_4$, and a cubic polynomial $f_3(x) \in K[x]$ with $(f_3)_0 = P_1 + P_2 + P_3$. Then the (Weierstraß) equation

$$y^2 = f_3(x)$$

defines an elliptic curve E/K and a double cover $\pi : E \to \mathbb{P}^1$ which is ramified at $P_1, \ldots, P_4$ and which maps the zero-point $0_E \in E(K)$ to P_4. Thus, the group of 2-torsion points of E is $E[2] = \{0_E, P'_1, P'_2, P'_3\}$, where $P'_i \in \pi^{-1}(P_i)$ is the (unique) point above P_i. Note that P'_i is K-rational if and only if P_i is K-rational.

Let $C_0 = E \times_{\mathbb{P}^1} \mathbb{P}^1$ be the fibre product of E and $\mathbb{P}^1$ over $\mathbb{P}^1$ (with respect to the morphisms π and φ). It is immediate that C_0 is an irreducible curve with function field $F = K(X, Y)$, with X and Y satisfying a hyperelliptic equation of the form

$$Y^2 = f_3(X) \cdot g_3(X)$$

where g_3 is a polynomial of degree 3 corresponding to the 3 unramified extensions of P_4 in the cover φ (cf. Convention 1.2). Thus, the normalization $C = \widetilde{C_0}$ of C_0 is a curve of genus 2 with morphisms $f : C \to E$ and $\pi' : C \to \mathbb{P}^1$, and $f_*(W_C) = 3 \cdot P'_4 + P'_1 + P'_2 + P'_3$.

We thus obtain the following diagram of morphisms:

The point P_5 has two distinct extensions P and $P' = [-1]_E P$ to E, and there is exactly one point Q resp. $Q' = \omega Q$ over P resp. P' which is ramified of order 2. Hence the discriminant divisor of f is equal to $\pi^*(P_5)$ and the different divisor of f is equal to $Q + Q'$.

If we assume that φ is primitive, then it follows that f is also primitive (or minimal). Thus, we have

PROPOSITION 2.3. *Let* $\varphi : \mathbb{P}_K^1 \to \mathbb{P}_K^1$ *be a primitive cover of odd degree* n *satisfying* (∗) *and Convention 1.2 with respect to the points* $P_1, \ldots, P_5$, *and let* E/K *be an elliptic curve with a morphism* $\pi : E \to \mathbb{P}^1$ *of degree 2 which is ramified at* $P_1, \ldots, P_4$ *such that* $\pi(0_E) = P_4$. *Then the normalization* C *of the fibre product* $\mathbb{P}^1 \times_{\mathbb{P}^1} E$ *is a curve of genus 2 defined over* K *and the morphism*

$$f = \varphi_{(\pi)} : C = (\mathbb{P}^1 \times_{\mathbb{P}^1} E)^\sim \to E$$

is normalized and has discriminant $\mathrm{Disc}(f) = \pi^*(P_5)$.

Conversely, assume that $f_0 : C \to E$ is a minimal cover of an elliptic curve E by a curve of genus 2 of odd degree n defined over K. After an appropriate translation (Lemma 2.2) we get a normalized cover

$$f : C \to E.$$

Since f satisfies (2.2), we can pass to the respective quotients and obtain a primitive cover

$$\varphi : C/\langle \omega \rangle = \mathbb{P}^1 \to E/\langle -id_E \rangle = \mathbb{P}^1$$

of degree n such that $\varphi \circ \pi' = \pi \circ f$.

The Weierstraß points of C are mapped onto $E[2]$ and hence to the ramification points of $\pi : E \to \mathbb{P}^1$. By our hypothesis, three of the Weierstraß points lie over the zero-point of E whose image in $\mathbb{P}^1(K)$ is denoted by P_4. The three others are mapped one-to-one to the points of order 2 of E. Denote by $P_1, ..., P_3$ the images of these points. The discriminant divisor of f is the conorm (or pullback) of a divisor of $\mathbb{P}^1$ with support equal to one point $P_5 \in \mathbb{P}^1(K)$. It follows that φ is unramified outside of $\{P_1, ..., P_5\}$. Summarizing, we have

PROPOSITION 2.4. *Let* $f_0 : C \to E$ *be a minimal genus 2 cover of odd degree* n *of an elliptic curve* E/K. *Then there is a unique translation* $\tau : E \to E$ *such that* $f = \tau \circ f_0$ *is normalized. Moreover,* $\pi \circ f$ *factors over the hyperelliptic cover* $\pi' :$ $C \to C/\langle \omega \rangle = \mathbb{P}^1$, *i.e. there is a unique cover* $\varphi : \mathbb{P}^1 \to \mathbb{P}^1$ *such that* $\varphi \circ \pi' = \pi \circ f$. *(In particular,* φ *is primitive and has degree* n.) *Let* $P_1, \ldots, P_3$ *denote the images of the non-zero 2-torsion points of* E *under* π *and put* $P_4 = \pi(0_E)$. *In addition, let* $P_5 \in \mathbb{P}^1(K)$ *be the unique point such that* $\mathrm{Disc}(f) = \pi^*(P_5)$. *Then* $\varphi : \mathbb{P}^1 \to \mathbb{P}^1$ *satisfies* (∗) *and Convention 1.2 if and only if* $\mathrm{Disc}(f)$ *is reduced, i.e. if and only if* $P_5 \notin \{P_1, \ldots, P_4\}$.

PROOF. Everything except the last statement has been explained above. Since $\mathrm{Disc}(f)$ is a divisor of degree 2 and invariant under $-id_E$ it is of the form $P + (-id_E(P))$ and hence it is reduced if and only if P_5 is unramified, i.e. not contained in $\{P_1, \ldots, P_4\}$.   □

Note that the above construction guarantees that $P_1, \ldots, P_4$ are distinct points, but not necessarily $P_1, \ldots, P_5$. Assume now that P_5 is different from P_i, $i = 1, ..., 4$ (the "generic" case, see Subsection 3.2). Then φ defines a point in the Hurwitz space H_n^*.

The previous propositions suggest that the moduli space H_n^* is closely related to another moduli space H_n, which represents the moduli problem $\mathcal{H}_n$ classifying isomorphism classes of pairs (C, f) where C is a curve of genus 2 and f is a *normalized* covering map from C to an elliptic curve E of degree n. More precisely, let us look at the subproblem $\mathcal{H}'_n$ of $\mathcal{H}_n$ which classifies the covers for which $\mathrm{Disc}(f)$ is reduced. Then we shall see that $\mathcal{H}_n$ can be (coarsely) represented by a surface H_n, and $\mathcal{H}'_n$ by an open subset H'_n of H_n. Moreover, over fields K of characteristic 0, H'_n turns out to be isomorphic to the the Hurwitz space H^*_n; cf. Theorem 4.9 below.

The advantage of the spaces H'_n and H_n is that we can embed them as open subsets into an explicitly given surface.

To describe this embedding we shall use torsion structures on elliptic curves.

2.3. The "Basic Construction".
For details of the following discussion we refer to [**FK**].

Let (C, f) be a (normalized) genus 2 cover of E, and let J_C denote the Jacobian variety of C. Then the map f induces homomorphisms $f_* : J_C \to E$ and $f^* : E \to J_C$. Because of the minimality of f, the morphism f^* is injective, and the kernel E_* of f_* is an elliptic curve intersecting $f^*(E)$ exactly in the points of order n. It follows that there is a G_K-module isomorphism

$$\alpha : E[n] \xrightarrow{\sim} E_*[n]$$

which turns out to be anti-isometric with respect to the Weil-pairings on $E[n]$ and on $E_*[n]$. The abelian variety J_C together with its natural principal polarization is isomorphic to $(E \times E_*)/\Gamma_\alpha$ where $\Gamma_\alpha \leq E[n] \times E_*[n]$ is the graph of α.

Conversely, take a K-rational triple (E, E', α) with E, E' elliptic curves over K and α a K-rational isomorphism from $E[n]$ to $E'[n]$ whose graph Γ_α is isotropic in $E[n] \times E'[n]$. Then $(E \times E')/\Gamma_\alpha$ is a principally polarized abelian variety, and there is a K-rational effective divisor C in the class of $n(0_E \times E' + E \times 0_{E'})/\Gamma_\alpha$ which is either a curve of genus 2 *or* a union of two elliptic curves intersecting in one point. If C is irreducible, then the projection p_1 of $E \times E'$ to E induces a cover

$$f_\alpha : C \to E$$

of degree n and hence we get a K-rational solution of the moduli problem $\mathcal{H}_n$ defined above.

Moreover, if the discriminant divisor $\mathrm{Disc}(f_\alpha)$ is reduced, then $f_\alpha \in \mathcal{H}'_n(K)$, and the associated cover $\varphi : \mathbb{P}^1 \to \mathbb{P}^1$ gives rise to a point in the moduli space H^*_n.

The triples (E, E', α) are parameterized by an open part of the *modular diagonal quotient surface* $Z_{n,-1}/K$ (cf. [**Ka7**]) and Subsection 4.1. If K contains a primitive n-th root of unity $\zeta_n \in K$, then $Z_{n,-1}$ is a quotient of the product surface $X(n) \times X(n)$, where $X(n)$ is the modular curve parameterizing elliptic curves with level-n-structure (of fixed determinant); explicitly,

$$Z_{n,-1} = (X(n) \times X(n))/\Delta_{-1}(\mathrm{SL}_2(\mathbb{Z}/n\mathbb{Z})),$$

where $\Delta_{-1}(G_n)$ is the "twisted diagonal subgroup" of $G_n = \mathrm{SL}_2(\mathbb{Z}/n\mathbb{Z})$ with respect to $\tau_{-1} \in \mathrm{Aut}(G_n)$, i.e. $\Delta_{-1}(G_n) = \{(g, \tau_{-1}(g)) : g \in G_n\}$ where τ_{-1} is the automorphism of G_n given by conjugation with an element $\beta \in \mathrm{GL}_2(\mathbb{Z}/n\mathbb{Z})$ of determinant -1; cf. [**KS**] and [**Fr2**]. If $\zeta_n \notin K$, then the construction of $Z_{n,-1}$ is more complicated; in this case $Z_{N,-1}$ is still a quotient of $X(n) \times X(n)$, (where now $X(n)/K$ denotes Shimura's canonical model of $X(n)/\mathbb{C}$), but the quotient has to be taken with respect to an étale group scheme rather than a (constant) group (scheme); cf. [**Ka7**] for more detail.

For us it is important that we have found a very explicit variety which is isomorphic to the Hurwitz space H_n^* (as we shall see), and this allows us to study its geometric and Diophantine properties. So we want to compare the moduli spaces H_n^* and H_n' which were introduced in the previous subsections. Although the discussion in there shows that there is a relation between their sets of K-rational points, this does not suffice to show that their coarse moduli spaces are isomorphic. For this, we shall study the associated moduli functors and/or stacks in more detail. We begin to do this "fiberwise".

3. Covers with Fixed Base Curve

3.1. The Hurwitz spaces $H_{E/K,N}$ and $H'_{E/K,N}$. Before we discuss the moduli problem $\mathcal{H}_n$ (with moduli space H_n) in more detail, let us consider the simpler problem of classifying the (normalized) genus 2 covers $f : C \to E$ over a *fixed* elliptic curve E/K. Thus, in the "dictionary" of the previous subsection, this corresponds (roughly) to case that we study covers $\varphi : \mathbb{P}^1 \to \mathbb{P}^1$ in which the points $P_1, \ldots, P_4$ are fixed.

In this discussion we can drop the condition that the degree of the cover is odd, and we shall indicate this by switching from n to N. Note, however, that in the case that N is even, the normalization condition (2.1) has to be replaced by the condition

$$(3.1) \qquad\qquad f_* W_C = 2(P_1' + P_2' + P_3').$$

Then, as in the case of odd degree covers, for each minimal cover $f_0 : C \to E$ there is a unique translation $\tau : E \to E$ such that $f = \tau \circ f_0$ is normalized (cf. [**Ka5**], Proposition 2.2). Moreover, f satisfies condition (2.2).

For any extension field L of K, let $E_L = E \otimes L$ denote the elliptic curve E lifted to L. We now consider the set

$$\mathrm{Cov}_{E/K,N}(L) := \{f : C \to E_L : \begin{array}{l} f \text{ is a normalized genus 2 cover} \\ \text{defined over } L \text{ with } \deg(f) = N\}/ \simeq \end{array}$$

of isomorphism classes of normalized covers. As is explained in [**Ka5**], the assignment $L \mapsto \mathrm{Cov}_{E/K,N}(L)$ can be extended in a natural way to a functor $\mathcal{H}_{E/K,N} : \underline{Sch}_{/K} \to \underline{Sets}$, and by Theorem 1.1 of [**Ka5**] we have:

THEOREM 3.1. *If $N \geq 3$, then the functor $\mathcal{H}_{E/K,N}$ is finely represented by a smooth, affine and geometrically connected curve $H_{E/K,N}/K$ with the property that $H_{E/K,N} \otimes K_s$ is an open subset of the modular curve $X(N)_{/K_s}$.*

REMARK 3.2. The fact that the curve $H := H_{E/K,N}$ *finely* represents the functor $\mathcal{H}_{E/K,N}$ means that there exists a *universal* normalized genus 2 cover $f_H : \mathcal{C}_H \to E \times H$ of degree N with the property that every normalized genus 2 cover $f : C \to E \times S$ of degree N (where S is any K-scheme) is obtained uniquely from f_H by base-change. In particular, the set $\mathrm{Cov}_{E/K,N}(K)$ of covers can be identified with the set of fibres $f_x := (f_H)_x : \mathcal{C}_x \to E_x = E$ of f_H, where $x \in H(K)$.

The main idea for proving this is to use the "basic construction" of genus 2 curves which was presented in [**FK**], [**Ka3**] (and which was sketched in Subsection

2.3). In [**Ka5**], this construction is generalized to families of covers of elliptic curves, and this yields an (open) embedding of functors

$$(3.2) \qquad\qquad \Psi = \Psi_{E/K,N} : \mathcal{H}_{E/K,N} \hookrightarrow \mathcal{X}_{E/K,N,-1},$$

where $\mathcal{X}_{E/K,N,-1}$ is the functor which classifies (isomorphism classes) of pairs (E', ψ) consisting of an elliptic curve E'/K and an anti-isometry $\psi : E[N] \overset{\sim}{\to} E'[N]$ (with respect to the e_N-pairings). Since this latter functor is finely representable by an affine curve $X_{E/K,N,-1}$, it follows that $\mathcal{H}_{E/K,N}$ is represented by an open subset $H_{E/K,N} \subset X_{E/K,N,-1}$. Note that if $E[N] \simeq (\mathbb{Z}/N\mathbb{Z})^2$ as a G_K-module, then the curve $X_{E/K,N,-1}$ is isomorphic to the usual affine modular curve $X'(N) = X(N) \setminus \{\text{cusps}\}$ of level N whose K-rational points correspond bijectively to isomorphism classes (E', α), where E'/K is an elliptic curve and $\alpha : E'[N] \overset{\sim}{\to} (\mathbb{Z}/N\mathbb{Z})^2$ is a level-N-structure (of fixed "determinant" ζ_N). Since this is always true over a finite separable extension of K, we see that $H_{E/K,N} \otimes K_s$ is (isomorphic to) a subset of $X'(N)_{/K_s}$.

REMARK 3.3. (a) It is interesting to note that this representability result is obtained by purely algebraic techniques and hence does not use (not even implicitly) the Riemann Existence Theorem (RET).

(b) Note that while the moduli scheme $X_{E/K,N,-1}$ depends only on the structure of the K-group scheme $E[N]$, the subscheme $H_{E/K,N}$ *does depend* on the choice of E/K; cf. [**Ka5**]. Moreover, if we replace E/K by a non-isomorphic twist E'/K of E/K, then $E[N] \not\simeq E'[N]$ (as G_K-modules), and hence the functors $\mathcal{X}_{E/K,N,-1}$ and $\mathcal{X}_{E'/K,N,-1}$ are not isomorphic (as functors on $\underline{Sch}_{/K}$), and the same is true for $\mathcal{H}_{E/K,N}$ and $\mathcal{H}_{E'/K,N}$. This shows that the moduli problem $\mathcal{H}_{E/K,N}$ is more refined than the corresponding one for covers $\varphi : \mathbb{P}^1 \to \mathbb{P}^1$ satisfying condition (∗) with fixed ramification points $P_1, \ldots, P_4$.

In view of the correspondence between genus 2 covers of elliptic curves and covers of $\mathbb{P}^1$ (cf. Proposition 2.4), we shall also be interested in the subfunctor $\mathcal{H}'_{E/K,N}$ of $\mathcal{H}_{E/K,N}$ which classifies those (normalized) genus 2 covers $f : C \to E$ for which $\text{Disc}(f)$ is reduced. More precisely, for any K-scheme S, let

$$\mathcal{H}'_{E/K,N}(S) = \{(C \overset{f}{\to} E_S) \in \mathcal{H}_{E/K,N}(S) : \text{Disc}(f_s) \text{ is reduced}, \forall s \in S\}.$$

It is clear that this defines a subfunctor of $\mathcal{H}_{E/K,N}$, and so we should expect that it is represented by an open subscheme $H'_{E/K,N}$ of $H_{E/K,N}$. This follows from the following more precise assertion.

PROPOSITION 3.4. *The rule* $f \mapsto \text{Disc}(f)$ *defines a morphism*

$$\delta_{E/K,N} : H_{E/K,N} \to \mathbb{P}^1$$

such that $\text{Disc}(f_x) = \pi^*(\delta_{E/K,N}(x))$, *for every* $x \in H_{E/K,N}(K)$. *In particular, for every* $\bar{P} \in \mathbb{P}^1(K)$ *we have*

$$\delta_{E/K,N}^{-1}(\bar{P}) = \text{Cov}_{E/K,N,\bar{P}}(K) := \{f \in \text{Cov}_{E/K,N}(K) : \text{Disc}(f) = \pi^*(\bar{P})\},$$

and hence the functor $\mathcal{H}'_{E/K,N}$ *is (finely) represented by the open subset* $H'_{E/K,N} := \delta_{E/K,N}^{-1}(\mathbb{P}^1_K \setminus \pi(E[2]))$ *of* $H_{E/K,N}$.

PROOF. The first two assertions are proved in [**Ka6**], Proposition 12, and the last one is an immediate consequence of this and Theorem 3.1 □

REMARK 3.5. Let $\underline{P} = (P_1, \ldots, P_5) \in (\mathbb{P}^1_K)^5(K)$ be a tuple of 5 distinct points with $P_4 = \infty$. For an odd integer n, let $\mathrm{Cov}_{\underline{P},n}(K)$ denote the set of isomorphism classes of K-covers $\varphi : \mathbb{P}^1_K \to \mathbb{P}^1_K$ of $\deg(\varphi) = n$ satisfying condition $(*)$ and Convention 1.2. If E/K is an elliptic curve such that $\pi_E(E[2]) = \{P_1, \ldots, P_4\}$ (and $\pi_E(0_E) = P_4$), then the discussion of subsection 2.2 shows that we have a bijection

$$\mathrm{Cov}_{\underline{P},n}(K) \xrightarrow{\sim} \mathrm{Cov}_{E/K,n,P_5}(K).$$

Thus, by allowing the point P_5 to move, we obtain a bijection

$$H_{P_1,\ldots,P_4,n}(K) \xrightarrow{\sim} H'_{E/K,n}(K),$$

where $H_{P_1,\ldots,P_4,n} \subset H^{in}(S_n, \mathbf{C})$ is the Hurwitz space with fixed ramification points $P_1, \ldots, P_4$. We therefore see that the set

$$H^\partial_{E/K,n} := H_{E/K,n} \setminus H'_{E/K,n} = \delta^{-1}_{E/K,n}(\pi(E[2]))$$

is a "boundary" of the Hurwitz space parametrizing $\mathbb{P}^1$-covers, even though it is contained in the Hurwitz space $H_{E/K,N}$ parametrizing genus 2 covers of E.

3.2. The relative boundary $H^\partial_{E/K,n} = H_{E/K,n} \setminus H'_{E/K,n}$. We now describe the genus 2 covers which correspond to the points of the "relative boundary" $H^\partial_{E/K,n} = H_{E/K,n} \setminus H'_{E/K,n}$ (where n is odd) in more detail. By Theorem 3.1 and Proposition 3.4 we know that each such point corresponds to a unique normalized genus 2 cover $f : C \to E$ of degree n such that its discriminant has the form $\mathrm{Disc}(f) = 2P'_k$, where $P'_k \in E[2]$; such covers are classified as Type II in [**Fr1**], p. 85.

By Proposition 2.4 we still have an associated $\mathbb{P}^1$-cover $\varphi : \mathbb{P}^1 \to \mathbb{P}^1$, but now the ramification structure of φ no longer satisfies the "generic" condition $(*)$ of Subsection 1.3. Indeed, since these special covers may be viewed as the "limits" of the generic case when we let the point P_5 move to coincide with one of the points $P_1, \ldots, P_4$, we obtain new ramification types, i.e. new cycle decompositions (σ_i), where, as before, $\sigma_i \in G_\varphi = \mathrm{Gal}(\overline{\varphi})$ denotes a generator of a ramification group above P_i.

Here we shall work out all the ramification types for these special covers. (For comparison purposes we also include the generic case as case 0.) It turns out that the ramification type $\mathbf{C} = \mathbf{C}_\varphi$ of φ is completely determined by the *different* $D_f = \mathrm{Diff}(f)$ of f.

PROPOSITION 3.6. *Put $m = \frac{n-1}{2}$ and let $i, k \in \{1, 2, 3\}$. Then the ramification structure of φ is given by the following table:*

Case	Condition	$(\sigma_i), i \neq k$	(σ_k)	(σ_4)	(σ_5)
0	$\mathrm{Disc}(f) \neq 2P$	$(2)^m$	$-$	$(2)^{m-1}$	(2)
1	$\mathrm{Disc}(f) = 2P'_k, D_f \neq 2P$	$(2)^m$	$(2)^{m-2}(4)$	$(2)^{m-1}$	$-$
2	$\mathrm{Disc}(f) = 2P'_k, D_f = 2P$	$(2)^m$	$(2)^{m-1}(3)$	$(2)^{m-1}$	$-$
3	$\mathrm{Disc}(f) = 2P'_4, D_f \neq 2P$	$(2)^m$	$-$	$(2)^{m-3}(4)$	$-$
4	$\mathrm{Disc}(f) = 2P'_4, D_f = 2P$	$(2)^m$	$-$	$(2)^{m-2}(3)$	$-$

PROOF. Put $f_1 = \pi \circ f = \varphi \circ \pi' : C \to \mathbb{P}^1$. In each of the cases, we shall first work out the ramification structure of the points in $f_1^{-1}(P_j)$, for $j = 1, \ldots, 5$, and then deduce that of $\varphi^{-1}(P_j)$ from this. From this we can then read off the cycle structure of $\sigma_j \in G_\varphi$. (Here G_φ is the *monodromy group* of φ, i.e. the Galois group of the Galois closure $\overline{\varphi}$ of φ.)

Case 0. Here $\mathrm{Disc}(f) = P_5' + [-1]P_5'$, and $P_5' \notin E[2] = \{P_1', P_2', P_3', P_4'\}$, where $P_4' = 0_E$. Thus, $P_1', \ldots, P_4'$ are unramified in f, but are ramified of order 2 for π_E. It follows that there are n points in $f_1^{-1}(P_j)$, $j = 1, \ldots, 4$, and all have ramification index 2 with respect to f_1. Now for $j = 1, 2, 3$, precisely one point (the Weierstrass point) $W_j \in f_1^{-1}(P_j)$ is ramified in π', whereas the others are not. Thus, $\pi'(W_j)$ is unramified wrt. φ, whereas the other $m = \frac{n-1}{2}$ points in $\varphi^{-1}(P_j)$ are each ramified of order 2. It follows that σ_j is a product of m transpositions, as claimed. On the other hand, since $f_1^{-1}(P_4)$ contains 3 Weierstrass points, there are 3 unramified points in $\varphi^{-1}(P_4)$, and the other $m - 1 = \frac{n-3}{2}$ points are ramified of order 2. Thus, $(\sigma_4) = (2)^{m-1}$. Finally, since $f_1^{-1}(P_5)$ contains precisely two points with ramification index 2 (which are interchanged by the hyperelliptic involution), whereas the others are unramified, it follows that $\varphi^{-1}(P_5)$ has a unique point of index 2, and so σ_5 is a transposition.

Case 1. Here $D_f = P_{k1} + P_{k2}$ with $P_{k1} \neq P_{k2}$ and $f(P_{k1}) = f(P_{k2}) = P_k'$, so the ramification structure above P_j for $j = 1, \ldots, 4$, $j \neq k$ is as in case 0. Now above P_k' the ramification structure is $(2)^2$ (wrt. f), so above P_k the ramification is $(2)^{n-4}(4)^2$. Now P_{kj} cannot be a Weierstrass point, for otherwise the ramification indices wrt. π' and f are both 2, which is impossible since all ramification groups are cyclic; cf. Abhyankar's Lemma. Thus, the unique Weierstrass point $W_k \in f_1^{-1}(P_k)$ has index 2 wrt. f_1 and hence $\pi(W_k)$ is unramified wrt. φ. Thus, σ_k has type $(2)^{\frac{n-5}{2}}(4)$.

Case 2. Here $D_f = 2W_k$, the unique Weierstrass point above P_k'. Thus, above P_k' the ramification structure is (3) and hence above P_k it is $(2)^{n-3}(6)$ wrt. f_1. It follows that σ_k has type $(2)^{\frac{n-3}{2}}(3)$.

Case 3. This is similar to case 1. Here $D_f = P_{41} + P_{42}$ with $P_{41} \neq P_{42}$ and $f(P_{41}) = f(P_{42}) = P_4' = 0_E$, so the ramification structure above P_j for $j = 1, \ldots, 3$, is as in case 0. Now above 0_E the ramification structure is $(2)^2$ (wrt. f), so above P_4 the ramification wrt. f_1 is $(2)^{n-4}(4)^2$. As in case 1, P_{4j} cannot be a Weierstrass point, so we see that σ_4 has type $(2)^{\frac{n-7}{2}}(4)$ because the images of the 3 Weierstrass points are unramified wrt. φ.

Case 4. Here $D_f = 2W_k'$, where W_k' is one of the 3 Weierstrass points above 0_E. Thus, above 0_E the ramification structure is (3) and hence above P_k it is $(2)^{n-3}(6)$ wrt. f_1. It follows (cf. case 3) that σ_4 has type $(2)^{\frac{n-5}{2}}(3)$. $\qquad\square$

COROLLARY 3.7. *(a) In cases 0 and 2, the monodromy group of φ and f is $G_\varphi = G_f = S_n$.*

(b) In case 4, we have $G_f = A_n$. Moreover, $G_\varphi = S_n$ when $n \equiv 1 \pmod 4$, and $G_\varphi = A_n$ when $n \equiv 3 \pmod 4$.

PROOF. Let $\tilde{F}_0$ (respectively, $\tilde{F}$) be the Galois hull of $\kappa(\mathbb{P}^1)/\varphi^*\kappa(\mathbb{P}^1)$ (respectively, of $\kappa(C)/f^*\kappa(E)$). Since $\tilde{F} = \tilde{F}_0\kappa(E)$ by Galois theory, we see that $G_\varphi = G_f$ if and only if $\kappa(E) \not\subset \tilde{F}_0$.

Moreover, since f is minimal, we see that $G_\varphi = \mathrm{Gal}(\tilde{F}_0/\kappa(\mathbb{P}^1))$ and $G_f = \mathrm{Gal}(\tilde{F}/\kappa(E))$ are primitive permutation groups of degree n; cf. [**Hup**], II.1.4. In case 0, σ_5 is a 2-cycle, so $G_\varphi = S_n$ by [**Hup**], II.4.5b). In case 2 (respectively, case

4) we have from Proposition 3.6 that σ_4^2 (respectively, σ_k^2) is a 3-cycle, so $G_\varphi \geq A_n$ by [**Hup**], II.4.5c.

(a) Since Proposition 3.6 shows that $G_\varphi \not\leq A_n$, it follows that $G_\varphi = S_n$ in these cases. Moreover, suppose $\kappa(E) \subset \tilde{F}_0$. Then $\kappa(E) = F_1 := (\tilde{F}_0)^{A_n}$, because A_n is the unique subgroup of index 2 of S_n. Now in case 0 and m even we see from Proposition 3.6 that σ_4 and σ_5 are odd and the others are even permutations, so $F_1/\kappa(\mathbb{P}^1)$ is ramified precisely at P_4 and P_5. Thus, $F_1 \neq \kappa(E)$ (which is ramified at $P_1, \ldots P_4$). Similarly, we have in the other cases that $F_1 \neq \kappa(E)$, and so $G_f = G_\varphi$.

(b) If m is even (i.e. $n \equiv 1(4)$), then $G_\varphi \not\leq A_n$ by Proposition 3.6 and so $G_\varphi = S_n$. Moreover, $F_1/\kappa(\mathbb{P}^1)$ is ramified precisely at P_4, P, so $F_1 \neq \kappa(E)$ and $G_f = G_\varphi = S_n$.

If m is odd, then by Proposition 3.6 we have $\sigma_i \in A_n$, for all $i = 1, \ldots, 4$, and so $G_\varphi \leq A_n$ (because G_φ is generated by the σ_i's and their conjugates). Thus $G_\varphi = A_n$. Since A_n has no subgroup of index 2, it follows that $\kappa(E) \not\subset \tilde{F}_0$ and so $G_f = G_\varphi$. $\square$

REMARK 3.8. (a) The cases 2 and 4 have been investigated in detail from a different perspective by authors studying problems in *billiard dynamics* and *square tilings*; cf. [**Mc1**], [**HL1**] and the references therein. More precisely, if $K = \mathbb{C}$, and if $f : C \to E$ is a genus 2 cover of type 2 or 4, and ω_E is a holomorphic differential on E, then $(C, f^*\omega_E)$ defines a point in the moduli space $\mathcal{H}(2)$ of genus 2 curves with a differential 1-form having a double zero (cf. [**Mc1**]).

By [**Mc1**], these two cases can be distinguished by their "spin invariant": those in case 2 have spin 0, and those in case 4 have spin 1. Indeed, in case 2 (respectively, case 4) we have 1 (respectively, 3) Weierstrass points above the ramification point P_k' (respectively, P_4'), so the assertion follows from [**Mc1**], Theorem 6.1. (Note that [**Mc1**] and [**HL1**] assume that the ramification takes place over the origin of E, so in case 2 they consider the cover $\tau_{P_k'} \circ f$ in place of the cover f.)

(b) Matthew Ingle has pointed out to us that Theorem III on p. 138 of E. Netto's book, *The Theory of Substitutions*, shows that $G_f \geq A_n$ in the other cases as well. (He also filled the gaps in Netto's proof of this theorem.)

3.3. The boundary of $H_{E/K,N}$. Since the moduli space $H := H_{E/K,N}$ is a smooth, irreducible (affine) curve, it has a unique normal compactification X. Here we shall study the "boundary" $\partial H = X \setminus H$ of H and how the correspondence between covers $f_x : C_x \to E$ and points of H (cf. Remark 3.2) extends to ∂H.

The first step towards this goal is to observe that $\partial H = S(C)$, the set of places of *bad reduction* of the curve C/F which is the generic fibre of the relative universal curve $\mathcal{C}_H/H$ (cf. Remark 3.2). As a consequence, ∂H decomposes as

$$\partial H = S(C) = S_0 \cup S_1$$

where S_0 is the set of points of X where the Jacobian J_C of C/F has bad reduction and where S_1 is the set of places where C has bad reduction yet J_C had good reduction.

To explain how the points of ∂H correspond to (degenerate) covers, we shall use the theory of minimal models to construct a *canonical compactification* $f : \mathcal{C} \to E \times X$ of the universal cover $f_H : \mathcal{C}_H \to E \times H$ (cf. Remark 3.2) and then study its fibres.

To explain this in more detail, assume for simplicity that $K = \overline{K}$ is algebraically closed. In this case we know that H is an open subset of the modular curve $X(N)$, which is therefore its canonical compactification, i.e. $X = X(N)$. In addition, we shall assume that $\text{char}(K) \nmid N!$ (i.e. that either $\text{char}(K) = 0$ or that $\text{char}(K) > N$); in the contrary case the structure of the boundary (and of the corresponding covers) is much more complicated.

Let $p : \mathcal{C} \to X$ denote the *minimal model* of C/F over X. Thus, $\mathcal{C}$ is a projective, smooth surface over K, and $p : \mathcal{C} \to X$ is a genus 2 fibration whose fibres do not contain any rational (-1)-curves. Moreover, the restriction of p to $p^{-1}(H)$ is canonically isomorphic to $p_H := pr_2 \circ f_H : \mathcal{C}_H \to H$ (which is the minimal model of C/F over H) and so we can identify $\mathcal{C}_H$ with the open subscheme $p^{-1}(H)$ of $\mathcal{C}$. Thus, $\mathcal{C}$ is a canonical compactification of $\mathcal{C}_H$.

It turns out that f_H extends to a *finite* morphism $f : \mathcal{C} \to E_X = E \times X$. Moreover, the precise structure of the fibres of $\mathcal{C}/X$ is as follows:

THEOREM 3.9. *If* $\text{char}(K) \nmid N!$, *then* $\mathcal{C}/X(N)$ *is a stable curve which has precisely one singular point in each singular fibre* $\mathcal{C}_x$, *and the universal cover* f_H *extends to a* finite *morphism* $f : \mathcal{C} \to E_{X(N)} := E \times X(N)$. *Furthermore, a fibre* $\mathcal{C}_x$ *is singular if and only if* $x \in \partial H$, *i.e.* $S = \partial H$, *and the set* S_0 *(as defined above) coincides with the set* $X(N)_\infty$ *of cusps of* $X(N)$. *Moreover, the structure of the singular fibres of* $\mathcal{C}/X(N)$ *is as follows:*

(a) If $x \in S_0$, *then the fibre* $\mathcal{C}_x$ *is an irreducible curve whose normalization is a curve of genus 1. Furthermore,* $\mathcal{C}_x$ *has a unique singular point* P_x, *and the restriction* $f_x = f_{\mathcal{C}_x} : \mathcal{C}_x \to (E_X)_x \simeq E$ *is a cover of degree* N *such that* $f_x(P_x) = 0_E$.

(b) If $x \in S_1$, *then* $\mathcal{C}_x = E_{x,1} \cup E_{x,2}$ *is the union of two curves of genus 1 which meet transversely in a unique point* P_x. *Furthermore,* $f_x(P_x) \in E[2]$ *and the cover* $f_x : \mathcal{C}_x \to E$ *has degree* N *in the sense that* $d_{x,1} + d_{x,2} = N$, *where* $d_{x,i} = \deg(f_{x,i})$ *is the degree of* $f_{x,i} := (f_x)_{|E_{x,i}} : E_{x,i} \to E$.

Moreover, the modular height of $\mathcal{C}/X(N)$ *is* $h_{\mathcal{C}/X} = \frac{1}{24} sl(N)$, *where* $sl(N) = \#\text{SL}_2(\mathbb{Z}/N\mathbb{Z})$, *and the self-intersection number of the relative dualizing sheaf* $\omega^0_{\mathcal{C}/X}$ *is given by*

$$(3.3) \qquad (\omega^0_{\mathcal{C}/X(N)})^2 = \frac{7}{5}\#S_1 + \frac{1}{5}\#S_0 = \frac{1}{24N}(7N - 6)sl(N).$$

REMARK 3.10. The proof of this proposition is based on the ideas and techniques of [**FK**], pp. 161–166. Note, however, that the conclusions obtained here are much stronger than those of [**FK**], for here we are able to determine the number of irreducible components of each fibre. In addition, in this situation one of the cases (i.e. the case β_2) of [**FK**] does not occur.

The reason for obtaining such strong results is due to two facts. First of all, by using the results of [**Ka3**], [**Ka4**] (cf. Step 8 below), we are able to determine precisely the total number of singular fibres in each case. Secondly, we can use a result of Mumford to determine certain invariants which give information about the individual degeneration types. It is interesting to note that such a formula does not seem to be available in the arithmetic case; indeed, a considerable portion of the paper [**FK**] was devoted to deriving a weak version of such a formula. (This was pointed out to us by Barry Mazur.)

PROOF OF THEOREM 3.9 (Sketch). For convenience of the reader, we outline the main ideas of the proof of Theorem 3.9 and refer to [**Ka6**] for details. (The above theorem is contained in Corollary 16, Theorem 17, Proposition 19, Theorem 22, Corollary 24 and Corollary 25 of [**Ka6**].)

Step 1: f_H *extends to a morphism* $f : C \to E_X$.

Since E_X is a smooth proper curve over $X = X(N)$, this follows from Zariski's Main Theorem; cf. [**Ka6**], Proposition 15 and Corollary 16.

Step 2: C/X *is semistable and* $S_0 = X(N)_\infty$.

By the basic construction, $J_C \sim E_F \times E'$, where $E_F = E \otimes F$ and E'/F is an elliptic curve. Now since $E'[N] \simeq E[N] \simeq (\mathbb{Z}/N\mathbb{Z})^2$ (as G_F-modules), it follows that both E' and J_C are semi-stable abelian varieties (as E has good reduction), and so C is a semi-stable curve (cf. [**Ka6**], proof of Theorem 17).

Clearly, J_C has bad reduction at $x \in X$ if and only if E' does. Since $E'/X(N)$ is the universal elliptic curve with level-N-structure (of fixed determinant), it follows that $S_0 = X(N)_\infty$.

Step 3: $S(C) = \partial H$.

Since $p_H : C_H \to H$ has smooth fibres, it is clear that $S(C) \subset \partial H$. The opposite inclusion follows from the construction of H in [**Ka5**]; cf. [**Ka6**], Proposition 19 for a more detailed explanation.

Step 4: *The preliminary structure of* C_x *for* $x \in S_0$.

Here we are in case β) of [**FK**], p. 163ff. By the argument given there (or by the one presented in [**Ka6**], proof of Proposition 19), we conclude that C_x has a unique component $C_{x,0}$ whose normalization has genus 1, and that there is at least one singular point $P_x \in C_x$ such that $C_x \setminus \{P_x\}$ is connected.

Step 5: *The preliminary structure of* C_x *for* $x \in S_1$.

Here we are in case α) of [**FK**], p. 161ff. Thus, if δ_x denotes the number of singular points of C_x, then C_x has $\delta_x + 1$ irreducible components: two elliptic curves $E_{x,1}$, $E_{x,2}$ which are connected by a chain of $\mathbb{P}^1$'s.

Step 6: *Computation of the modular height* $h_{C/X}$.

Recall that if $p : C \to X$ is any semi-stable curve of genus $g \geq 1$, then its *modular height* is defined by $h_{C/X} = \deg(\lambda)$, where $\lambda = \lambda_{C/X} = \wedge^g p_* \omega^0_{C/X} \in \mathrm{Pic}(X)$. Note that we also have $\lambda \simeq \lambda_{\mathcal{J}/X} := s^* \Omega^1_{\mathcal{J}/X}$, where $\mathcal{J}/X$ denote the Néron model of the Jacobian J_C and $s : X \to \mathcal{J}$ denotes the zero-section of $\mathcal{J}/X$; this is a formula due to Parshin and Arakelov (cf. [**Ka6**], proof of Theorem 17).

Since $J_C \sim E_F \times E'_F$ (by an étale isogeny), and since E_F/F is constant, we see (by using the second description of λ) that $\lambda_{C/X} \simeq \lambda_{\mathcal{E}'/X}$, where $\mathcal{E}'$ denotes the minimal model of E' over X. Now for any (semi-stable) elliptic curve $\mathcal{E}'/X$ we have (by Noether's formula) that $h_{\mathcal{E}'/X} = \frac{1}{12}\delta_{\mathcal{E}'/X}$. But $\delta_{\mathcal{E}'/X} = N\#S_0$ (because each singular fibre $\mathcal{E}'_x$ of $\mathcal{E}'/X$ lies over $X(N)_\infty = S_0$ (cf. Step 2) and is a Néron polygon of length N), and so we obtain

$$(3.4) \qquad h_{C/X} = h_{\mathcal{E}'/X} = \frac{N}{12}\#S_0 = \frac{1}{24}sl(N),$$

where the latter equation follows from the first equality of (3.9) below.

Step 7: *The invariants* δ_0, δ_1, ω^2 *and Mumford's formula*.

Let δ_1 (respectively, δ_0) denote the number of singular points of the fibres of C/X which disconnect (respectively, which do not disconnect) the fibre. By Steps 4 and

5 we see that

(3.5) $$\delta_0 \geq \#S_0 \quad \text{and} \quad \delta_1 \geq \#S_1.$$

Indeed, if $x \in S_0$, then by Step 4 there is at least one singular point P_x of $\mathcal{C}_x$ which does not disconnect the fibre, and if $x \in S_1$, then by Step 5 each of the $\delta_x \geq 1$ singular points in $\mathcal{C}_x$ disconnects the fibre.

There is a remarkable relation which connects the invariants $h = h_{\mathcal{C}/X}$, δ_0, and δ_1:

(3.6) $$12h = \frac{6}{5}\delta_0 + \frac{12}{5}\delta_1.$$

This equation is an immediate consequence of the following two relations which also involve the invariant $\omega^2 = (\omega_{\mathcal{C}/X}^0)^2$, the self-intersection number of the relative dualizing sheaf $\omega_{\mathcal{C}/X}^0$ of $\mathcal{C}/X$:

(3.7) $$12h = \omega^2 + \delta_0 + \delta_1, \quad \text{and} \quad \omega^2 = \frac{1}{5}\delta_0 + \frac{7}{5}\delta_1,$$

which are called *Noether's formula* and *Mumford's formula* respectively; cf. [**Ka6**], proof of Theorem 22 for the relevant references.

By combining equation (3.6) with the inequalities (3.5), we obtain the inequality

(3.8) $$\#S_1 \leq \frac{1}{12}(5N - 6)\#S_0.$$

Indeed, since $h = \frac{N}{12}\#S_0$ by (3.4), we thus obtain from (3.6) and (3.5) that $N\#S_0 = 12h \geq \frac{6}{5}\#S_0 + \frac{12}{5}\#S_1$, which is (3.8).

Step 8: *Computation of $\#S_0$ and $\#S_1$.*

Whereas the computation of $\#S_0$ is easy (because $S_0 = X(N)_\infty$ by step 2), the determination of $\#S_1$ is much harder. Indeed, as is explained in the proof of [**Ka6**], Theorem 21 (see also [**Ka5**], Theorem 6.2), this is essentially the "mass formula" of [**Ka3**], [**Ka4**]. (The hypothesis $\text{char}(K) \nmid N!$ is crucial here.) One obtains:

(3.9) $$\#S_0 = \frac{sl(N)}{2N} \quad \text{and} \quad \#S_1 = \frac{1}{24N}(5N - 6)sl(N).$$

Step 9: $\delta_x = 1$, *for all $x \in S = S_0 \cup S_1$.*

By equation (3.9) we see that we have equality in (3.8), and so we must have equality in (3.5) as well. Thus, $\delta_0 = \#S_0$ and $\delta_1 = \#S_1$, which means that $\delta_x = 1$, for all $x \in S = S_0 \cup S_1$, i.e. that $\mathcal{C}_x$ has a unique singular point. From this (together with Steps 4, 5) it is clear that the structure of $\mathcal{C}_x$ for $x \in S_0 \cup S_1$ is as claimed in parts (a), (b), and so $\mathcal{C}/X$ is a stable curve.

Step 10: *f is finite.*

Since f is proper, it is enough to show that f is quasi-finite, i.e. that none of the components of a fibre $\mathcal{C}_x$ of p is mapped to a point under f. For the points $x \notin S_1$, this is easy (since $\mathcal{C}_x$ is irreducible). If $x \in S_1$, it is clear that at least one of the two components is not mapped to a point. To see that both have this property is a bit more difficult, particularly if N even; see [**Ka6**], Proposition 20, for the proof.

Since the rest of the assertions (concerning $f_x(P_x)$) are verified easily, and since (3.3) follows from (3.7) (because $\delta_0 = \#S_0$ and $\delta_1 = \#S_1$ by step 9), this concludes the proof (sketch) of the theorem. $\qquad\square$

3.4. Application: the Rigidity Number. As an application of the above geometric ideas and constructions, we shall now compute (following [**Ka6**]) the number $\#\mathrm{Cov}_{E/K,N,\bar{P}}(K)$ of covers with fixed discriminant $\pi^*(\bar{P})$; this number may be viewed (via the identification of Remark 3.5) as a *measure of non-rigidity* of the Hurwitz space H_n^* because for rigid systems this number is 1; cf. [**V1**], p. 39.

Although this "rigidity number" can be defined via group theory, it seems very difficult to compute this number directly by counting tuples of conjugacy classes (cf. Remark 3.17). Here instead we shall use the geometric results of the previous subsections to calculate this number.

THEOREM 3.11. *Let $N \geq 3$ be an integer, and suppose that K is an algebraically closed field. If $\mathrm{char}(K) \nmid N!$, then for every $\bar{P} \in \mathbb{P}^1(K) \setminus \pi(E[2])$ we have*

$$\#\mathrm{Cov}_{E/K,N,\bar{P}}(K) = \frac{1}{12}(N-1)sl(N),$$

where, as before, $sl(N) := \#\mathrm{SL}_2(\mathbb{Z}/N\mathbb{Z})$.

We briefly sketch the main ideas involved in the proof of Theorem 3.11. (For more detail, cf. [**Ka6**].)

The basic idea is to relate the non-rigidity number $\#\mathrm{Cov}_{E/K,N,\bar{P}}(K)$ to the degree of the discriminant map $\delta_{E/K,N}$ (cf. Proposition 3.4) which in turn can be computed by using the intersection theory on compactification $\mathcal{C}$ of the universal genus 2 cover $f : \mathcal{C} \to E \times X(N)$. The key ingredients for this latter computation are the precise knowledge of the degenerate fibres of $\mathcal{C}$ (cf. Theorem 3.9) and the formula (3.3) for the self-intersection number $(\omega_{\mathcal{C}/X}^0)^2$ of the relative dualizing sheaf.

Step 1: $\#\mathrm{Cov}_{E/K,N,\bar{P}}(K) = \deg(\delta_{E/K,N})$.

By Proposition 3.4 we know that $\#\mathrm{Cov}_{E/K,N,\bar{P}}(K) = \#\delta_{E/K,N}(\bar{P}))$. On the other hand, since $\delta_{E/K,N}$ is finite and étale over $\bar{P} \notin \pi(E[2])$ (cf. [**Ka6**], Theorem 13), the assertion follows.

Step 2: *Analysis of the different divisor D on the compactification $\mathcal{C}$.*

We next want to interpret $\deg(\delta)$ as an intersection number. For this, we shall study the (Kähler) different divisor $\mathrm{Diff}(f)$ of the compactification $f : \mathcal{C} \to E \times X(N)$ of the universal cover f_H; cf. subsection 3.3.

Let D_F denote the different divisor of the generic (universal) cover $f_F : C = \mathcal{C}_F \to E_F$, and let D be its closure in the compactification $\mathcal{C}$. By using the explicit structure of the fibres of $p : \mathcal{C} \to X(N)$ (cf. Theorem 3.9) together with the Riemann-Hurwitz formula (for relative curves), we obtain

PROPOSITION 3.12. *The divisor D is an irreducible curve on $\mathcal{C}$ which is a degree 2 cover of $X(N)$ via the map $\pi_D := pr_2 \circ f_{|D}$. If $\delta_D := pr_1 \circ f_{|D} : D \to E$ denotes the projection onto the first factor, then*

$$\pi_E \circ \delta_D = \bar{\delta}_{E/K,N} \circ \pi_D.$$

In addition, $D = \mathrm{Diff}(f)$, the different divisor of $f : \mathcal{C} \to E_X$, and hence we have $\omega_{\mathcal{C}/X(N)}^0 \sim D$. Thus, for any $P \in E(K)$ we have

(3.10) $\deg(\delta_{E/K,N}) = \deg(\delta_D) = (\omega_{\mathcal{C}/X(N)}^0 . f^*(P \times X)).$

PROOF. [**Ka6**], Theorem 26. Note that this makes heavy use of the Structure Theorem 3.9. □

Step 3: *Analysis of the Weierstraß divisor W.*

By using the above proposition we see that the theorem will follow once we have computed the intersection number $(\omega^0_{\mathcal{C}/X} \cdot f^*(P \times X))$. As we shall see next, this number is closely related to the self-intersection number of the *Weierstraß divisor* W on $\mathcal{C}$. This divisor W is the closure in $\mathcal{C}$ of the (usual) Weierstraß divisor W_F on $C = C_F$.

PROPOSITION 3.13. *We have $6D \sim 2W + p^*A$, for some divisor $A \in \mathrm{Div}(X)$ of degree*

$$(3.11) \qquad \deg(A) = \#X(N)_\infty - \frac{4}{3}W^2 = \frac{1}{6}(9(\omega^0_{\mathcal{C}/X})^2 - W^2) = \frac{1}{2N}(N-1)sl(N),$$

and hence

$$(3.12) \qquad \deg(\delta_D) = \frac{N}{6}\deg(A) = \frac{1}{12}(N-1)sl(N).$$

PROOF. (Sketch; cf. [**Ka6**], Proposition 29). Since $6D_F$ and $2W_F$ are both 6-canonical divisors on C, we have $6D_F \sim 2W_F$. Moreover, since the fibres over S_1 are the only reducible fibres of $\mathcal{C}/X$ and since for $x \in S_1$, both $6D$ and $2W$ meet each component $E_{x,i}$ of $\mathcal{C}_x$ with multiplicity 6 (cf. the diagram on p. 162 of [**FK**]), we can conclude that $6D - 2W \sim p^*(A)$, for some $A \in \mathrm{Div}(X)$.

To calculate the degree of A, we shall use formula (34) of [**Ka6**]:

$$(3.13) \qquad (\omega^0_{\mathcal{C}/X(N)} \cdot W) + W^2 = \#X(N)_\infty = \frac{1}{2N}sl(N).$$

This formula is established by applying the adjunction formula to the divisor W (after base-changing to $X(2N)$).

From (3.13) and the equivalence $6D \sim 2W + p^*A$, the first two equalities of (3.11) follow readily. To deduce the third, note that the first two yield that $6\deg(A) = (9\omega^2 - W^2) = 6\#X(N)_\infty - 8W^2$, and so

$$(3.14) \qquad W^2 = \frac{6}{7}\#X(N)_\infty - \frac{9}{7}\omega^2 = -\frac{3}{8N}(N-2)sl(N),$$

the latter by the important formula (3.3). From this, the last equation of (3.11) is immediate.

Finally, since it easy to see that $f_*(2W) \sim 12(0_E \times X)$ (and since $\omega \sim D$ by Proposition 3.12) we have $6f_*\omega \sim 12(0_E \times X) + N(E \times A)$, and so by (3.10) and the projection formula we obtain

$$6\deg(\delta_D) = 6(\omega \cdot f^*(0_E \times X)) = ((12(0_E \times X) + N(E \times A)).0_E \times X) = N\deg(A),$$

which proves the first equality of (3.12). The second follows immediately from (3.11). $\qquad \square$

Step 4: *Conclusion.*

Combining the formula of Step 1 with equations (3.10) and (3.12) yields the desired formula of Theorem 3.11.

From the above result (and its proof) we can also compute the number of points in the "interior boundary" $H^\partial_{E/K,N}$ (cf. Remark 3.5(b)).

COROLLARY 3.14. *In the above situation we have*

$$(3.15) \qquad \#H^{\partial}_{E/K,N} = \frac{(4N-3)(N-2)}{24N} sl(N).$$

PROOF. We apply the Riemann-Hurwitz formula to the cover $\overline{\delta} = \overline{\delta}_{E/K,N}$: $X = X(N) \to \mathbb{P}^1$. Since $\overline{\delta}$ is tamely ramified (cf. [**Ka6**], Proposition 31) and since $\overline{\delta}$ is unramified outside of $\pi(E[2])$ (cf. Step 1 of the proof of Theorem 3.11), we have by Riemann-Hurwitz

$$2g_X - 2 = \deg(\overline{\delta})(-2) + 4\deg(\overline{\delta}) - \#(\overline{\delta}^{-1}(\pi(E[2]))).$$

It is easy to see that $\overline{\delta}^{-1}(\pi(E[2])) = \partial H \cup H^{\partial}_{E/K,N}$. Thus, since $2g_X - 2 = \frac{N-6}{12N} sl(N)$, since $\#\partial H = \frac{1}{24N}(5N+6)sl(N)$ by (3.9), and since $\deg\overline{\delta} = \frac{1}{12}(N-1)sl(N)$ (cf. (3.12) and (3.10)), we obtain that $\#H^{\partial}_{E/K,N} = 2\deg(\overline{\delta}) - \#\partial H - (2g_X - 2) = \frac{1}{6}(N-1)sl(N) - \frac{1}{24N}(5N+6)sl(N) - \frac{N-6}{12N} sl(N) = \frac{4N^2-11N+6}{24N} sl(N)$, and so (3.15) follows. $\qquad\square$

REMARK 3.15. If we compare the above formula (3.15) with (3.9), we see that there are more points in the relative boundary $H^{\partial}_{E/K,N}$ than in the boundary $\partial H_{E/K,N}$; more precisely, we have

$$\#H^{\partial}_{E/K,N} - \#\partial H_{E/K,N} = \frac{1}{6}(N-6)sl(N).$$

As yet another application of Theorem 3.11, we compute the number $c_{N,D} = \#\mathrm{Cov}^{all}_{E,N,D}$ of *all* genus 2 covers of E of fixed degree $N \geq 1$ and fixed discriminant divisor $D \in \mathrm{Div}(E)$. (Thus, here we no longer assume that the cover is normalized.) Since this number is closely related to the *weighted number* $\overline{c}_{N,D} := \sum_{f \in \mathrm{Cov}^{all}_{E,N,D}} \frac{1}{|\mathrm{Aut}(f)|}$ of such covers and since the latter leads to simpler formulae, we determine $\overline{c}_{N,D}$ as well.

COROLLARY 3.16. *If* char$(K) \nmid N!$ *and* $D \in \mathrm{Div}(E)$ *is an effective divisor of degree 2, then*

$$(3.16) \qquad \overline{c}_{N,D} = \frac{N}{3\mu_D}\left(\sigma_3(N) - N\sigma_1(N)\right) - \frac{\mu_D - 1}{24}\left(7\sigma_3(N) - (6N+1)\sigma_1(N)\right)$$

where $\mu_D = 1$ *if* D *is reduced and* $\mu_D = 2$ *otherwise, and where* $\sigma_k(n) = \sum_{d|n} d^k$ *denotes the sum of the kth powers of the divisors of n. Moreover, if we put* $\sigma_1(N/2) = 0$ *if* N *is odd, then the total number of genus 2 covers is given by*

$$(3.17) \qquad c_{N,D} = \overline{c}_{N,D} + \left(\frac{N}{\mu_D} - (\mu_D - 1)\right)\sigma_1(N/2).$$

PROOF. [**Ka6**], Theorem 1. As is explained in section 2 of that paper, this theorem can be deduced from Theorem 3.11. $\qquad\square$

REMARK 3.17. It is interesting to note that the number $\overline{c}_{N,D}$ (for D reduced) can be computed by a method that is essentially group-theoretic (and which goes back to Hurwitz[**Hu**]); cf. [**Dij**] and also [**Ma**]. Thus, by reversing the reasoning of the proof of the above corollary (by using a Moebius inversion formula), one can obtain a group-theoretical proof of Theorem 3.11.

4. Comparison of the Moduli Spaces

4.1. The Moduli Spaces H_N, H'_N **and** $Z_{N,-1}$. We now turn to the study of the moduli problems $\mathcal{H}_N$ and $\mathcal{H}'_N$ which were (briefly) introduced at the end of Subsection 2.2. For this, we need to extend the previous Hurwitz functor $\mathcal{H}_{E/K,N}$ (cf. Subsection 3.1) for a fixed elliptic curve E/K to a functor over a *variable* elliptic curve E/K.

A convenient framework for studying such questions is the concept of a *moduli problem for elliptic curves* as introduced by Katz-Mazur[**KM**], p. 107: by definition, such a moduli problem is a contravariant functor

$$\mathcal{P} : \underline{Ell}_{/R} \to \underline{Sets}$$

from the category $\underline{Ell}_{/R}$ to the category of sets. Here, as in [**KM**], p. 122, the *moduli stack $\underline{Ell}_{/R}$ of all elliptic curves* over a ring R is the category whose objects are (relative) elliptic curves E/S where S is any R-scheme, and whose morphisms from E/S to E'/S' consist of cartesian diagrams (with vertices E/S and E'/S') over R.

As is explained in [**KM**], pp. 108 and 125, each moduli problem $\mathcal{P}$ on $\underline{Ell}_{/R}$ gives rise to a contravariant functor

$$\widetilde{\mathcal{P}} : \underline{Sch}_{/R} \to \underline{Sets}$$

which classifies isomorphism classes of $\mathcal{P}$-structures. This functor $\widetilde{\mathcal{P}}$ gives us a more naive interpretation of our moduli problem. However, although the study of $\widetilde{\mathcal{P}}$ suffices for the discussion of coarse moduli spaces, it is the study of $\mathcal{P}$ which yields the most powerful (and natural) results.

Here we shall be interested in two (or three) specific moduli problems on $\underline{Ell}_{/R_N}$, where throughout $R_N = \mathbb{Z}[1/(2N)]$; these are defined by the rules

$$\mathcal{H}_N(E/S) = \mathcal{H}_{E/S,N}(S) \quad \text{and} \quad \mathcal{Z}_{N,-1}(E/S) = \mathcal{X}_{E/S,N,-1}(S),$$

for $E/S \in ob(\underline{Ell}_{/R})$, where the functors $\mathcal{H}_N(E/S)$ and $\mathcal{Z}_{N,-1}$ are as in Subsection 3.1. Clearly, each morphism in $\underline{Ell}_{/R}$ induces a natural map (by base-change) between the relevant sets, so that we obtain a moduli problem on $\underline{Ell}_{/R}$ as claimed.

In addition, we shall be interested in the moduli problem $\mathcal{H}'_N$ which is defined by rule $\mathcal{H}'_N(E/S) = \mathcal{H}'_{E/S,N}(S)$ (where $\mathcal{H}'_{E/S,N}$ is the subfunctor of $\mathcal{H}_{E/S,N}$ defined at the end of Subsection 3.1). Since this extra condition is obviously compatible with base change, we see that $\mathcal{H}'_N$ is a subfunctor of $\mathcal{H}_N$ and hence a moduli problem on $\underline{Ell}_{/R}$.

Note that the functor $\widetilde{\mathcal{H}}_N : \underline{Sch}_{/R} \to \underline{Sets}$ associated to $\mathcal{H}_N$ has the following natural interpretation:

$$\widetilde{\mathcal{H}}_N(S) = \{(C \xrightarrow{f} E) \in \mathcal{H}_N(E/S) : E/S \text{ is a (relative) elliptic curve}\}/\sim .$$

Moreover, $\widetilde{\mathcal{H}}'_N$ and $\widetilde{\mathcal{Z}}_{N,-1}$ have similar interpretations. Thus, although the moduli problems $\mathcal{H}_N$ and $\mathcal{H}'_N$ "classify" *isomorphism classes* of covers, their associated functors $\widetilde{\mathcal{H}}_N$ and $\widetilde{\mathcal{H}}'_N$ naturally "classify" *equivalence classes* of covers (in the sense of Definition 1.1).

THEOREM 4.1. *If $N \geq 3$, then the moduli problems $\mathcal{H}_N$, $\mathcal{H}'_N$ and $\mathcal{Z}_{N,-1}$ are relatively representable, (quasi-) affine, smooth and geometrically connected of relative* *dimension* 1 *over* $\underline{Ell}_{/R_N}$. *Furthermore, we have open embeddings (of functors)*

$$\Psi_N : \mathcal{H}_N \hookrightarrow \mathcal{Z}_{N,-1} \quad and \quad \Psi'_N : \mathcal{H}'_N \hookrightarrow \mathcal{H}_N \hookrightarrow \mathcal{Z}_{N,-1}.$$

PROOF. For a fixed elliptic curve E/S, the "fibre functor" of $\mathcal{H}_N$ at E/S is by construction just the functor $\mathcal{H}_{E/S,N}$, i.e.

$$(\mathcal{H}_N)_{E/S} = \mathcal{H}_{E/S,N} : \underline{Sch}_{/S} \to \underline{Sets}.$$

Now by Theorem 5.18 of [**Ka5**] the functor $\mathcal{H}_{E/S,N}$ is representable for every E/S, and this means precisely that $\mathcal{H}_N$ is relatively representable (cf. [**KM**], p. 108). (Actually, in [**Ka5**] the representability of $\mathcal{H}_{E/S,N}$ is asserted only for an affine base S, but from this the assertion follows (by gluing) for any base S.) Moreover, since the scheme $H_{E/S,N}$ representing $\mathcal{H}_{E/S,N}$ is quasi-affine, smooth and geometrically connected of relative dimension 1 over S, this means that functor $\mathcal{H}_N$ also has these properties by definition (cf. [**KM**], p. 109).

By construction, the fibre functor $(\mathcal{H}'_N)_{E/S} = \mathcal{H}'_{E/S,N}$ is a subfunctor of $(\mathcal{H}_N)_{E/S}$, and it is easy to see that $(\mathcal{H}'_N)_{E/S}$ is represented by an open subscheme $H' = H'_{E/S,N}$ of $H = H_{E/S,N}$. If S is a field, then this was proven in Proposition 3.4. In the general case we have $H' = H \setminus Z$, where Z is the closed subscheme of H which is universal for the relation $\mathrm{Disc}(f_H) \leq E_H[2]^{\#}$ in the sense of [**KM**], Key Lemma 1.3.4. Here, $f_H : C \to E_H$ denotes the universal normalized genus 2 cover of degree N over E_H and $E[2]^{\#} = E[2] - 0_{E_H}$ (viewed as a Cartier divisor as in [**Ka5**]).

The proof of the assertions about $\mathcal{Z}_{N,-1}$ are entirely analogous to those for $\mathcal{H}_N$; here we use [**Ka5**], Corollary 4.3 in place of Theorem 5.18. (Note that $\mathcal{X}_{E/S,N,-1}$ is the fibre functor of $\mathcal{Z}_{N,-1}$ at E/S.) Finally, the open embedding Ψ_N is induced by the (open) embeddings $\Psi_{E/S,N} : \mathcal{H}_{E/S,N} \to \mathcal{X}_{E/S,N,-1}$; cf. (3.2) or [**Ka5**], Theorem 5.18.   □

REMARK 4.2. The above moduli problems $\mathcal{H}'_N, \mathcal{H}_N$ and $\mathcal{Z}_{N,-1}$ are in fact affine and not just quasi-affine. For $\mathcal{Z}_{N,-1}$, this is immediate from the above construction, but not for $\mathcal{H}'_N$ and $\mathcal{H}_N$. Since we don't need this here, we omit the proof.

It is clear that none of the functors $\mathcal{H}_N$, $\mathcal{H}'_N$ and $\mathcal{Z}_{N,-1}$ is representable, for none is rigid (because the $[-1]$ map acts trivially on each). However, since all are (quasi-) affine and relatively representable, there exists a *coarse moduli scheme* $H_N := M(\mathcal{H}_N)$, and $H'_N := M(\mathcal{H}'_N)$ and $Z_{N,-1} = M(\mathcal{Z}_{N,-1})$ for each; cf. [**KM**], p. 224. Note that $Z_{N,-1}$ (and hence also H_N and H'_N) comes equipped with a natural morphism

$$p_N : Z_{N,-1} \to M_1 := \mathrm{Spec}(R_N[j]) = M([\Gamma(1)]),$$

to the j-line over R_N which is the coarse moduli space $M_1 = M([\Gamma(1)])$ attached to the moduli problem $[\Gamma(1)]$ classifying isomorphism classes of elliptic curves; cf. [**KM**], p. 229. Thus we have:

COROLLARY 4.3. *The coarse moduli scheme* $H_N := M(\mathcal{H}_N)$ *of* $\mathcal{H}_N$ *is an open subscheme of the coarse moduli scheme* $Z_{N,-1} = M(\mathcal{Z}_{N,-1})$ *of* $\mathcal{Z}_{N,-1}$. *The latter is normal and affine of relative dimension* 1 *over* $M_1 = \mathrm{Spec}(R_N[j])$. *In addition, the coarse moduli scheme* $H'_N := M(\mathcal{H}'_N)$ *of* $\mathcal{H}'_N$ *is an open subscheme of* H_N *and hence also of* $Z_{N,-1}$.

REMARK 4.4. As a first approximation, we can view the previously considered moduli spaces $H_{E/K,N}$ as the fibres of the morphism $p_N : H_N \to M_1$. Indeed, since each E/K gives rise to a point $x_E \in M_1$ such that $\kappa(x_E) = K$ and $j(E) = j(x_E)$, we might expect that the fibre $(H_N)_x$ over x is "essentially" (i.e. up to twists) the curve $H_{E/K,N}$. While this is true "in general", it is not always true. For example, at a (geometric) point x where $\mathrm{Aut}(E_x) \neq \pm 1$, the fibre of p_N is an affine curve whose compactification has lower genus than those of the generic case.

By the above corollary we see that H_N and $Z_{N,-1}$ are (quasi-) affine normal surfaces over $\mathrm{Spec}(R_N)$. We next want to identify the surfaces in terms of the modular diagonal quotient surfaces introduced in [**KS**].

Even though these surfaces can be defined over R_N (cf. [**Ka7**]), we shall explain their construction only over the ring $\tilde{R}_N = R_N[\zeta_N] = \mathbb{Z}[\frac{1}{2N}, \zeta_N]$, where ζ_N is a primitive N-th root of unity. (This situation suffices for geometric purposes, but not for arithmetic applications.) Over $\tilde{R}_N$ we have the smooth affine curve $X(N)'_{/\tilde{R}_N}$ which represents the moduli problem $[\Gamma(N)]^{can}$ of [**KM**], p. 283. It comes equipped with a natural action of the group $G_N = \mathrm{SL}_2(\mathbb{Z}/N\mathbb{Z})/\{\pm 1\}$; thus $G_N \times G_N$ acts on the product $Y_N = X(N)'_{/\tilde{R}_N} \times_{\tilde{R}_N} X(N)'_{/\tilde{R}_N}$. For any $\varepsilon \in (\mathbb{Z}/N\mathbb{Z})^\times$, let

$$\Delta_\varepsilon = \{(g, \tau_\varepsilon(g)) : g \in G_N\} \leq G_N \times G_N$$

denote the "twisted diagonal subgroup", i.e. the graph of the automorphism $\tau_\varepsilon \in \mathrm{Aut}(G_N)$ defined by $\tau_\varepsilon(g) = \sigma_\varepsilon g \sigma_\varepsilon^{-1}$, where $\sigma_\varepsilon = \left(\begin{smallmatrix} \varepsilon & 0 \\ 0 & 1 \end{smallmatrix}\right) \in \mathrm{GL}_2(\mathbb{Z}/N\mathbb{Z})$. Then the quotient

$$Z_{N,\varepsilon} = Y_N/\Delta_\varepsilon$$

is called the *modular diagonal quotient surface* of type (N, ε).

PROPOSITION 4.5. *The modular diagonal quotient surface $Z_{N,\varepsilon}$ is the coarse moduli space of the moduli problem $\mathcal{Z}_{N,\varepsilon}$, i.e. $M(\mathcal{Z}_{N,\varepsilon}) = Z_{N,\varepsilon}$, and hence the coarse moduli spaces H_N and H'_N are open subsets of $Z_{N,-1}$.*

PROOF. (Sketch.) As was mentioned above, the moduli problem $[\Gamma(N)]^{can}_{/\tilde{R}_N}$ is represented by E_N/X, where $X = X(N)'_{\tilde{R}_N}$, Thus, since $\mathcal{Z}_{N,\varepsilon}$ is affine and relatively representable by Theorem 4.1, it follows that the simultaneous moduli problem $\mathcal{P}_{N,\varepsilon} := [\Gamma(N)]^{can} \times \mathcal{Z}_{N,\varepsilon}$ is represented by $p^* E_N/Y_\varepsilon$ where $p : Y_\varepsilon \to X$ represents the functor $\mathcal{X}_{E_N/X,N,\varepsilon}$ on $\underline{Sch}_{/X}$; cf. [**KM**], p. 109. Thus, by [**KM**], p. 224, we see that the coarse moduli space of $\mathcal{Z}_{N,\varepsilon}$ is $M(\mathcal{P}_{N,\varepsilon})/G_N$ because $[\Gamma(N)]^{can}_{/\tilde{R}_N}$ is étale with group G_N.

Now $Y_\varepsilon \simeq X \times_{\tilde{R}_N} X$ because X also represents the functor $\mathcal{X}_{E_N/\tilde{R}_N,N,\varepsilon}$ on $\underline{Sch}_{\tilde{R}_N}$, and via this identification the group G_N is mapped to $\Delta_\varepsilon \leq G_N \times G_N$. Thus $M(\mathcal{Z}_{N,\varepsilon}) = Z_{N,\varepsilon}$, as claimed. The last assertion follows from Corollary 4.3. $\square$

While the above result gives a concrete interpretation of the coarse moduli space $Z_{N,-1}$ and hence of the subspaces H_n and H'_n, it has the disadvantage that its interpretation as a coarse moduli space has nothing to do with covers of curves. As a result, it seems somewhat mysterious how the spaces H_n and H'_n are embedded in $Z_{n,-1}$. To remedy this, we now explain how $Z_{n,-1}$ is the coarse moduli space of a suitable Hurwitz functor.

As a first step towards this, let $\mathcal{A}_N$ denote the moduli problem on $\underline{Ell}$ which, for a fixed elliptic curve E/S, is the set

$$\mathcal{A}_N(E/S) = \mathcal{A}_{E/S,N}(S) = \{(J, \lambda, h)\}/\simeq,$$

of isomorphism classes of triples (J, λ, h) consisting of a principally polarized abelian surface (J, λ) over S and an injective homomorphism $h : E \to J$ of degree N in the sense of [**Ka5**], §5.1 (p. 24). We then have:

PROPOSITION 4.6. *There is a natural isomorphism* $\mathbf{\Psi}' : \mathcal{A}_N \xrightarrow{\sim} \mathcal{Z}_{N,-1}$ *of moduli problems, and hence* $Z_{N,-1}$ *is also a coarse moduli scheme of* $\mathcal{A}_N$.

PROOF. For each $E/S \in ob(\underline{Ell})$ we have by Theorem 5.10 of [**Ka5**] an isomorphism $\mathbf{\Psi}'_{E/S} : (\mathcal{A}_N)_{E/S} = \mathcal{A}_{E/S,N} \xrightarrow{\sim} (\mathcal{Z}_{N,-1})_{E/S} = \mathcal{X}_{E/S,N,-1}$ of the corresponding fibre functors, and these fit together to define the desired isomorphism. The last assertion follows from Proposition 4.5. $\square$

We next need to identify $\mathcal{A}_N$ with a Hurwitz moduli problem. This, however, is more problematic. Consider first the moduli problem H_N^{st} on $\underline{Ell}$ which, for a given E/S, classifies finite flat covers $f : C \to E$ of degree N where C/S is a *stable* curve of genus 2 (in the sense of [**DM**]), i.e.

$$\mathcal{H}_N^{st}(E/S) = \{f : C \to E \text{ finite, flat}, C/S \text{ stable of genus } 2, \deg(f) = N\}/\simeq.$$

Next, let $\mathcal{H}_N^{\S}$ be the subfunctor of $\mathcal{H}_N^{st}$ consisting of the those S-covers $f : C \to E$ such that $(J_{C_s}, \theta_{C_s}, f_s^*) \in \mathcal{A}_N(\mathrm{Spec}(\kappa(s)))$, for every $s \in S$; here $J_{C_s}/\kappa(s)$ is the Jacobian of the fibre C_s and θ_s its canonical polarization. Since the Jacobian exists for any stable curve C/S (cf. [**BLR**], Th. 9.4/1), the rule $(f : C \to E) \mapsto (J_C, \theta, f^*)$ defines a morphism of moduli problems

$$j_N^{\S} : \mathcal{H}_N^{\S} \to \mathcal{A}_N.$$

However, $j_N^{\S}$ fails to be injective since no normalization condition was imposed. To remedy this, we might like to impose (if $N = n$ is odd) the normalization condition (2.1), but this will not work if the cover $f_s : C_s \to E_s$ is reducible (i.e. C_s is a union of two elliptic curves) because the hyperelliptic involution σ_{C_s} now has 7 (in place of 6) fixed points. Instead, we could require that (in all cases) $f^{-1}(0_{E_s}) \cap \mathrm{Fix}(\sigma_{C_s})$ has at least 3 points, and (as will be shown elsewhere) this does lead to a moduli problem $\mathcal{H}_n^{\dagger}$ which is isomorphic to $\mathcal{A}_n$ and which contains H_n as a natural open subproblem.

4.2. Hurwitz Moduli Stacks. We now analyze the moduli space H_n^* which was introduced in Subsection 1.3. It turns out the corresponding moduli problem $\mathcal{H}_n^*$ can be studied by a formalism which is very similar to that used for the moduli problem $\mathcal{H}_n$. Here, however, we have to replace the stack $\underline{Ell}_{/R}$ of elliptic curves over R by the stack $(\underline{M_0})_{/R}$ of curves of genus 0 over R. The latter is defined completely analogous to $\underline{Ell}_{/R}$, i.e. $(\underline{M_0})_{/R}$ is the category fibred in groupoids over $\underline{Sch}_{/R}$ whose objects are relative (smooth, proper) curves $X \to S$ of genus 0, and whose morphisms are cartesian diagrams.

We shall now interpret $\mathcal{H}_n^*$ as a *moduli problem on* $(\underline{M_0})_{/R}$, i.e. as a contravariant functor $\mathcal{H}_n^* : (\underline{M_0})_{/R} \to \underline{Sets}$; here R is any ring in which $n!$ is invertible.

For this, let $X/S \in ob((\underline{M}_0)/R)$ be a relative genus 0 curve, and consider the set

$$\mathrm{Cov}^*_{X/S,n} := \{\varphi : Y \to X : \quad \varphi \text{ is a finite cover of degree } n$$
$$\text{whose fibres } \varphi_t \text{ satisfy } (*) \}/ \simeq_X .$$

Then the rule $\mathcal{H}^*_n(X/S) = \mathrm{Cov}^*_{X/S,n}$ defines a moduli problem on $(\underline{M}_0)/R$, i.e. a functor $\mathcal{H}^*_n : (\underline{M}_0)/R \to \underline{Sets}$.

Similar to the case of moduli problems on $\underline{Ell}/R$, each moduli problem $\mathcal{P}$ on $(\underline{M}_0)/R$ gives rise to a functor $\widetilde{\mathcal{P}} : \underline{Sch}/R \to \underline{Sets}$ by considering isomorphism classes of $\mathcal{P}$-structures. In particular, for $\mathcal{P} = \mathcal{H}^*_n$ we obtain:

$$\widetilde{\mathcal{H}}^*_n(S) = \{(Y \to X) \in \mathrm{Cov}^*_{X/S,n} : X/S \text{ is a (relative) curve of genus } 0\}/\sim,$$

where, as before, the indicated equivalence relation is the weak equivalence of covers. Note that if $S = \mathrm{Spec}(K)$ is a field, then the set $\widetilde{\mathcal{H}}^*_n(S)$ coincides with the set $\widetilde{\mathcal{H}}^*_n(K)$ defined in Remark 1.5(a).

REMARK 4.7. (a) Although we don't need this here, we observe that the moduli problem $\mathcal{H}^*_n \simeq \tilde{H}^*_n$ is *relatively representable*, i.e. for each $X/S \in ob((\underline{M}_0)/R)$, the fibre functor $(\mathcal{H}^*_n)_{X/S} : \underline{Sch}/S \to \underline{Sets}$ defined by $T \mapsto \mathcal{H}^*_n(X_T/T)$ is representable. Indeed, if $X = \mathbb{P}^1_K$, where K is a field, then the results of Fried-Völklein and Wewers show that this functor is represented by the scheme $H^{in}(S_n, \mathbf{C}) \otimes K$, as was explained in subsection 1.3. For a general scheme X/S, this follows from the results of Wewers [**We**].

(b) For later usage we observe here that the concept of a moduli problem on $(\underline{M}_0)/R$ can be reformulated in terms of *moduli stacks* over $(\underline{M}_0)/R$: the latter are categories $p : \underline{C} \to (\underline{M}_0)/R$ over $(\underline{M}_0)/R$ which are fibred in groupoids. Indeed, given a moduli problem $\mathcal{P} : (\underline{M}_0)/R \to \underline{Sets}$, we can define its *classifying stack* $\underline{\mathcal{P}}$ by the rule that its fibre category over $X/S \in ob((\underline{M}_0)/R)$ is $\underline{\mathcal{P}}_{X/S} = \mathcal{P}(X/S)$ (where the morphisms are the identities). Conversely, each moduli stack $\underline{C}$ over $(\underline{M}_0)/R$ gives rise to a moduli problem $\mathcal{P}_{\underline{C}}$ on $(\underline{M}_0)/R$.

(c) In the case of the moduli problem $\mathcal{H}^*_n$, it is immediate that the associated classifying stack is equivalent (in the sense of categories) to the fibred category $\underline{\mathcal{H}}^*_n$ whose objects are triples $Y \to X \to S$ with $X/S \in ob((\underline{M}_0)/R)$ and $Y/X \in \mathcal{H}^*_n(X/S)$, and whose morphisms are (extended) cartesian diagrams. Thus, by the discussion of Subsection 1.3, we see that we can identify $\underline{\mathcal{H}}^*_n$ with the stack $\mathcal{H}_{0,0,m}$ of [**BR**], Theorem 6.22, for a suitable monodromy $m = (G, H, \xi)$. (More precisely, $G = S_n$ and $H \simeq S_{n-1}$ is the stabilizer of an element of the set $\{1, \ldots, n\}$ (on which S_n acts), and ξ is the Hurwitz data given by the class $\mathbf{C}$.)

It is clear that the moduli problem $\mathcal{H}^*_n$ cannot be representable because the group scheme $\mathrm{Aut}(\mathbb{P}^1)$ acts non-trivially. However, we have:

PROPOSITION 4.8. *The moduli problem $\mathcal{H}^*_n$ has a coarse moduli space which is given by*

$$M(\mathcal{H}^*_n) = H^*_n = H^{in}(S_n, \mathbf{C})/\mathrm{Aut}(\mathbb{P}^1).$$

PROOF. This follows easily from [**BR**]. Indeed, via identification $\underline{\mathcal{H}}^*_n = \mathcal{H}_{0,0,m}$ of Remark 4.7(c) it follows from [**BR**], Théorème 6.22 that $\underline{\mathcal{H}}^*_n \simeq \mathcal{H}_{g,S_n,\xi}/\Delta(m)$ for a suitable integer g. But since here $N_G(H) = H$, we see that $\Delta(m) = 1$, and so $\underline{\mathcal{H}}^*_n \simeq \mathcal{H}_{g,S_n,\xi}$. Now by formula (6.17) of [**BR**] we know that $\mathcal{H}_{g,S_n,\xi} \simeq$

$\mathcal{F}H_{g,S_n,\xi}/\mathrm{Aut}(\mathbb{P}^1)$. Since the results of Fried/Völklein/Wewers[**We**] show that the Fried-Völklein stack $\mathcal{F}H_{g,S_n,\xi}$ is representable by $H^{in}(S_n, \mathbf{C})$, the assertion about the coarse moduli spaces follows from [**BR**], Théorème 6.3 (3). □

4.3. The Main Result. We now want to relate the moduli problem $\mathcal{H}'_n$ (on $\underline{Ell}_{/R}$) to the moduli problem $\mathcal{H}^*_n$ (on $(\underline{M}_0)_{/R}$). In order to be able to compare them, we shall pass to their classifying stacks $\underline{\mathcal{H}}'_n$ and $\underline{\mathcal{H}}^*_n$ (over $\underline{Ell}$ and over $\underline{M}_0$, respectively), and view both of these as fibred categories over $\underline{Sch}_{/R}$; cf. Remark 4.7(b). We then have:

THEOREM 4.9. *The rule* $(C \xrightarrow{f} E) \mapsto (C/\langle\omega_C\rangle \to E/\langle[-1]\rangle)$ *defines a functor* $q = q_n : \underline{\mathcal{H}}'_n \to \underline{\mathcal{H}}^*_n$, *and the induced map*

$$M(q) : H'_n = M(\mathcal{H}'_n) \to H^*_n = M(\mathcal{H}^*_n)$$

on the coarse moduli schemes is surjective and radical. Thus, if $R = K$ is a field of characteristic 0, then $M(q)$ is an isomorphism and hence $H'_n \simeq H^*_n$ is an irreducible, normal affine surface.

PROOF. Let $(C \xrightarrow{f} E) \in ob(\underline{\mathcal{H}}'_n)$, where C and E are S-curves. Then, since f is normalized, we have $f\omega_C = [-1]f$, and so we have an induced morphism $\varphi : Y := C/\langle\omega_C\rangle \to X := E/\langle[-1]\rangle$. (The quotients exist and are smooth S-curves by Lønsted-Kleiman[**LK**], Theorem 4.12.) Moreover, if $s \in S$, then $\mathrm{Disc}(f_s)$ is reduced (by hypothesis), and by Proposition 2.4 we see that φ_s is a cover of type (∗), which means that $(Y \xrightarrow{\varphi} X) \in ob(\underline{\mathcal{H}}^*_n)$. Since this construction is compatible with base-change (because 2 is invertible in S), we see that this defines the desired functor q.

The functor q induces a natural transformation $\widetilde{q} : \widetilde{\mathcal{H}}'_n \to \widetilde{\mathcal{H}}^*_n$ between the associated functors on $\underline{Sch}_{/R}$. Thus, by the universal property of coarse moduli spaces, there is unique morphism $M(q) = M(\widetilde{q}) : H'_n = M(\mathcal{H}'_n) \to H^*_n = M(\mathcal{H}^*_n)$ such that $\mu'\widetilde{q} = h_{M(q)}\mu^*$, where $\mu' : \mathcal{H}^*_n \to h_{M(\mathcal{H}'_n)}$ and $\mu^* : \mathcal{H}^*_n \to h_{M(\mathcal{H}^*_n)}$ are the natural transformations associated to the coarse moduli spaces and, as usual ([**EGA**], $(0_I, 1.1.2)$), h_X denotes the functor associated to the object X. (Recall that Propositions 4.5 and 4.8 show that we have the identifications $M(\mathcal{H}'_n) = H'_n$ and $M(\mathcal{H}^*_n) = H^*_n$.)

Now the discussion of Subsection 2.2 shows:

Claim: If $S = \mathrm{Spec}(k)$, where k is an algebraically closed field, then

$$\widetilde{q}_S : \widetilde{\mathcal{H}}'_n(S) \to \widetilde{\mathcal{H}}^*_n(S)$$

is a bijection.

Thus, since $\mu'_S : \widetilde{\mathcal{H}}'_n(S) \xrightarrow{\sim} h_{M(\mathcal{H}'_n)}(S) = H'_n(k)$ and $\mu^*_S : \widetilde{\mathcal{H}}^*_n(S) \xrightarrow{\sim} \widetilde{\mathcal{H}}^*_n(S) = H^*_n(k)$ are bijections, it follows that $M(q)_k : H'_n(k) \to H^*_n(k)$ is a bijection. Thus, $M(q)$ is surjective and radical; cf. [**EGA**], (I,3.6.3) and (I,3.7.1). Moreover, if $R = K$ is a field of characteristic 0, then H'_n is normal and irreducible (cf. Proposition 4.5), and so we see that H^*_n is also irreducible, and that hence $M(q)$ is a birational map. In addition, $M(q)$ is separated because H'_n is quasi-affine and H^*_n is affine. Thus, since H^*_n is normal (cf. Remark 1.5(b)), it follows from Zariski's Main Theorem that $M(q)$ is a local isomorphism; cf. [**EGA**], (Err_{IV}, 30). By [**EGA**], (I, 4.4.8) we thus have that $M(q)$ is an open immersion, and hence an isomorphism since $M(q)$ is surjective. □

REMARK 4.10. (a) Even though $M(q)$ is an isomorphism, the functor

$$q : \underline{\mathcal{H}}'_n \to \underline{\mathcal{H}}^*_n$$

is not an isomorphism (of stacks). Indeed, q cannot be an equivalence (of categories) because of the presence of twists of elliptic curves. More precisely, if $K \supset R$ is any field for which $K^*/(K^*)^2 \neq 1$, then the discussion of Subsection 2.2 shows that $\widetilde{q}_K : \widetilde{\mathcal{H}}'_n(K) \to \widetilde{\mathcal{H}}^*_n(K)$ cannot be injective, and so in particular q cannot be an equivalence of categories.

(b) In his preprint, Fried[**Fri**] proves the irreducibility of $H^{in}(S_n, \mathbf{C})$ by the method of Nielsen classes. This gives a different proof of the irreducibility of H^*_n.

4.4. Connection with Humbert surfaces. If H is any Hurwitz scheme which classifies curve covers $f : C \to C_1$ with fixed genus $g = g_C$, then the forget map $(f : C \to C_1) \mapsto C$ induces a natural morphism

$$\mu_H : H \to M_g$$

to the moduli space M_g which classifies isomorphism classes of genus g curves. The image $\mu_H(H)$ is thus a subscheme of M_g which frequently carries interesting information. For example, if $H = H_{d,g}$ is the Hurwitz space of simple genus g covers of degree d of $\mathbb{P}^1$, then the $\mu_H(H_{d,g})$'s $(d \geq 2)$ define cycles on M_g which are important for the study of the geometry of M_g; cf. [**HM**], p. 32, 175, for more detail.

Specializing this to our case $H = H_N$ which classifies normalized genus 2 covers, we thus obtain a morphism

$$\mu_N = \mu_{H_N} : H_N \to M_2.$$

It turns out that the image of μ_N is essentially a *Humbert surface*. To make this more precise, recall (cf. [**Hum**] or [**vdG**], section IX.2) that for each positive integer $D \equiv 0, 1(4)$, there is an irreducible surface Hum_D contained in the space A_2 of principally polarized abelian surfaces; Hum_D is called the *Humbert surface* with *Humbert invariant* (or discriminant) D.

Since each point of $\mu_N(H_N)$ corresponds to a genus 2 curve with an elliptic differential, it follows from Humbert[**Hum**], Théorème 15, [3] that $\mu_N(H_N)$ is contained in some Humbert surface Hum_D with square invariant $D = d^2$; here we view M_2 as an open subset of A_2 via the (birational) map $M_2 \to A_2$ (which take a curve to its polarized Jacobian). More precisely:

PROPOSITION 4.11. *We have* $\mu_N(H_N) = \mathrm{Hum}_{N^2} \cap M_2$. *More precisely, if* τ *denotes the involution of* $Z_{N,-1}$ *induced by the automorphism of* $X(N) \times X(N)$ *which interchanges the two factors, then* μ_N *factors over the quotient* $H_N^{sym} := H_N/\langle \tau \rangle$, *and the induced map*

$$\mu_N^{sym} : H_N^{sym} \to \mathrm{Hum}_{N^2} \cap M_2$$

is the normalization map of the Humbert surface $\mathrm{Hum}_{N^2} \cap M_2$.

PROOF. The first assertion follows from Theorem 1.9 of [**Ka2**] (together with the fact that every minimal cover can be normalized; cf. [**Ka5**], Proposition 2.2). The second assertion follows from Corollary 1.8 of [**Ka2**] because H_N is precisely

---

[3]In fact, this result was already proven by Biermann in 1883 (cf. [**Kr**], p. 485), long before the discovery of Humbert surfaces.

the (open) part of $Z_{N,-1}$ whose points correspond to genus 2 curve covers; cf. Corollary 4.3, together with the discussion of the next section. $\hfill\square$

REMARK 4.12. If D is not a square, then the Humbert surface H_D is birationally isomorphic to the *symmetric Hilbert modular surface* Hilb_D^{sym}; the latter is the quotient of the *Hilbert modular surface* $\mathrm{Hilb}_D = \mathrm{SL}_2(\mathfrak{O}_D)\backslash(\mathfrak{H}\times\mathfrak{H})$ by the involution τ induced by the involution of $\mathfrak{H}\times\mathfrak{H}$ which interchanges the two factors; cf. [**vdG**], Proposition (IX.2.5). Thus, the Corollary 1.8 of [**Ka2**] may be viewed as an extension of this fact to square discriminants by viewing the modular diagonal quotient surfaces $Z_{N,\varepsilon}$ as "degenerate Hilbert modular surfaces". This is the point of view taken by Hermann, who investigated the geometry of $Z_{N,\varepsilon}$ and of $Z_{N,\varepsilon}^{sym} = Z_{N,\varepsilon}/\langle\tau\rangle$ as analogues of Hilbert modular surfaces; cf. [**He1**], [**He2**].

This point of view is also adopted by McMullen in his papers. Indeed, he uses a more general definition of Hilbert modular surfaces Hilb_D in which also square discriminants D are allowed, and then shows in all cases that Hum_D is birationally isomorphic to Hilb_D^{sym}; cf. [**Mc2**], Theorem 4.5.

In the case that $D = N^2$, this result may be stated as follows: if $C/\overline{K}$ is a curve of genus 2 with Jacobian J_C, then

$$\langle C\rangle \in \mathrm{Hum}_{N^2}(\overline{K}) \;\Leftrightarrow\; \mathrm{End}(J_C) \text{ has real multiplication by } \mathfrak{O}_{N^2};$$

the latter means that we have an embedding $\rho : \mathfrak{O}_{N^2} := \mathbb{Z}[x]/(x^2-Nx) \hookrightarrow \mathrm{End}(J_C)$ which is *primitive*, i.e. $\rho(\mathfrak{O}_{N^2})\mathbb{Q}\cap\mathrm{End}(J_C) = \rho(\mathfrak{O}_{N^2})$. [Indeed, if $\langle C\rangle \in \mathrm{Hum}_{N^2}(\overline{K})$, then by Proposition 4.11 we have a (normalized) subcover $f : C \to E$ of degree N to some elliptic curve E, and then the rule $x \mapsto f^*f_*$ induces such an embedding. Conversely, if we have such an embedding, then $\varepsilon = \frac{1}{N}\rho(x) \in \mathrm{End}^0(J_C)$ is a nontrivial idempotent, and one concludes (using Lange[**Lan**]) that $\rho(x) = N\varepsilon = f^*f_*$, for a suitable minimal cover $f : C \to E$.]

5. The Compactification of H_n' and of H_n^* and Related Boundary Curves

5.1. Compactifications: an Overview. As the constructions of the previous section show, the moduli spaces H_n' and H_n^* are not compact. It is thus of interest to construct natural compactifications of these spaces and to investigate whether or not the boundary components have a modular interpretation in terms of covers of curves.

Since H_n' was constructed as an open subset of the affine surface $Z_{n,-1}$, the natural compactification $\overline{Z}_{n,-1}$ of $Z_{n,-1}$ (which was studied in [**KS**]) also serves as a compactification of H_n'. While this compactification has the advantage of being very explicit, its disadvantage is that it does not readily lead to a modular interpretation in terms of covers. However, such an interpretation can be obtained by studying the fibres of the canonical compactification $\mathcal{C}$ of the universal cover $C \to E_H$ over $H = H_{E/K,n}$ for each fixed elliptic curve E/K; cf. Subsection 3.3.

More precisely, we saw in Theorem 3.9 that the study of the fibres of $\mathcal{C}$ lead to the following four types of covers $f_x : \mathcal{C}_x \to E$ of E:

I. $\mathcal{C}_x$ is a smooth curve of genus 2 and $\mathrm{Diff}(f_x)$ is reduced;
II. $\mathcal{C}_x$ is smooth curve of genus 2 and $\mathrm{Diff}(f_x) = 2P$ with $P \in E[2]$;
III. $\mathcal{C}_x$ is the union of two elliptic curves meeting at a point;
IV. $\mathcal{C}_x$ is a singular irreducible curve of arithmetic genus 2.

These cases correspond to the stratification

$$X(N) = H'_{E/K,N} \stackrel{.}{\cup} H^{\partial}_{E/K,N} \stackrel{.}{\cup} S_1 \stackrel{.}{\cup} S_0$$

which was constructed in Section 3. By varying E, we should expect that these induce a similar stratification for $\overline{Z}_{N,-1}$. This, however, is not quite correct since we also have to allow the case that E degenerates. Thus, we have a fifth case:

V. $C \to E$ is a cover of a stable curve E of arithmetic genus 1.

The precise nature of the covers of type V will be discussed in the next section.

On the other hand, the theory of Wewers [**We**] gives a recipe for an (abstract) compactification of $H^{in}(S_n, \mathbf{C})$ in terms of covers of $\mathbb{P}^1$; here the boundary components correspond to covers in which the monodromy group, the ramification type and the curve are (suitable) degenerations of the generic situation. This, therefore, also gives a compactification of $H_n^* = H^{in}(S_n, \mathbf{C})/\mathrm{Aut}(\mathbb{P}^1) \simeq H'_n$, and so one might expect that these boundary components match up with those of H'_n defined by the first method. This is indeed the case. Thus, in both interpretations the boundary curves of the Hurwitz spaces correspond to interesting degenerations of covers.

5.2. The Boundary Curves of H'_n. In the case of the Hurwitz space $H'_n \subset Z_{n,-1}$ which parameterizes normalized genus 2 covers of elliptic curves, the above overview (§5.1) shows that boundary curves naturally split into four main types. As we shall see, these correspond to the stratification induced by the inclusions

$$H'_n \subset H_n \subset Z_{n,-1} \subset \overline{Z}^{(1)}_{n,-1} \subset \overline{Z}_{n,-1}.$$

Here $\overline{Z}^{(1)}_{n,-1} = \psi_1^{-1}(X'(1))$ is the inverse image of $X'(1) \simeq \mathbb{A}^1$ of the morphism

$$\psi_1 = pr_1 \circ \psi : \overline{Z}_{n,-1} \to X(1) \times X(1) \to X(1) \simeq \mathbb{P}^1,$$

where, as in [**KS**], $\psi : \overline{Z}_{n,-1} \to X(1) \times X(1)$ is the map induced by the inclusion $\Delta_{-1}(G_n) \leq G_n \times G_n$. Note that the fibres of ψ_1 are geometrically irreducible (but not necessarily reduced); cf. [**KS**], p. 353.

We have two tasks: (i) to interpret (if possible) these boundary curves as the (coarse) moduli curves of suitable Hurwitz functors; and (ii) to identify the components of the boundary curves and to study their geometric properties.

Type I: the generic case (the points in H'_n):

By the modular description of H'_n mentioned in subsection 4.1, the points of H'_n classify genus 2 covers of type I; i.e. normalized genus 2 covers $f : C \to E$ of some elliptic curve E with reduced discriminant; cf. Corollary 4.3. We also know that H'_n is an open subset of the irreducible projective normal surface $\overline{Z}_{N,-1}$.

Type II: The points in $\partial_{II} := H_n \setminus H'_n$.

Since H_n is a coarse moduli scheme for the moduli problem $\mathcal{H}_n$ (cf. Corollary 4.3), we would expect that ∂_{II} is the coarse moduli space for the (closed) sub-problem $\partial \mathcal{H}_n$ of $\mathcal{H}_n$ which classifies normalized genus 2 covers with non-reduced discriminants. This is almost correct. It is easy to see (using Theorem 3.1 and Proposition 3.4) that $\partial \mathcal{H}_n$ is relatively representable and closed in $\mathcal{H}_n$. Thus, we have a coarse moduli space $M(\partial \mathcal{H}_n)$ (of dimension 1) and a canonical morphism $M(\partial \mathcal{H}_n) \to M(\mathcal{H}_n) = H_n$ whose image is ∂_{II}. However, without further work we only know that this map is *radical*, but not necessarily that it is a closed immersion (or that $M(\partial \mathcal{H}_n)$ is reduced).

Recall from subsection 3.2 that the covers of Type II break up into four sub-cases, which we shall label here as types $II_1, \ldots, II_4$. (Thus, for $i = 1, \ldots, 4$, type II_i corresponds to case i of subsection 3.2.) These give rise to subproblems $\partial\mathcal{H}_n^{(i)}$ of $\mathcal{H}_n$, and the corresponding coarse moduli schemes induce curves $\partial_{II_i} = \partial H_n^{(i)}$ on H_n. One might expect these curves to be irreducible, and to have large genus for large n.

REMARK 5.1. The work of [**Mc1**] and [**HL1**] (and others) shows that this expectation is correct for the curves ∂_{II_2} and ∂_{II_4}. More precisely, they show that their (birational) images in M_2 with respect to $\mu_n : H_n \to M_2$ (as defined in Subsection 4.4) can be interpreted as *Teichmüller curves*. This means in particular that for $i = 0, 1$ there exist subgroups $\Gamma_{i,n} \leq \mathrm{SL}_2(\mathbb{Z})$ of finite index and a birational map

$$\Gamma_{i,n}\backslash\mathfrak{H} \;\to\; \mu_n(\partial H_n^{(2+2i)}) \subset M_2, \quad \text{for } i = 0, 1.$$

Here i is the value of the *spin* as defined in [**Mc1**]. As Hubert/Lelièvre show, $\Gamma_{i,n}$ has index $d_{i,n} = \frac{3}{16n}(n - 1 - 2i)sl(n)$ in $\mathrm{SL}_2(\mathbb{Z})$, at least if n is prime; note that the case $i = 0$ corresponds to their type A. Furthermore, they show that the genus of both curves is asymptotic to $\frac{1}{64}n^3$ as $n \to \infty$; cf. [**HL1**], Theorem 1.2. (Moreover, one can show, using the results of [**Mc1**], that their genus is ≥ 3 for $n \geq 11$, n prime.) It is interesting to note that in general $\Gamma_{i,n}$ is not a congruence subgroup; cf. [**HL2**].

Type III: The points in $\partial_{III} := Z_{n,-1} \setminus H_n$.

By the discussion at the end of Subsection 4.1 we know that $Z_{n,-1}$ is a coarse moduli space for the functor $\mathcal{A}_n$ and so the points in ∂_{III} correspond to singular (stable) genus 2 curves whose Jacobian is smooth, i.e. to curves C that are the union of two elliptic curves meeting in a single point. In [**Ka3**], Theorem 2.3, the set ∂_{III} was interpreted in terms of conditions involving the functor $\mathcal{Z}_{N,-1}$, i.e. in terms of triples $(E_1, E_2, \psi) \in \mathcal{Z}_{N,-1}(K)$. Roughly speaking, the condition is that ψ should be induced by an isogeny $h : E_1 \to E_2$ of a specific type. In the next section we will see in general that the set of ψ's induced by an isogeny lie on so-called *Hecke curves* $T_{m,k} \subset Z_{n,-1}$, and that $T_{m,k}$ is birationally equivalent to $X_0(m)$, provided that $mk^2 \equiv -1 \pmod{n}$; cf. Subsection 6.2 below. Reinterpreting [**Ka3**], Corollary 2.4 and Remark 2.5 in terms of Hecke curves, we obtain:

PROPOSITION 5.2. *If $m = s(n - s)/t^2$ and $k \equiv ts^* \ (n)$, where $1 \leq s \leq n - 1$, $(s, n) = 1$, and $s^*s \equiv 1 \ (n)$, then $T_{m,k}$ is an irreducible component of ∂_{III}. Furthermore, if $n = p$ is prime, then every component of ∂_{III} has this form, and hence*

$$\partial_{III}(H_p) \;=\; \bigcup_{s=1}^{\frac{p-1}{2}} \bigcup_{t^2 | s(p-s)} T_{\frac{s(p-s)}{t^2}, ts^*}.$$

Type IV: The points in $\partial_{IV} := \bar{Z}_{n,-1}^{(1)} \setminus Z_{n,-1}$.

We first work out the geometric structure of ∂_{IV}. For this we note that since $X(1) = X'(1) \cup \{P_\infty\}$, we have $(X(1) \times X(1)) \setminus (X'(1) \times X'(1)) = P_\infty \times X(1) \cup X(1) \times P_\infty = pr_1^{-1}(P_\infty) \cup pr_2^{-1}(P_\infty)$, and so

$$\overline{Z}_{n,-1} \setminus Z_{n,-1} = \psi^{-1}(pr_1^{-1}(P_\infty) \cup pr_2^{-1}(P_\infty))_{red} = \psi_1^{-1}(P_\infty)_{red} \cup \psi_2^{-1}(P_\infty)_{red},$$

where ψ and ψ_i are as defined at the beginning of this subsection. By the discussion preceding Proposition 2.5 of [KS] we know that $C_{\infty,i} := \psi_i^{-1}(P_\infty)_{red}$ is an irreducible curve isomorphic to $X_1(n)$. We thus see that

$$\overline{Z}_{n,-1} \setminus Z_{n,-1}^{(1)} = \psi_1^{-1}(P_\infty)_{red} = C_{\infty,1},$$

and that hence

$$\partial_{IV} = Z_{n,-1}^{(1)} \setminus Z_{n,-1} = C_{\infty,2} \setminus C_{\infty,1}$$

is an open subset of $C_{\infty,2} \simeq X_1(n)$.

In order to see how the points of ∂_{IV} give rise to curve covers, we first note that the surface $\overline{Z}_{n,-1}$ has the following modular interpretation (as is shown in [Ka7]): it is the coarse moduli space of the functor $\overline{\mathcal{Z}}_{n,-1}$ which classifies triples (E_1, E_2, ψ) where E_1 and E_2 are generalized elliptic curves "of type n" over a scheme S (i.e. the fibres of E_i/S are either smooth elliptic curves or Néron polygons of length n) and $\psi : E_1[n] \to E_2[n]$ is an isomorphism of group schemes of determinant -1.

Via this interpretation, the points of $\partial_{IV}(\overline{K})$ can be identified with triples (E_1, E_2, ψ) where $E_1/\overline{K}$ is an elliptic curve and $E_2/\overline{K}$ is the Néron polygon of length n. By an extension of the "basic construction" (cf. section 2.3) we can construct a stable curve C_ψ on $J_\psi = (E_1 \times E_2)/\text{Graph}(\psi)$ together with morphism $f : C_\psi \to E_1$ (induced by the projection onto the first factor). Note that the Classification Theorem 3.9 shows that C_ψ is an irreducible singular curve whose normalization is an elliptic curve and which has a unique singular point.

Type V: The points in $\partial_V := \overline{Z}_{n,-1} \setminus \overline{Z}_{n,-1}^{(1)}$.

By definition, $\partial_V = \psi_1^{-1}(P_\infty)_{red} = C_{\infty,1} \simeq X_1(n)$. For purposes of discussing the associated covers, it is useful to write

$$\partial_V = \partial_V^* \,\dot\cup\, \partial_{VI}, \quad \text{where } \partial_{VI} = C_{\infty,1} \cap C_{\infty,2}.$$

Via the above modular description (see type IV), the points of $\partial_V^*(\overline{K})$ (respectively, of ∂_{VI}) correspond to triples (E_1, E_2, ψ) where $E_1/\overline{K}$ is the Néron polygon of length n and $E_2/\overline{K}$ is an elliptic curve (respectively, $E_2/\overline{K}$ is the Néron polygon of length n).

If $(E_1, E_2, \psi) \in \partial_V^*(\overline{K})$, then by the same method as for type IV we obtain an irreducible singular curve C_ψ on J_ψ. Here, however, the projection onto the first factor only induces a rational map $f_0 : C_\psi \cdots \to E_1$. Nevertheless, if we compose f_0 with the map $c_n : E_1 \to \bar{E}_1$ which contracts all components of E_1 (and hence $\bar{E}_1$ is the irreducible Néron polygon of length 1), then $f = c_n \circ f_0 : C_\psi \to \bar{E}_1$ is a morphism and hence defines a cover of degree n.

If $(E_1, E_2, \psi) \in \partial_{VI}$, then the situation is more complicated (and has not been fully worked out). Here we need to look at stable curves C_ψ (of genus 2) whose Jacobian is totally degenerate; thus, C_ψ is the reduction of a suitable Mumford curve (of genus 2). In addition, we should have (as above) a morphism $f : C_\psi \to \bar{E}_1$ of degree n.

Note that ∂_{VI} is a finite set; in fact, one can show that its cardinality is $\#\partial_{VI}(\overline{K}) = \frac{\phi(n)}{2}$, where ϕ is the Euler totient function. It is interesting to observe that this count agrees with the intersection formula (19) of [KS].

5.3. The Boundary Curves of H_n^*. According to Wewers[**We**], p. 65, the space $H_r^N(S_n) \supset H^{in}(S_n, \mathbf{C})$ has a natural compactification $\overline{H}_r^N(S_n)$ whose boundary components classify "admissible m-covers" (as defined on p. 60 of [**We**]). As Wewers shows on p. 68ff, these components can be further analyzed in terms of "degeneration structures" of conjugacy classes.

Here we will not directly follow this description; instead we will only use it as a motivation for the ideas leading to the compactification of $H^{in}(S_n, \mathbf{C})$ given below. In particular, we shall use the following two key ideas for the construction of the compactification: 1) We need to allow covers of stable curves (which are degenerations of smooth curves) and 2) we need to consider other ramification data (which is obtained by degeneration of the generic data).

In our case, this means that we need to analyze the different types of admissible covers (and their ramification data) that can arise as a degeneration of covers $\varphi : \mathbb{P}^1 \to \mathbb{P}^1$ of type $(*)$.

For this, let us first consider the degenerations of (normalized) covers $f : C \to E$. Then we have the following possibilities:

1) E and C are smooth, but the ramification data "degenerates";

2) E is smooth and C is a stable reducible curve (of arithmetic genus 2);

3) E is smooth and C is a stable singular irreducible curve;

4) E is a stable singular curve of arithmetic genus 1.

Note that the above cases correspond to the boundary components of the compactification of H_n' of Subsection 5.2; indeed, the cases 1) to 4) correspond to the boundary components ∂_{II} to ∂_V, respectively.

Each degenerate cover $f : C \to E$ gives rise to a cover $\varphi : \overline{C} := C/\langle \omega_C \rangle \to \mathbb{P}^1$ because f is still quasi-normalized. (More precisely, C and E come equipped with involutions ω_C and $[-1]_E$, respectively, such that $[-1]_E f = f\omega_C$.)

Now one has to analyze the ramification structure of the cover φ. We shall restrict ourselves to the cases 1) - 3). So the base E is an elliptic curve, and the branch points $P_1, ..., P_4$ remain distinct (because they are also the branch points of $\pi_E : E \to \mathbb{P}^1$). Thus, in these cases P_5 will move into one of $P_1, ..., P_4$, and because of symmetry, it is enough to look at the case that P_5 moves to P_4 or to P_3.

To see how the associated tuple σ changes if we coalesce two branch points P, P' of the cover φ, we choose a homotopy basis of $\mathbb{P}^1$ minus the branch points in the following way: Take a small disc D containing P and P', but none of the other branch points, and choose the loops around P and P' such that they agree outside of D and their product is homotopic (in $\mathbb{P}^1$ minus the branch points) to a loop that winds once around D and agrees with the other two loops outside of D. If $(\sigma_1, ..., \sigma_r)$ is the tuple associated with φ relative to this homotopy basis, and we coalesce the last two branch points, then the resulting cover is associated with the tuple $(\sigma_1, ..., \sigma_{r-2}, \sigma_{r-1}\sigma_r)$. In other words, coalescing the last two branch points means replacing the last two entries of the tuple by one consisting of their product.

Since we are interested in the cases that P_5 moves either to P_3 or P_4 we put $\tau = \sigma_3\sigma_5$ and $\rho = \sigma_4\sigma_5$. Define $H := \langle \sigma_1, \sigma_2, \sigma_3 \rangle = \langle \sigma_1, \sigma_2, \sigma_3, \rho \rangle$ and $F = \langle \sigma_1, \sigma_2, \sigma_4 \rangle = \langle \sigma_1, \sigma_2, \tau \rangle$.

Case 1: C is smooth. The ramification types occurring in this case were already considered above and are listed in the table of Proposition 3.6.

Case 2: C is reducible. A very interesting case for arithmetical investigations is that C is reducible ([**FK**]). It is the union of two *isogenous* elliptic curves $\hat{E}_1$ and $\hat{E}_2$ linked at one point.

The cover $f : C \to E$ of degree n induces two finite covers $f_i = f_{|\hat{E}_i} : \hat{E}_i \to E$ of degree n_i with $n_1 + n_2 = n$. We can assume that n_1 is even and n_2 is odd.

This case is classified in the world of Galois representations on torsion points of elliptic curves in [**Ka3**]: the isomorphism α on points of order n is induced by restriction of this isogeny and satisfies an additional condition.

Here we give a classification in terms of degeneration of ramification types.

One possible ramification type of τ is $(4)^1(1)^1(2)^{(n-3)/2}$. In this case $F = N\langle\sigma_1\rangle$ where N is abelian of order n and σ_1 acts by inversion on N. F has exactly two orbits of length n_1 and n_2 with $n_1 + n_2 = n$.

One possible ramification type of ρ is $(1)^5(2)^{(n-5)/2}$. Then H is not a dihedral group, has two orbits of length n_1 and n_2 and necessarily $4 \mid n_2$.

Case 3: C is singular and irreducible. Here C is obtained from a genus one curve $\hat{E}$ by identifying two points. The induced cover $\tilde{f} : \hat{E} \to E$ is unramified by the Riemann-Hurwitz formula. Hence it is an isogeny of degree n for suitable choice of the zero points of E and $\hat{E}$. The two points of $\hat{E}$ that coincide in C are then n-division points. This situation is well-understood, and the corresponding boundary curve of the diagonal surface is a curve which is an open subset of $X_1(n)$, as we saw above in the discussion of type IV.

The characterization in terms of covers of $\mathbb{P}^1$ is given as follows: P_5 coalesces with P_4, and the ramification type of ρ is $(1)^1(2)^{(n-1)/2}$. The monodromy group $H = D_{2n}$ is the dihedral group naturally embedded into S_n.

By using group theory and Riemann's existence theorem, Völklein[**V2**] proves the following.

PROPOSITION 5.3 (Völklein). *The above list covers all possibilities for the cycle types of ρ and τ. All these cases do actually occur, for any n, n_1, n_2 satisfying the given conditions.*

6. Rational Points Related to Isogenies

As we have seen in Section 4, rational points on the Hurwitz space H_n^* correspond to rational points on H_n'. Because of the large rigidity number it may be preferable to use the modular interpretation of such points as corresponding to pairs of elliptic curves with isomorphic level-n-structures and then to exploit arithmetical properties of $Z_{n,-1}$ rather than to use group theoretical methods. At the same time this will shed some light to the conjectures in Subsection 1.2. On the other hand, the interpretation of points by covers will give information about the questions asked in Subsection 1.1 (see Section 7).

6.1. Construction of Points by Isogenies. By the above-mentioned modular interpretation we know that if E and E' are elliptic curves over K and if $\psi : E[n] \to E'[n]$ is a G_K-isomorphism which is an anti-isometry with respect to the Weil pairings, then the triple (E, E', ψ) determines a K-rational point on $Z_{n,-1}$. By imposing further restrictions, one can ensure that this point lies on the open subset $H_n' \subset Z_{n,-1}$.

One way to construct such G_K-isomorphisms is by considering K-isogenies of elliptic curves. Thus, we *assume* in the following that there exists a *cyclic* K-isogeny

$$\eta : E \to E'.$$

Its degree will be denoted by $d = \deg(\eta)$. To avoid trivial cases *we always assume that $d > 1$ and that η is separable.*

Let α_n denote the restriction of η to $E[n]$. We then get G_K-isomorphisms $z \cdot \alpha_n$ from $E[n]$ to $E'[n]$ for all n prime to d and $z \in \mathbb{Z}$ prime to n.

Of course, there may be other G_K-isomorphisms between $E[n]$ and $E'[n]$. We call the triples $(E, E', z \cdot \alpha_n)$ "generic" because of the following evident observation.

LEMMA 6.1. *Assume that the centralizer of G_K in $\mathrm{Aut}(E_n)$ is $\mathbb{Z} \cdot \mathrm{id}_{E[n]}$ and that n is prime to d. Then every G_K-isomorphism between $E[n]$ and $E'[n]$ is of the form $z \cdot \alpha_n$ with $z \in \mathbb{Z}$ prime to d. Moreover, if E has no complex multiplication, then there is a number c (depending on E and K) such that this property holds for every n with $(n, c) = 1$.*

PROPOSITION 6.2. *For all n prime to d and all $z \in \mathbb{Z}$ prime to n the abelian variety $J_{z,n} := (E \times E')/\mathrm{Graph}(z \cdot \alpha_n)$ is isomorphic to $E \times E'$.*

PROOF. (cf. [**DiFr**]). The isogeny

$$\Phi : E \times E' \longrightarrow E \times E'$$

given by the matrix $\begin{pmatrix} n \circ \mathrm{id}_E & 0 \\ -z \cdot \eta & \mathrm{id}_{E'} \end{pmatrix}$ has kernel $\mathrm{Graph}(z \cdot \alpha_n)$. $\square$

In order to get points on H_n via the "basic construction" we need two additional properties:

Firstly, the graph of $\psi = z \cdot \alpha_n$ has to be isotropic with respect to the Weil pairing and secondly, one has to verify that the resulting curve C_ψ is irreducible.

The first condition means that

$$\deg(z \cdot \eta) = z^2 \cdot d \equiv -1 \bmod n.$$

The second condition was analyzed in [**Ka3**], Theorem 2.3. In the special case $n = p$ is prime, this criterion states that C_ψ is irreducible if and only if there is no isogeny $\theta : E \to E'$ of degree $k(p - k)$ with $1 \leq k \leq p - 1$ such that $\theta_{|E[n]} = k\psi$ $(= kz\alpha_n)$; cf. [**Ka3**], Theorem 3.

We state the result in the simplest case (cf. [**Fr1**]).

PROPOSITION 6.3. *Assume that d and n are distinct odd primes. In addition, assume that E has no complex multiplication.[4] Then there is an element $z \in \mathbb{Z}$ such that $z \cdot \alpha_n$ induces a covering $C \to E$ of degree n if and only if n is split into two non principal prime ideals in $\mathbb{Z}[\sqrt{-d}]$.*

REMARK 6.4. (a) There is a similar (but more complicated) criterion in the case of a general d; cf. [**Fr1**], Proposition 3.1.

(b) If d is odd and if $E' \not\simeq E$, then there always exists a cover $C \to E$ of degree 2 with $J_C \simeq E \times E'$; cf. [**DiFr**]. In particular, $E \times E'$ is the Jacobian of a curve of genus 2.

(c) On the other hand, if $d = 1, 2, 4, 6, 10, 12, 18, 22, 28, 30, 42, 58, 60, 70, 78, 102, 130, 190, 210, 330$ or 462 and E does not have CM, then there is *no* (irreducible)

---

[4]For the case that E has CM see [**Fr1**].

curve $C/\overline{K}$ such that $J_C \simeq E \times E'$; cf. [**Ka8**]. In fact, the above list of values of d are conjectured to be *all* the values for which no such curve exists. As is explained in [**Ka8**], this would follow from an old *Conjecture of Gauss* on the structure of class groups of imaginary quadratic fields (which is known to be true under the Generalized Riemann Hypothesis).

The following result may be viewed as a converse to Proposition 6.2.

PROPOSITION 6.5. *Let C/K be a curve with Jacobian $J_C \simeq E \times E'$. If* $\mathrm{Hom}_K(E, E') = \mathrm{Hom}_{\overline{K}}(E, E')$, *then there is a cyclic isogeny $\eta : E \to E'$, and integers $n \geq 2$ and z with $\deg(\eta)z^2 \equiv -1(\mathrm{mod}\ n)$ such that $C \simeq C_\psi$, where $\psi = z\eta_{|E[n]}$. Thus, there is a minimal cover $f : C \to E$ of degree n such that $f_* \circ \alpha = pr_E$, for a suitable isomorphism $\alpha : E \times E' \to J_C$.*

PROOF. (cf. [**Ka8**]). Fix an isomorphism $\alpha : A := E \times E' \overset{\sim}{\to} J_C$ and let $\lambda' = \hat{\alpha}\lambda_C\alpha : A \overset{\sim}{\to} \hat{A}$ be the principal polarization induced by $\lambda_C : J_C \overset{\sim}{\to} \hat{J}_C$. The given hypothesis implies that the Néron-Severi group $\mathrm{NS}(A \otimes \overline{K})$ of $A \otimes \overline{K}$ is K-rational, i.e. $\mathrm{NS}(A \otimes \overline{K}) = \mathrm{NS}(A)$, and so there exists a divisor $D \in \mathrm{Div}(A)$ which defines the polarization λ', i.e. we have $\lambda' = \phi_D$. By Riemann-Roch we can choose D to be effective, and then we have that $C \otimes \overline{K} \simeq D \otimes \overline{K}$. By a Galois descent argument similar to that for Torelli's theorem (cf. Milne[**Mi**], p. 203) one shows that in fact $C \simeq D$. Thus, we have an embedding $j : C \to J_C$ such that $j(C) = \alpha(D)$.

Now every $D \in \mathrm{Div}(A)$ has the form $D \equiv a\theta_1 + b\theta_2 + \Gamma_h$, where $a, b \in \mathbb{Z}$ and $h \in \mathrm{Hom}(E, E')$; here $\theta_i = pr_i^*(0_{E_i})$ (where $E_1 = E$, $E_2 = E'$), and Γ_h denotes the graph of $h \in \mathrm{Hom}(E, E')$. Since D defines a principal polarization (so D is ample and $D^2 = 2$), it follows that there are positive integers n and r such that $b = n - 1$, $a = r - \deg(h)$ and $nr - \deg(h) = 1$. Thus, $\deg(h) \neq 0$ for else $D \equiv \theta_1 + \theta_2$, which is impossible since D is is irreducible. Thus, $h = k\eta$, for some cyclic isogeny $\eta : E \to E'$ and $k \neq 0$. We thus have

$$(6.1) \qquad D \equiv (r - k^2 d)\theta_1 + (n - 1)\theta_2 + \Gamma_{k\eta}, \quad \text{where } d = \deg(\eta),\ rn - k^2 d = 1.$$

Let $\psi = -k\eta_{|E[n]} : E[n] \to E'[n]$. Since $d(-k)^2 \equiv -1 \mod n$, we see that ψ is an anti-isometry. Let $\Phi : A \to A$ be as defined in the proof of Proposition 6.2 (with $z = -k$). We thus obtain an isomorphism $\beta : J_\psi \overset{\sim}{\to} A$ such that $\beta\pi_\psi = \Phi$. Moreover, one checks that $\Phi^*(D) \equiv n^2(\theta_1 + \theta_2)$, and so $C_\psi \equiv \beta^* D \equiv \beta^*\alpha^*(j(C))$.

Now the basic construction gives a normalized cover $f_\psi : C_\psi \to E$ of degree n such that $(f_\psi)_* \circ \pi_\psi = [n]_E \circ pr_E$. Put $\gamma = (f_\psi)_* \circ \beta^{-1} \circ \alpha^{-1} : J_C \to E$, and let $f = \gamma \circ j : C \to E$. Then $f_* = \gamma$ and $\deg(f) = \deg(f_\psi) = n$ because $\beta^*\alpha^* j(C) \equiv C_\psi$.

From the definition of Φ we see that $pr_E \circ \Phi = [n] \circ pr_E = (f_\psi)_* \circ \pi_\psi = f_* \circ \alpha \circ \beta \circ \pi_\psi = f_* \circ \alpha \circ \Phi$, and hence $f_* \circ \alpha = pr_E$ because Φ is an isogeny. □

COROLLARY 6.6. *In the above situation put $d = \deg(\eta)$ and $r = \frac{dz^2+1}{n}$. Then for any pair of integers $x, y \in \mathbb{Z}$ with $(dx, y) = 1$, there is a minimal elliptic subcover $f_{x,y} : C \to E$ such that*

$$(6.2) \qquad (f_{x,y})_* \circ \alpha = ypr_E + x\hat{\eta}pr_{E'} \quad \text{and} \quad \deg(f_{x,y}) = drx^2 - 2dzxy + ny^2,$$

and hence there exist infinitely many minimal subcovers $C \to E$. Furthermore, if E does not have complex multiplication, then every minimal cover $f : C \to E$ is of this form (up to translates), and hence $(d, \deg(f)) = 1$.

PROOF. Choose integers $a, c \in \mathbb{Z}$ such that $ay - cdx = 1$, so $g := \left(\begin{smallmatrix} a & x \\ cd & y \end{smallmatrix}\right) \in \Gamma_0(d)$. Put $\alpha_g = \left(\begin{smallmatrix} a & -x\widehat{\eta} \\ -c\eta & y \end{smallmatrix}\right) \in \mathrm{End}(A)$, where $A = E \times E'$. Note that $\alpha \in \mathrm{Aut}(A)$ because $\left(\begin{smallmatrix} y & x\widehat{\eta} \\ c\eta & x \end{smallmatrix}\right)$ is the inverse of α_g. Now $g^t \left(\begin{smallmatrix} dr & dk \\ dk & n \end{smallmatrix}\right) g = \left(\begin{smallmatrix} dr_g & dk_g \\ dz_g & n_g \end{smallmatrix}\right)$, for some integers $r_g, n_g, k_g \in \mathbb{Z}$ with $n_g r_g - k_g^2 d = 1$ and $n_g, r_g > 0$. Moreover, $n_g = drx^2 + 2dkxy + ny^2 = drx^2 - 2dzxy + ny^2$. Define $D_g \in \mathrm{Div}(A)$ by (6.1), using (r_g, n_g, k_g) in place of (r, n, k). If $D = \alpha^*(j(C))$ is as in Proposition 6.5, then we have $\alpha_g^*(D) \equiv D_g$ (cf. [**Ka8**]), so $\alpha\alpha_g(D_g) \equiv j(C)$. By (the proof of) Proposition 6.5 we thus have a minimal cover $f_g : C \to E$ of degree n_g such that $(f_g)_* \circ \alpha \circ \alpha_g = pr_E$, and so $(f_g)_* \circ \alpha = pr_E \circ \alpha_g^{-1} = pr_E \circ \left(\begin{smallmatrix} y & x\widehat{\eta} \\ c\eta & x \end{smallmatrix}\right) = ypr_E + x\widehat{\eta}pr_{E'}$. Thus, $f_{x,y} := f_g$ satisfies (6.2).

Now suppose that E has no complex multiplication. Then $\mathrm{End}(E) = \mathbb{Z}1_E$ and $\mathrm{Hom}(E', E) = \mathbb{Z}\widehat{\eta}$, so $\mathrm{Hom}(A, E) = \mathbb{Z}pr_E + \mathbb{Z}\widehat{\eta}pr_{E'}$. Let $f : C \to E$ be a minimal cover. Then $h := f_* \circ \alpha \in \mathrm{Hom}(A, E)$, so $\exists x, y \in \mathbb{Z}$ such that $h = ypr_E + x\widehat{\eta}pr_{E'}$. We have $(x, y) = 1$ for otherwise h factors over the isogeny $[g]_E$, where $g = (x, y)$, and then f cannot be minimal. Moreover, we have $q := (d, y) = 1$ for otherwise we can factor η as $\eta = \eta_2\eta_1$, where $\ker(\eta_1) = \ker(\eta)[q]$, and then h factors over the isogeny $\widehat{\eta}_1$ of degree q because $h := \widehat{\eta}_1(\frac{y}{q}\eta_1 pr_E + x\widehat{\eta}_2 pr_{E'})$. Thus, $(dx, y) = 1$, and so f is the same as $f_{x,y}$ up to a translation because $f_* = (f_{x,y})_*$. Note that the formula (6.2) for the degree shows that $(\deg(f), d) = 1$ because $(n, d) = 1$ (as $dz^2 \equiv -1 \mod n$). $\qquad\square$

6.2. The Universal Construction. Let j be transcendental over the prime field K_0 of K and let $F_d = K_0(j, j_d)$, where j_d is the j-invariant of an elliptic curve $E_{j,d} = E_j/\mathrm{Ker}(\eta_d)$ obtained by applying a cyclic isogeny η_d of degree d to E_j, the elliptic curve defined over $K_0(j)$ (let's say with Hasse invariant -1/2). Hence F_d is the function field of $X_0'(d)/K_0$, the affine modular curve which parameterizes elliptic curves with cyclic isogenies of degree d. For any n and z such that $dz^2 \equiv -1 \mod n$ we thus obtain an F_d-rational point $P_{d,z} = (E_j, E_{j,d}, (z\eta_d)_{|E[n]}) \in Z_{n,-1}(F_d)$ or, equivalently, a rational map $\tau_{d,z} : X_0'(d) \to Z_{n,-1}$ (which is defined over K_0).

The above construction has the following natural modular interpretation. Let $\mathcal{X}_0(d)$ denote the moduli functor which classifies triples (E, E', η), where $\eta : E \to E'$ is a *cyclic* isogeny of degree d; thus, as was mentioned above, $\mathcal{X}_0(m)$ is coarsely represented by the modular curve $X_0'(d)$. Then the rule $(E, E', \eta) \mapsto (E, E', z\eta_{|E[n]})$ defines a morphism of functors and hence of moduli schemes

$$\tau_{d,z} = \tau_{d,z}^{(n)} : \mathcal{X}_0(d) \to \mathcal{Z}_{n,-1} \quad \text{and} \quad \tau_{d,z} = \tau_{d,k}^{(n)} : X_0'(d) \to Z_{n,-1},$$

Thus, the image $T_{d,z} = T_{d,z}^{(n)} := \tau_{d,k}^{(n)}(X_0'(d))$ is a curve lying on $Z_{n,-1}$ with generic point $P_{d,z} \in Z_{n-1}(F_d)$. We call $T_{d,z}$ a *Hecke curve* because it is induced by a certain Hecke correspondence on $X(n)$; cf. [**KS**], p. 364 or [**Ka7**]. Note that

$$T_{d,z} = T_{d',z'} \quad \Leftrightarrow \quad d = d' \text{ and } z \equiv \pm z' \mod n,$$

as is easy to see. (In fact, if $T_{d,k} \neq T_{d',k'}$, then these curves can intersect only at *CM-points*, i.e. at points $(E, E', \psi) \in Z_{n,-1}(\overline{K})$, where E and E' are elliptic curves with complex multiplication (or are supersingular).) Note also that $\tau_{d,z}$ is easily seen to be a birational equivalence (so $T_{d,z}$ is birationally equivalent to $X_0'(d)$), but in general $\tau_{d,z}$ cannot be an isomorphism because $T_{d,z}$ has singularities (at certain CM-points) whereas $X_0'(d)$ does not.

We now want to find conditions to ensure that $P_{d,z}$ lies on H_n and on H'_n or, equivalently, that the curve $T_{d,z}$ meets H_n and H'_n. Note that there are values of d and z such that this is not the case, as we saw above in Proposition 5.2.

For this, assume that $\mathbb{Q}(\sqrt{-d})$ does not have class number 1 and that d is prime. Then by Chebotartev there exist infinitely many prime ideals which are not principal, and so by Proposition 6.3 there are infinitely many prime numbers n such that there is a curve $C_{j,n}$ of genus 2 defined over $F_d = K_0(j, j_d)$ covering E_j of degree n. Hence we constructed a non-constant point on $H_n(F_d)$.

However, it seems more difficult to decide whether or not such a point lies in H'_n. Here we prove:

PROPOSITION 6.7. *Let d and n be odd integers such that $dz^2 \equiv -1 \bmod n$ for some $z \in \mathbb{Z}$. If $T_{d,z} \cap H_n \neq \emptyset$, then either $T_{d,z} \cap H'_n \neq \emptyset$ or $T_{d,z}$ is a component of $\partial_{II_3}(H'_n)$.*

PROOF.   By the basic construction, the hypothesis means that $(z \cdot \eta_d)_{|E_j[n]}$ induces a cover f_n of $E = E_j$ by an absolutely irreducible curve C of genus 2 over $F_d = K_0(j, j_d)$.

Suppose that $T_{d,z} \cap H'_n = \emptyset$. Then the ramification type of f_n is one of the cases (1)–(4) listed in Proposition 3.6. Note that if we are in case (3), then we are done, for then $T_{d,z}$ is a component of $\partial_{II_3}(H'_n)$.

In cases (1) and (2) we see that E_j has an F_d-rational point of order 2 because the discriminant divisor of f_n is rational over F_d. This, however, contradicts the fact E_j has no such point over F_d (because otherwise $F_{2d} \subset F_d$, which is impossible).

Finally, suppose that we are in case (4). Since the different divisor $\mathrm{Diff}(f_n) = 2W'_k$ is rational, it follows that the Weierstraß point W'_k is rational. Let $f'_n : C \to E' = E_{j,d}$ be the complementary cover. Since $f_n^{-1}(0_E) \cap (f'_n)^{-1}(0_E)$ is a subgroup of odd order, we see that $f'_n(W'_k) = P'$, for some $P' \in E'[2] \setminus \{0_{E'}\}$. Thus, E' has an F_d-rational point of order 2, and hence so does E_j (because $d = \deg(\eta)$ is odd). Thus, we obtain the same contradiction.   □

REMARK 6.8. In Corollary 7.8 below we shall show that for a given d with $(d, 30) = 1$, there are only finitely many n's such that $T_{d,z_n}^{(n)}$ is a component of $\partial_{II_3}(H'_n)$. This is in marked contrast to the components of $\partial_{III}(H_n)$: if there is one (prime) n such that $T_{d,z}^{(n)}$ is a component of $\partial_{III}(H_n)$, then there are infinitely many primes n_i such that $T_{d,z_i}^{(n_i)}$ is a component of $\partial_{III}(H_{n_i})$. Indeed, if d is prime, then this follows from Proposition 6.3 and Chebotarev, and the general case is similar.

EXAMPLE 6.9. Take $d = 13$. Then $X_0(d)$ has genus 0 and $\mathbb{Q}(\sqrt{-d})$ has class number 2, so for infinitely many n's we get a rational curve in H_n which is defined over K_0. Moreover, by Corollary 7.8 below we see that infinitely many of these also lie in H'_n (and hence also in H_n^*).

6.3. The Answer to Q4 is Yes. Recall Question **Q4** which asks in our context: For given $n_0 \in \mathbf{N}$ and finitely generated field K, are there curves C of genus 2 defined over K such that there are projective absolutely irreducible Galois covers of C of degree $\geq n_0$?

By results of [**Fr1**] and [**Ki**] we know already that the answer is positive for finite fields K. If K is infinite we can do even better.

By Example 6.9 we find a prime $n \geq n_0$ and a K-cover $\varphi : \mathbb{P}^1 \to \mathbb{P}^1$ of degree n which is ramified in points $P_1, \ldots, P_5$ (say) of order ≤ 2 with monodromy group S_n.

Now choose $P_6 \in \mathbb{P}^1(K)$ different from $P_1, \ldots, P_5$ and let $\pi : C_\varphi \to \mathbb{P}^1$ be a genus 2 cover of degree 2 which is ramified at $P_1, \ldots, P_5, P_6$.

Then the normalization of the fibre product of $\varphi : \mathbb{P}^1 \to \mathbb{P}^1$ and $\pi : C_\varphi \to \mathbb{P}^1$ over $\mathbb{P}^1$ is unramified over C_φ (Abhyankar's Lemma) and its Galois closure $\tilde{C}$ has Galois group S_n. Since P_6 is unramified in φ and ramified in P_6 it follows that $\tilde{C}$ is regular.

So we get

PROPOSITION 6.10. *For all K and all n_0 there is a curve C of genus 2 defined over K which has a regular unramified Galois cover of degree $> n_0$.*

Moreover, if K is not a finite field, there are infinitely many such curves C which are not a twist of a curve defined over a finite field.

6.4. Finiteness Conjectures. We now discuss the Diophantine background of the conjectures presented in Subsection 1.2.

In view of (the reverse of) the basic construction, Conjecture 3 predicts that for given elliptic curve E and for large enough natural numbers N, the twisted modular curves $H_{E/K,N}$ have only "obvious" K-rational points coming from iso-genies. This is a conjecture of the type of Fermat's Last Theorem: We have an infinite family of curves of growing genus depending on a fixed "parameter" (here the fixed ramification points $P_1, \ldots, P_4$) and we look at the union of all solutions over K of these curves. So it is not surprising that a proof of Conjecture 3 requires not only Faltings' theorem about the finiteness of rational solutions on *each* of the curves but also additional arithmetical information about these solutions. As was mentioned in Subsection 1.2, the height conjecture for elliptic curves (see [**Fr1**]) delivers such an information.

If one looks at small N (less or equal to 5), the related modular curves are of genus 0 and so one can explain why there are so many examples for non-isogenous elliptic curves with Galois-isomorphic N-torsion structures. But by experiments one finds easily many examples for $N = 7$ (and not so easily) examples for $N = 11$ for $K = \mathbb{Q}$. The examples become very rare for $N = 13$ and just recently Cremona has found examples for $N = 17$. As far as we know this is the record for the ground field $\mathbb{Q}$. These phenomena cannot be explained by properties of modular curves, and in fact, a question of Mazur concerning these examples was one of the main motivations for the investigations described in this paper (and in [**Ka7**]).

In Mazur's question, as in Conjectures 1 and 2, the elliptic curve E/K is no longer fixed, and so this leads to the study of *pairs* (E, E') of elliptic curves with G_K-isomorphic N-torsion structures. Any such isomorphism $\psi : E[N] \to E'[N]$ has a "determinant" $\varepsilon \in (\mathbb{Z}/N\mathbb{Z})^\times$, and the triple (E, E', ψ) determines a unique K-rational point on the modular diagonal quotient surface $Z_{N,\varepsilon}$ (which is a natural generalization of the surface $Z_{N,-1}$ which was considered above).

By the work of Hermann [**He1**], Kani and Schanz [**KS**], one knows that for $N \leq 10$, the (desingularization of the) compactification $\overline{Z}_{N,\varepsilon}$ of $Z_{N,\varepsilon}$ is either a rational surface, a K3-surface or an elliptic surface. Thus, we can find infinitely many curves of genus ≤ 1 lying on $Z_{N,\varepsilon}$, and each of these will have infinitely many points which are rational over a suitable finite extension of K. But for $N \geq 13$ (or for $N = 11, 12$ and for some ε's such as $\varepsilon = -1$), it turns out that these surfaces are of *general type*, so one expects that there are only finitely many such curves on these surfaces.

Indeed, *Lang's conjecture* predicts that for surfaces S/K of general type almost all of the K-rational points lie on the curves of genus ≤ 1 on S, and that there are only finitely many such curves (even over $\overline{K}$). This conjecture plays the role of Faltings' theorem (Mordell conjecture) for curves of genus ≥ 2.

Even if we believe this conjecture, the Conjectures 1 and 2 would not be immediate consequences. As in the case of families of curves, we would need to have additional information to find all curves of low genus on the surfaces $Z_{n,\varepsilon}$.

In Subsection 6.2 we saw that the Hecke curves $T_{m,k}$ (for m small) give rise to such curves on $Z_{N,-1}$, and the same construction yields curves on the surfaces $Z_{N,\varepsilon}$ whenever $mk^2 \equiv \varepsilon \mod N$. One might hope that every curve is of this form:

CONJECTURE 4 (Kani). *For prime numbers $n \geq 23$, all curves on $Z_{n,\varepsilon}$ of genus ≤ 1 are Hecke curves.*

One might expect that Lang's Conjecture and Conjecture 4 imply Conjecture 2. This is indeed the case, but is not completely obvious, due to the fact that different (non-isomorphic) triples (E_i, E_i', ψ_i) give rise to the same K-rational point on $Z_{N,\varepsilon}$. However, if $j(E_i) \neq 0, 1728$, then (as shown in [**Ka7**]) this can happen if and only if the points are *simultaneous twists* of each other, i.e. if there is a (quadratic) character χ on G_K such that $(E_2, E_2', \psi) \simeq ((E_1)_\chi, (E_1')_\chi, \psi_\chi)$ where $(E_1)_\chi$ and $(E_1')_\chi$ denote the χ-twists of E_1 and E_1', respectively and $\psi_\chi (E_1)_\chi[N] \to (E_1')_\chi[N]$ is the isomorphism induced by ψ. (In particular, we see that if K isn't finite, then there are infinitely many non-isomorphic triples which give rise to the same point on $Z_{N,\varepsilon}$.) We thus obtain:

PROPOSITION 6.11. *Lang's Conjecture and Conjecture 4 imply Conjecture 2.*

7. Towers

We now want to discuss the questions concerning the existence of towers of unramified regular Galois covers of curves of genus 2. As in the proof of Proposition 6.10, we want to use Abhyankar's Lemma to construct composites of covers of a given curve C ramified at the same places to get unramified covers. To get an infinite tower by this strategy, we have to find infinitely many covers $f_n : C \to E$ with the same discriminant divisor. Hence, we have to analyze ramification points (i.e. the discriminant) of covers resulting from the universal construction in Subsection 6.2.

7.1. Ramification points. We continue to assume that E and E' are isogenous curves with a K-rational isogeny η of minimal degree d. From now on we shall assume in addition that this degree is odd and that the j-invariant of E is not contained in an algebraic extension of the prime field K_0 of K. Thus, E does not have complex multiplication and we have $\mathrm{Hom}_K(E, E') = \mathrm{Hom}_{\overline{K}}(E, E') = \mathbb{Z}\eta$.

Let C be a curve over defined over K with $J_C \simeq E \times E'$. Then by Proposition 6.5 we know that C has a subcover $f : C \to E$ which is induced by an isogeny; in fact, C has infinitely many such subcovers.

Let $I = I(C, E, E')$ denote the set of integers $n > 1$ for which there exists a minimal subcover

$$f_n : C \to E$$

of degree n which is induced by an isogeny. (Fix one for each $n \in I$.) Let $\Delta_n = P_1^n + P_2^n$ denote the discriminant divisor (branch locus) of f_n and let $\text{Diff}(f_n) = Q_1^n + Q_2^n$ its different divisor (ramification locus). Then by Proposition 6.7 we know that either $P_1^n \neq P_2^n$ or that $P_1^n = P_2^n = 0_E$ and $Q_1^n \neq Q_2^n$ (Case (3)). We denote the set of n's for which the first case holds by I_1 and the rest by I_0.

We remark that if $P_1^n \neq 0_E$ is K-rational then Q_1^n is K-rational.

LEMMA 7.1. *If the set of points $\{P_j^n \in E(\overline{K}) : n \in I_1\}$ is finite, then the set $\{Q_j^n \in C(\overline{K}) : n \in I_1\}$ is finite.*

PROOF. Assume that $\{P_j^n : n \in I_1\}$ is finite. Thus all points in this set are K'-rational, for a finite extension K'/K, and hence $\{Q_j^n : n \in I_1\}$ is finite. Here we use the result of Faltings/Grauert/Samuel that the set of K'-rational points on C is finite. Note that C is not isotrivial over K_0 because $J_C \simeq E \times E'$ isn't isotrivial (as $j(E) \notin \overline{K}_0$). $\qquad\square$

In the sequel we shall use a criterion of [**DiFr**] about equality of ramification divisors. This is based on the following criterion for ramification points.

LEMMA 7.2. *Let $\iota : C \to J_C$ be an embedding, and let*
$$c : C \to E$$
be a minimal cover which maps Q to 0_E. Then c is unramified in Q if and only if $\ker(c_*)$ intersects transversally with $\iota(C)$ in $\iota(Q)$.

PROOF. The subscheme $\iota(c^{-1}(0_E))$ of $\iota(C)$ has degree $\deg c$, contains $\iota(Q)$ and is equal to $\ker(c_*) \cap \iota(C)$. Since Q isn't a ramification point if and only if Q occurs with multiplicity 1 in this scheme, the lemma follows. $\qquad\square$

Since translations on E are étale we can apply the above criterion to arbitrary minimal covers. In [**DiFr**] the following result is deduced from this:

PROPOSITION 7.3 ([**DiFr**]). *Assume that*
$$c_i : C \to E_i, \quad for\ i = 1, 2$$
are two minimal covers such that $\ker(c_{1*}) \cap \ker(c_{2*})$ is a finite group scheme. Then the ramification locus of c_1 is different (and hence disjoint) from the ramification locus of c_2 if and only if $\ker(c_{1*}) \cap \ker(c_{2*})$ is a reduced group scheme.

COROLLARY 7.4. *Let K be a field of characteristic 0, and let c_1 and c_2 be minimal covers from C to E with $c_1 \neq \pm c_2$. Then the ramification loci of c_1 and c_2 are disjoint.*

Thus, if $\{f_n; n \in I_1\}$ is an infinite set of minimal covers from C to E, then the set of ramification loci on C is infinite.

The case of positive characteristic behaves totally different.

PROPOSITION 7.5. *Assume that char$(K) = p > 0$. Then the ramification points of all minimal covers from C to E lie in a finite set of order $\leq p^{\dim_{\mathbb{Z}} Hom_K(J_C, E)}$.*

PROOF. Let c_1 and c_2 be two minimal covers with $c_{1*} \equiv c_{2*}$ modulo $p \cdot \text{Hom}_K(J_C, E)$. We claim that then $\text{Diff}(c_1) = \text{Diff}(c_2)$. Indeed, if $\ker(c_{1*}) = \ker(c_{2*})$, then this is clear since then $c_2 = g \circ c_1$ for some automorphism g, so assume that $\ker(c_{1*}) \neq \ker(c_{2*})$. Now the group schemes $\ker(c_{i*})[p]$ (of rank p^2)

are equal as closed subschemes of J_C and hence are contained in the finite group scheme $\ker(c_{1*}) \cap \ker(c_{2*})$. Thus, the latter is not étale, and hence $\mathrm{Diff}(c_1) = \mathrm{Diff}(c_2)$ by Proposition 7.3. $\qquad\square$

Note that the above two propositions apply to arbitrary genus 2 curve covers $C \to E$. We now return to the situation of the curve C as at the beginning of this subsection and prove:

PROPOSITION 7.6. *If* $(d, 15) = 1$, *then the set* I_0 *is finite.*

PROOF. Suppose first that $\mathrm{char}(K) = p > 0$. If I_0 were infinite, then by Proposition 7.5 there exists an infinite subset $I_0' \subset I_0$ such that $D := \mathrm{Diff}(f_n) = \mathrm{Diff}(f_m)$ for all $n, m \in I_0'$. Note that since C, η and the f_n's are all defined over $F_d = \mathbb{F}_p(j_E, j_{E,d})$, we can and will assume henceforth that $K = F_d$.

Write $D = Q_1 + Q_2$, and put $D' = Q_1 - Q_2$. Note that D' is not necessarily rational over $K = F_d$, but that $\sigma(D') = \pm D'$, for all $\sigma \in G_K$ (because D is K-rational). Then $cl(D') \in \ker(f_{n*}) \cap \ker(f_{m*})$ for all $n, m \in I_0'$ and hence $cl(D')$ has finite order. Thus, $H := \langle cl(D') \rangle$ is a finite, K-rational subgroup of $\ker(f_{n*}) \simeq E'$, and so $H \leq \ker(\eta^t)$ because all finite étale K-rational subgroup schemes of E' have this property. Let $q|r := \#H$ be a prime. Since $q|d$, we have that $H[q] = \ker(\eta^t)[q]$, and so $\frac{\phi(q)}{2} | [K(H[q]) : K]$. But by construction, $[K(H[q]) : K]|2$, so $\phi(q)|4$, i.e. $q \leq 5$. But since $(d, 30) = 1$, it follows that $H = 1$. This is impossible because $Q_1 \neq Q_2$, and so we conclude that I_0 must be finite.

Now suppose that $\mathrm{char}(K) = 0$, so $K_0 = \mathbb{Q}$. As above we can assume that $K = F_d = \mathbb{Q}(j_E, j_{E,d})$. Choose a prime $p \nmid 2d$. Since $X_0(d)$ has good reduction at p (Igusa), we can reduce K to $K_p = \mathbb{F}_p(j_p, j_{p,d})$ (where j_p is transcendental over $\mathbb{F}_p$). Then E, E' reduce to elliptic curves E_p, E_p'/K_p, and η reduces to a cyclic isogeny between E_p and E_p'. Thus, the G_K-isomorphisms $\psi_n : E[n] \to E'[n]$ reduce to G_{K_p}-isomorphisms $\psi_{n,p} : E_p[n] \to E_p'[n]$, and hence each cover $f_n : C \to E$ reduces to a cover $f_{n,p} : C_p \to E_p$ of the same degree. Note that C_p is again an irreducible curve because if not, there exists an isogeny diamond configuration (f, H_1, H_2) for $\psi_{n,p}$ (in the sense of [Ka3]), and this could be lifted to one of ψ_n because $\mathrm{Hom}(E, E') \to \mathrm{Hom}(E_p, E_p')$ is an isomorphism.

Thus, $I(C, E, E') \subset I(C_p, E_p, E_p')$, and hence we also have $I_0(C, E, E') \subset I_0(C_p, E_p, E_p')$ since the discriminants specialize properly. Now by the first part of the proof we know that $I_0(C_p, E_p, E_p')$ is a finite set and hence so is $I_0(C, E, E')$. $\qquad\square$

Combining the last proposition with Corollary 7.4 and Lemma 7.1 yields

COROLLARY 7.7. *If* $\mathrm{char}(K) = 0$, *then the set* $\{Q_j^n \in C(\overline{K}) : n \in I\}$ *and the set* $\{P_j^n \in E(\overline{K}) : n \in I\}$ *are infinite, provided that* $(d, 30) = 1$.

We can now finally prove the following fact which was mentioned in Remark 6.8.

COROLLARY 7.8. *If* $(d, 30) = 1$, *then there are only finitely many* n's *such that* $T_{d,z_n}^{(n)}$ *is a component of* $\partial_{II}(H_n')$, *for a suitable* $z_n \in \mathbb{Z}$.

PROOF. Let $\tilde{I}$ denote the set of integers $n \geq 2$ such that $T_{d,z_n}^{(n)}$ is a component of $\partial_{II}(H_n')$. Then for each n there exists an irreducible curve C_n and a minimal cover $f_n : C_n \to E$ defined over $K = F_d = K_0(X_0(d))$ with $J_{C_n} \simeq E \times E'$, where

E, E' are as at the beginning of Subsection 6.2. (In particular, E and E' satisfy the conditions of this subsection.)

By the proof of Proposition 6.5 we know that every C_n is isomorphic to an effective divisor D_n on $A = E \times E'$ which defines a principal polarization of A. Since there are only finitely many such (up to automorphisms), we conclude that there are finitely many curves $C_{n_1}, \ldots, C_{n_t}$ such that every curve C_n is K-isomorphic to one of these.

By Proposition 6.7 we know that $\tilde{I} \subset \cup_i I_0(C_{n_i}, E, E')$. Since the latter is a finite set by Proposition 7.6, we see that $\tilde{I}$ is finite as well. □

The above discussion shows that our strategy for building infinite unramified towers breaks down over fields of characteristic 0 because by Corollary 7.7 the ramification and branch loci of $f_n : C \to E$ are not contained in finite sets. Nevertheless, it turns out that the different divisors of these covers cannot be too "big" in the sense that they lie in a set of *bounded height*. Note that this gives another proof of the finiteness result of Proposition 7.5.

PROPOSITION 7.9. *The divisors* $\{\mathrm{Diff}(f_n) : n \in I\}$ *have bounded height over* $K = F_d$.

PROOF. Let F'_d be the function field of $X := X(d)/K_0$, and let $\mathcal{C}/X$ be the minimal model of $C \otimes F'_d/F'_d$. We will prove that there is an effective divisor B_0 on $\mathcal{C}$ such that for all $n \in I$ we have

$$(7.1) \qquad (\omega^0_{\mathcal{C}/X} . \overline{\mathrm{Diff}(f_n)}) \ \leq \ (\omega^0_{\mathcal{C}/X})^2 - \frac{1}{12} sl(d) - B_0^2 + 4(B_0.\mathcal{L}),$$

where $\overline{D}$ denotes the closure of the effective divisor $D = \mathrm{Diff}(f_n)$ in $\mathcal{C}$, and $\mathcal{L}$ is any ample invertible sheaf on $\mathcal{C}/X$.

To see that this implies the statement of the proposition, recall (cf. Lang[**La**], p. 58) that if $K = k(T)$ is a function field of a normal projective curve T/k, then the *height* of a point $P \in \mathbb{P}^n_K(K)$ is given by $h_K(P) = \deg_T(s^*_P(L))$, where $s_P : T \to \mathbb{P}^n_T$ is the section defined by P and $L = \mathcal{O}(1)$ is the usual very ample line bundle on $\mathbb{P}^n_T$. We can extend h_K to a function on $\mathbb{P}^n_K(\overline{K})$ by setting $h_K(P') := \frac{1}{[K':K]} h_{K'}$, for $P' \in \mathbb{P}^n_{K'}$, where K'/K is a finite extension. Thus, if V/K is any projective variety and $\varphi : V \hookrightarrow \mathbb{P}^n_K$ is a projective embedding, then $h_{K,\varphi} := h_K \circ \varphi$ is the height function on $V(\overline{K})$ determined by φ. A basic result is that, up to a bounded function, $h_{K,\varphi}$ depends only on $D = \varphi^*(\mathcal{O}_{\mathbb{P}^n_K}(1))$; cf. [**La**], p. 85. Note that if $\mathcal{V}/T$ is a projective model of V/K, and if φ extends to morphism $\tilde{\varphi} : \mathcal{V} \to \mathbb{P}^n_T$, then with $D = \tilde{\varphi}^*\mathcal{O}(1)$ we have

$$h_K, \varphi(P) = \frac{1}{\deg(\varphi(P))} \deg_{T_P}(s^*_P D) = \frac{1}{\deg(\varphi(P))} (\overline{P}.D)_{\mathcal{V}}, \quad \text{for all } P \in V(\overline{K});$$

here T_P is the normalization of T in the extenion $\kappa(P)$ of K, $s_P : T_P \to \mathcal{V}$ the induced T-morphism, and $\overline{P} = s_P(T_P)$ is the closure of P in $\mathcal{V}$ (if we view P as a closed point on V).

If $V = C$ is a curve, then the usual properties of height functions ([**La**], p. 94) show that if $S \subset C(\overline{K})$ has bounded height with respect to one height function h_{K,φ_0}, then it has bounded height with respect to every height function $h_{K,\varphi}$.

We now specialize the discussion to C/K as in the proposition (except that F_d has been replaced by F'_d). Since $C \sim E \times E$ and $E/k(X)$ has semi-stable reduction, the same is true for C; i.e. $\mathcal{C}/X$ is a semi-stable (cf. proof of Step 2 of

Theorem 3.9). Now although $\omega^0_{\mathcal{C}/X}$ isn't necessarily relatively ample over T, we do have that a sufficient large multiple $(\omega^0_{\mathcal{C}|X})^{\otimes m}$ defines a T-morphism $\varphi_m : \mathcal{C} \to \mathbb{P}^N_T$ for some $N = N_m$. (Indeed, if $\mathcal{C}_s/X$ denotes the associated stable model with contraction morphism $c : \mathcal{C} \to \mathcal{C}_s$, then $\omega^0_{\mathcal{C}_s/X}$ is relatively ample (cf. [**DM**]) and so the assertion follows since $\omega^0_{\mathcal{C}/X} \simeq c^* \omega^0_{\mathcal{C}_s/X}$.) Thus, by the above formula we see that (a multiple of) the left hand side of (7.1) can be interpreted as a height function, provided we restrict to those f_n's for which $\mathrm{Diff}(f_n)$ is irreducible. (The reducible ones can be dealt with in a similar manner, but they are anyways finite in number by Faltings/Grauert.)

Thus, the assertion of the proposition follows once we have proved (7.1). For this, let $q : \mathcal{E} \to X$ denote the minimal model of E/K. Since $\mathcal{E}/X$ is semi-stable, we have (cf. [**DR**], p. 175)

$$(7.2) \qquad \omega^0_{\mathcal{E}/X} \simeq q^* \mathfrak{A}, \quad \text{where } \mathfrak{A} = q_* \omega^0_{\mathcal{E}/X} \in \mathrm{Pic}(X).$$

Note that by (3.4) we know that $h := \deg(\mathfrak{A}) = h_{E/X} = \frac{1}{24} sl(d)$.

Now consider $f_n : C \to E$. By Proposition 8.1 below there exists an effective divisor $B_n \geq 0$ consisting entirely of components of the (reducible) fibres of the structure map $p : \mathcal{C} \to X$ such that

$$(7.3) \qquad \omega^0_{\mathcal{C}/X} \sim p^* \mathfrak{A} + \overline{D}_n + B_n,$$

where $D_n = \mathrm{Diff}(f_n)$. Here we have used the fact that $(f_n^*(\omega^0_{E/X}))_K \simeq p^* \mathfrak{A}$ which follows from (7.2) because $p_K = q_K \circ f_n$.

Fix one $n_0 \in I$ and write $D_0 = D_{n_0}$ and $B_0 = B_{n_0}$. Then from (7.3) (for n and n_0) we obtain

$$(7.4) \qquad B_n - B_0 \sim \overline{D}_0 - D_n.$$

Now for any component Γ of a fibre we have $0 \leq (\Gamma.\overline{D}_n) \leq (p^*(p(\Gamma)).\overline{D}_n) = 2$, and so $|(\Gamma.(D_0 - D_n))| \leq 4$. Thus, since $D_0 =: \sum b_\Gamma \Gamma \geq 0$, we obtain from (7.4) that $|(B_0.(B_n - B_0))| = \sum b_\Gamma |(\Gamma.(B_n - B_0))| \leq 4 \sum b_\Gamma \leq 4(B_0.\mathcal{L})$, the latter since $(\Gamma.\mathcal{L}) \geq 1$ for all Γ because $\mathcal{L}$ is ample. Thus

$$(7.5) \qquad -4(B_0.\mathcal{L}) + B_0^2 \leq (B_0.B_n) \leq B_0^2 + 4(B_0.\mathcal{L}).$$

Next, since $\overline{D}_0$ and B_n are effective and without common components we see that $\overline{D}_0.B_n \geq 0$. Thus, by (7.3) we obtain

$$(\omega.B_n) = (p^* \mathfrak{A} + \overline{D}_0 + B_0).B_n = \overline{D}_0.B_n + B_0.B_n \geq B_0.B_n,$$

where $\omega := \omega^0_{\mathcal{C}/X}$. From this, together with (7.3) and (7.5) we obtain

$$(\omega.\overline{D}_n) = \omega.(\omega - p^* \mathfrak{A} - B_n) \leq \omega^2 - 2h - B_0.B_n \leq \omega^2 - 2h - B_0^2 + 4(B_0.\mathcal{L});$$

here we have also used the fact $(\overline{D}_n.p^*\mathfrak{A}) = \deg(D_n)\deg(\mathfrak{A}) = 2h$. Since $h = \frac{1}{24} sl(d)$, this proves (7.1). $\qquad \square$

7.2. Towers of unramified extensions. In this subsection we discuss the question whether we can use different covers of C to E to compose them to towers of regular unramified Galois covers. First of all we will restrict ourselves to fields K of positive characteristic p, and we can assume that K is of transcendence degree 1 over K_0. Then we know by Proposition 7.5 that there are infinitely many covers

$$f_i : C \to E$$

with the same *different divisor*. In [**DiFr**] one finds examples of curves C which admit infinitely many subcovers $f_i : C \to E$ with the same *discriminant divisor*. We now want analyze the condition that (certain) genus 2 curve C admit subcovers with this property.

Recall the situation: We assume that we have an isogeny $\eta : E \to E'$ of degree d and that J_C is isomorphic to $E \times E'$. Moreover, we shall impose here the restriction that $d > 5$ is *prime*. Then we have that every elliptic curve E'' which is separably isogenous to E over $K = F_d = \mathbb{F}_p(j_E, j_{E'})$ is isomorphic to E or to E', and hence does not have any non-trivial torsion points in any quadratic extension of K. In particular, the ramification points of the f_i's on C are mapped to non-torsion points on E (or to 0_E).

In the following we denote by I the set of integers n such that there is a minimal cover $f_n : C \to E$ of degree n with $\mathrm{Disc}(f_n)$ reduced. For each n, fix such a cover f_n. Note that it follows from Corollary 6.6 and Proposition 7.6 that the set I is infinite. We also assume that the covers are compatible in the sense that there is an embedding $\iota : C \to J_C$ such that $(f_n)_* \circ \iota = f_n$, for all n.

As was mentioned above, it follows from Proposition 7.5 that there is an infinite subset I' of I such that the covers f_n with $n \in I'$ all have the same ramification points on C. We now want to find conditions which ensure that they also have the same branch point on E.

LEMMA 7.10. *Assume that Q is a ramification point on C for f_{n_1} and f_{n_2} with $n_1 \neq n_2$ and that $f_{n_1}(Q) = f_{n_2}(Q)$. Then there is a unique elliptic subgroup E_0 on J_C which contains $\iota(Q)$.*

PROOF. Let $h = f_{n_1,*} - f_{n_2,*} : J_C \to E$. Since $n_1 \neq n_2$, its kernel is a group scheme of dimension 1 and so the connected component of $\ker(h)$ is an elliptic curve E_0 defined over K. If $h_0 : J_C \to \overline{E}_0 := J_C/E_0$ denotes the quotient map, then $h = h_1 h_0$, for some isogeny $h_1 : \overline{E}_0 \to E$.

By hypothesis we have $\iota(Q) \in \ker(h)$, so $h_0(\iota(Q)) \in \ker(h_1)$ has finite order. Since $\overline{E}_0$ is an elliptic curve isogenous to E, it has no non-torsion points in quadratic extensions (by our hypothesis on K), and thus we have $h_0(\iota(Q)) = 0$. This means that $\iota(Q) \in E_0$, so $\iota(Q)$ lies on the elliptic subgroup E_0. Note that E_0 is the only elliptic subgroup containing $\iota(Q)$, for if $E_0' \neq E_0$ were another, then $\iota(Q)$ is in the finite subgroup scheme $E_0 \cap E_0'$, which contradicts the fact that $\iota(Q)$ has infinite order. $\square$

LEMMA 7.11. *If $E_1 \leq A := E \times E'$ is an elliptic subgroup, then there is another elliptic subgroup E_2 on A such that $(E_1.E_2) = 1$. In particular, if $i_{E_k} : E_k \to A$ denote the inclusion maps, them the map $\alpha_{E_1,E_2} = i_{E_1} + i_{E_2} : E_1 \times E_2 \xrightarrow{\sim} A$ is an isomorphism.*

PROOF. Write $\theta_1 = pr_E^*(0_E)$, $\theta_2 = pr_{E'}^*(0_{E'})$ and $\Gamma^* = -d\theta_1 - \theta_2 + \Gamma_{-\eta}$. Then $\theta_1, \theta_2, \Gamma^*$ is a basis of $\mathrm{NS}(A)$ (cf. [**Ka8**]), so $E_1 \equiv m_1\theta_1 + m_2\theta_2 + m_3\Gamma^*$, for some $m_i \in \mathbb{Z}$. Now since $(m_1\theta_1 + m_2\theta_2 + m_3\Gamma^*)^2 = 2(m_1m_2 - m_3^2 d)$, it follows that $m_1m_2 = m_3^2 d$ as $E_1^2 = 0$. Moreover, since E_1 is primitive (cf. [**Ka2**], Theorem 2.8) we have that $(m_1, m_2, m_3) = 1$. Since d is square free, it follows that $(m_1, m_2) = 1$. Thus $d = d_1d_2$ with $(d_1, d_2) = 1$ and $d_i | m_i$, and similarly $m_3^2 = (n_1n_2)^2$ with $(n_1, n_2) = 1$ and $n_i | m_i$. Thus $m_i = n_i^2 d_i$, for $i = 1, 2$, and $(n_1d_1, n_2d_d) = 1$, and hence there exist $k_1, k_2 \in \mathbb{Z}$ such that $k_1m_1d_1 - k_2m_2d_2 = 1$. Put $D = k_2^2d_2\theta_1 + k_1^2d_1\theta_2 + k_1k_2\Gamma^*$. Then $D^2 = 0$, and so there is an elliptic subgroup

E_2 with $D \equiv mE_2$ for some $m \in \mathbb{Z}$ (cf. [**Ka2**], Prop. 2.3). But D is primitive since $(k_2^2 d_2, k_1^2 d_1) = 1$, so $m = \pm 1$. Moreover, since $D.\theta_1 = k_1^2 d_1 > 0$, we must have $m = 1$, i.e. $E_2 \equiv D$. Now $(E_1.E_2) = \frac{1}{2}(E_1 + E_2)^2 = (n_1^2 d_1 + k_2^2 d_2)(n_2^2 d_2 + k_1^2 d_1) - (n_1 n_2 + k_1 k_2)^2 d = (n_1 k_1 d_1 - n_2 k_2 d_2)^2 = 1^2 = 1$, and so E_2 has the desired properties. The second assertion is an immediate consequence of the first.   □

COROLLARY 7.12. *If $E_0 \leq E \times E'$ is an elliptic subgroup, then ther is an automorphism $\alpha \in \mathrm{Aut}(E \times E')$ such that either $pr_E \circ \alpha \circ i_{E_0} = 0$, or $pr_{E'} \circ \alpha \circ i_{E_0} = 0$.*

PROOF. By Lemma 7.11 there is an elliptic subgroup $E_0' \leq A$ and an isomorphism $\beta : A_0 := E_0 \times E_0' \xrightarrow{\sim} A$ such that $\beta \circ i_{E_0 \times 0}^{A_0} = i_{E_0}^A$.

Suppose first that $E_0 \simeq E$. Then $E_0' \simeq A/E_0 \simeq A/E \simeq E'$, so we have isomorphisms $\alpha_0 : E_0 \xrightarrow{\sim} E$ and $\alpha_1 : E_0' \xrightarrow{\sim} E'$. Thus, $\alpha = \beta \circ (\alpha_0^{-1} \times \alpha_1^{-1}) \in \mathrm{Aut}(A)$ satisfies $\alpha \circ i_{E_0} = i_{E \times 0} \circ \alpha_0$. Thus, $pr_{E'} \circ \alpha \circ i_{E_0} = pr_{E'} \circ i_{E \times 0} \circ \alpha_0 = 0$.

Next, suppose $E_0 \not\simeq E$. Then (since d is prime) $E_0 \simeq E'$, and one has by a similar argument as above that $E_0' \simeq E$, so we have isomorphisms $\alpha_0 : E_0 \xrightarrow{\sim} E'$ and $\alpha_1 : E_0' \xrightarrow{\sim} E$. If $\tau : E \times E' \xrightarrow{\sim} E' \times E$ denotes the natural map, then $\alpha = \beta \circ (\alpha_0^{-1} \times \alpha_1^{-1}) \circ \tau \in \mathrm{Aut}(A)$ satisfies $\alpha \circ i_{E_0} = i_{0 \times E'} \circ \alpha_0$, and so $pr_E \circ \alpha \circ i_{E_0} = pr_E \circ i_{0 \times E'} \circ \alpha_0 = 0$.   □

PROPOSITION 7.13. *Let $d > 5$ be a prime and let E/K be an elliptic curve defined over $K = F_d$ with j-invariant $j = j_E$ transcendental over $\mathbb{F}_p \subset F_d$. Let $\eta : E \to E'$ be an isogeny of degree d to some curve E'/K, and let C/K be a genus 2 curve such that $J_C \simeq E \times E'$. Then the following conditions are equivalent.*

(i) There exist infinitely many minimal covers $f_n : C \to E$ with the same branch locus.

(ii) There exist two minimal covers $f_{n_k} : C \to E$ of different degrees with the same ramification locus and the same branch locus.

(iii) There is a minimal subcover $f : C \to E$ such that its ramification locus Q lies on some elliptic subgroup $E_0 \leq J_C$.

PROOF. (i) $\Rightarrow$ (ii): Clear by by Proposition 7.5.

(ii) $\Rightarrow$ (iii): Lemma 7.10.

(iii) $\Rightarrow$ (i): Fix an isomorphism $\beta : J_C \xrightarrow{\sim} E \times E'$. Applying Corollary 6.6 to $\beta(E_0) \leq E \times E'$, we have two possibilities:

Case 1: $pr_E \circ \alpha \circ i_{\beta(E_0)} = 0$, for some $\alpha \in \mathrm{Aut}(E \times E')$.

Then as in the proof of Corollary 6.6 there are integers $x_0, y_0 \in \mathbb{Z}$ with $(dx_0, y_0) = 1$ such that $f_* \circ \beta^{-1} \circ \alpha^{-1} = y_0 pr_E + x_0 \widehat{\eta} pr_{E'}$. Take $y \in \mathbb{Z}$ such that $y \equiv y_0$ mod p and $(y, x_0 d) = 1$, and put $f_y := (y pr_E + x_0 \widehat{\eta} pr_{E'}) \circ \alpha \circ \beta_{|\iota(C)}$. By (the proof of) Corollary 6.6 we know that $f_y : C \to E$ is minimal. Moreover, since $(f_y)_* = f_* + (y - y_0) pr_E \circ \alpha \circ \beta \equiv f_*$ mod $p\mathrm{End}(J_C)$, the proof of Proposition 7.5 shows that f and f_y have the same ramification point Q. Moreover, we have $(f_* - (f_y)_*)(Q) = (y - y_0) pr_E(\alpha(\beta(Q))) = 0$, and so f and f_y have the same branch point, and hence (i) holds in this case.

Case 2: $pr_{E'} \circ \alpha \circ i_{\beta(E_0)} = 0$, for some $\alpha \in \mathrm{Aut}(E \times E')$.

Again by Corollary 6.6 there are integers $x_0, y_0 \in \mathbb{Z}$ with $(dx_0, y_0) = 1$ such that $f_* \circ \beta^{-1} \circ \alpha^{-1} = y_0 pr_E + x_0 \widehat{\eta} pr_{E'}$. Take $x \in \mathbb{Z}$ such that $x \equiv x_0$ mod p and $(y_0, x) = 1$, and put $f_x := (y_0 pr_E + x \widehat{\eta} pr_{E'}) \circ \alpha \circ \beta_{|\iota(C)}$. By (the proof of) Corollary

6.6 we know that $f_x : C \to E$ is minimal. Here $(f_x)_* = f_* + (x - x_0)\widehat{\eta} \circ pr_{E'} \alpha \circ \beta \equiv f_*$ mod $p\mathrm{End}(J_C)$, so as before f and f_x have the same ramification point Q. Moreover, we have $(f_* - (f_y)_*)(Q) = (x - x_0)\widehat{\eta}(pr_{E'}(\alpha(\beta(Q)))) = 0$, and so f and f_x have the same branch point, and hence (i) holds here as well. $\square$

In [**DiFr**] it is shown that there always exists a cover $f : C \to E$ satisfying the condition (iii) of Proposition 7.13, the degree of f being 2. In addition, there is an explicit description of an infinite family of minimal covers $f_n : C \to E$ which have the same ramification and branch points.

Now we are nearly done. To end we have to overcome a technical difficulty. The monodromy groups occurring for the covers are the full symmetric groups S_{n_i}. These are only "nearly simple" groups and so when we take composites of covers (with normalization) it could happen that there occur constant field extensions with 2-elementary Galois groups. Hence we have to generalize the discussion and replace the ground field K by an appropriate affine scheme $\mathrm{Spec}(S)$ with function field K such that $\mathrm{Pic}(S)/\mathrm{Pic}(S)^2$ is finite, and then show that there are infinitely many minimal covers with the same *branch divisor* over S.

This implies that there is a finite extension K_1 of K such that the curve $C \times K_1$ has an infinite tower of K_1-rational regular unramified Galois extensions with Galois group being a product of alternating groups A_n. Hence we get

THEOREM 7.14. *Let K be a field of finite type of odd positive characteristic. Let E be an elliptic curve over K which has a K-rational isogeny η of prime degree $d > 5$. (For instance, take E belonging to a K-rational point on $X_0(13)$.)*

Then there is a curve C of genus 2 over K with $J_C \simeq E \times \eta(E)$ and a finite extension K_1 of K such that C has an infinite unramified regular Galois pro-cover defined over K_1.

If K is not a finite field, then the curve C can be chosen such that it is not a twist of constant curves, i.e. such that it is not isotrivial.

Note that in the above theorem we can take for the curve C/K a twist of the curve discussed in Proposition 7.13 and constructed in [**DiFr**].

COROLLARY 7.15. *Let K be a finitely generated field over $\mathbb{F}_p$. There is an algebraic subset of positive dimension in the moduli space of curves of genus 2 over K such that Question **Q3** has a positive answer for all curves in this subset.*

REMARK 7.16. It is easy to see that in the infinite towers constructed in Theorem 7.14 over the non-constant curve C, there is no point of C which splits totally. So we get no answer to Question **Q2** by this construction.

Question: Can one find a curve C/F_d which satisfies the condition (iii) of Proposition 7.13 which is not isomorphic to the one used in [**DiFr**]? (For a given d, this curve is uniquely characterized by the property that $J_C \simeq E \times E'$ and that it has a subcover $f : C \to E$ of degree 2.)

8. Appendix: The Riemann-Hurwitz Relation for Relative Curves

Let k be a field and let T/k be a smooth projective irreducible curve with function field $K = \kappa(T)$. Let X/T be a relative curve, i.e. X/k is a smooth projective irreducible surface with a surjective morphism $p_X : X \to T$. We denote by $X_K = X \otimes K$ the generic fibre of p_X (which is a smooth projective curve over K).

Now suppose Y/T is another relative curve and that we have a non-constant, separable morphism

$$f_K : X_K = X \otimes K \to Y_K = Y \otimes K$$

of their generic fibres. Then the usual Riemann-Hurwitz relation states that we have an isomorphism

$$(8.1) \qquad \omega_{X_K/K} \simeq f_K^*(\omega_{Y_K/K}) \otimes \mathcal{L}(D_{f_K}),$$

where $D_{f_K} \geq 0$ is an *effective* divisor (the *different divisor* of f_K). We now want to extend this relation to a similar one involving the *relative dualizing sheaves* of X/T and Y/T, which are given by

$$\omega_{X/T}^0 = \omega_{X/k} \otimes p_X^* \omega_{T/k}^{-1} \quad \text{and} \quad \omega_{Y/T}^0 = \omega_{Y/k} \otimes p_Y^* \omega_{T/k}^{-1},$$

where $\omega_{X/k} = \wedge^2 \Omega_{X/k}^1$ and $\omega_{Y/k} = \wedge^2 \Omega_{Y/k}^1$; cf. Kleiman[**Kl**].

In the case that f_K extends to a T-morphism $f : X \to Y$, such a formula was established in the Appendix of [**Ka6**]. Here we want to show that a similar formula holds without this assumption.

For this, note first that the map f_K naturally extends to a morphism $f : U \to Y$, where $U \subset X$ is an open set such that $X \setminus U$ has codimension 2, i.e. $X \setminus U$ consists of finitely many closed points. Thus, every divisor of U extends uniquely to a divisor on X, and hence every invertible sheaf $\mathcal{L}$ on U extends uniquely to an invertible sheaf $\mathcal{L}_X$ on X.

PROPOSITION 8.1. *Let $\overline{D}$ denote the schematic closure in X of the different divisor D_{f_K} of f_K. Then there exists an effective divisor $B = \sum b_\Gamma \Gamma \geq 0$ consisting entirely of fibre components of p_X such that if we write $D_f = \overline{D} + B$ then we have*

$$(8.2) \qquad \omega_{X/T}^0 \simeq (f^* \omega_{Y/T}^0)_X \otimes \mathcal{L}(D_f).$$

PROOF. First note that since $f^* \Omega_{Y/k}^1$ and $\Omega_{U/k}^1$ are locally free sheaves of rank 2 on U, we have the exact sequence

$$0 \to f^* \Omega_{Y/k}^1 \to \Omega_{U/k}^1 \to \Omega_{U/Y}^1 \to 0.$$

Let $\mathcal{D}(U/Y) = \mathcal{F}^0(\Omega_{U/Y}^1)$ denote the Kähler different of f which is by definition (cf. Kunz[**Ku**], p. 159) the 0-th Fitting ideal sheaf of the sheaf of differentials $\Omega_{U/Y}^1$. From the above exact sequence it follows that

$$f^* \omega_{Y/k} = \mathcal{D}(U/Y) \omega_{U/k}$$

(viewed as subsheaves of $\omega_{U/k} = \wedge^2 \Omega_{U/k}^1$), and hence $\mathcal{D}(U/Y)$ is invertible. Thus, if D_f denotes the associated effective divisor defined by the invertible ideal sheaf $\mathcal{D}(U/Y)$, then we have

$$\omega_{U/k} \simeq f^* \omega_{Y/k} \otimes \mathcal{L}(D_f),$$

from which equation (8.2) follows immediately. Finally we show that the divisor D_f has the form asserted in the proposition. Indeed, since $\mathcal{D}(U/Y)_{|X_K} = \mathcal{D}(X_K/Y_K)$ by general properties of the different (cf. [**Ku**], p. 160, property a)), we see that $D_f = \overline{D} + B$, where B is an effective divisor consisting entirely of fibre components. $\square$

References

[BR]   J. Bertin, M. Romagny, *Champs de Hurwitz.* Preprint, 151pp.

[BLR]   S. Bosch, W. Lütkebohmert, M.Raynaud, *Néron Models.* Springer-Verlag, Berlin, 1990.

[Da]   H. Darmon, *Serre's Conjectures.* In: *Seminar on Fermat's Last Theorem* (Toronto, 1993-4) CMS Conf. Proc. **17** (1995), pp. 135-153.

[DM]   P. Deligne, D. Mumford, *The irreducibility of the space of curves of given genus.* Publ. Math. I.H.E.S. **36** (1969), 75–109.

[DR]   P. Deligne, M. Rapoport, *Les schémas de modules de courbes elliptiques.* In: Modular functions of one variable, II, Lecture Notes in Math. **349**, Springer-Verlag, Berlin, 1973, pp. 143–316.

[Di]   C. Diem, *Families of elliptic curves with genus 2 covers of degree 2.* Collect. Math. **57** (2006), 1-25.

[DiFr] C. Diem, G. Frey, *Non-constant curves of genus 2 with infinite pro-Galois covers.* Israel Journal of Mathematics, **164** (2008), No. 1, 193–220.

[Dij]   R. Dijkgraaf, *Mirror symmetry and elliptic curves.* In: The Moduli Space of Curves (Dijkgraaf, Faber, van der Geer, eds.), Birkhäuser, Boston, 1995, pp. 149–164.

[EGA]   A. Grothendieck, J. Dieudonné, *Eléments de Géométrie Algébrique.* Chapter I: Springer-Verlag, 1971; Chapter II-IV: Publ. IHES 8, 11, 17, 20, 24, 28, 32 (1961-1967).

[Fr1]   G. Frey, *On elliptic curves with isomorphic torsion structures and corresponding curves of genus 2.* In: Conference on Elliptic Curves, Modular Forms and Fermat's Last Theorem (J. Coates, S.T. Yau, eds.), International Press, Boston, 1995, pp. 79-98.

[Fr2]   G. Frey, *On ternary equations of Fermat type and relations with elliptic curves.* In: Modular Forms and Fermat's Last Theorem, Boston, 1995 (G. Cornell, J. Silverman, G. Stevens, eds.), Springer-Verlag, New York, 1997, pp. 527–548.

[Fr3]   G. Frey, *Galois representations attached to elliptic curves and Diophantine problems.* In: Number Theory (M. Jutila, T. Metsänkylä, eds.), De Gruyter, Berlin, 2001, pp. 71-104.

[FJ]   G. Frey, M. Jarden *Horizontal isogeny theorems.* Forum Math. **14** (2002), 931-952.

[FK1]   G.Frey, E. Kani, *Projective p-adic representations of the K-rational geometric fundamental group.* Arch. Math. **77** (2001), 32-46.

[FK]   G. Frey, E. Kani, *Curves of genus 2 covering elliptic curves and an arithmetical application.* In: Arithmetic Algebraic Geometry (G. van der Geer, F. Oort, J. Steenbrink, eds.), Progress In Math. vol. 89, Birkhäuser, Boston, 1991, pp. 153-176.

[FKV]   G. Frey, E. Kani, H. Völklein, *Curves with infinite K-rational geometric fundamental group.* In: Aspects of Galois Theory, (H. Völklein et al., eds), LMS Lect. notes Ser. 256 (1999), 85-118.

[Fri]   M. Fried, *Frey-Kani covers.* Preprint.

[FV]   M. Fried and H. Völklein, *The inverse Galois problem and rational points on moduli spaces.* Math. Annalen **290** (1991), 771-800.

[Fu]   W. Fulton, *Hurwitz schemes and irreducibility of moduli of algebraic curves.* Ann. Math. (2) **90** (1969), 542–575.

[HM]   J. Harris, I. Morrison, *Moduli of Curves.* Springer Graduate Text **187**, Springer-Verlag, New York, 1998.

[He1]   C.F. Hermann, *Modulflächen quadratischer Diskriminante.* Manus. math. **77** (1991), 95–110.

[He2]   C.F. Hermann, *Symmetrische Modulflächen der Diskriminante p^2.* Manus. math. **76** (1992), 169–179.

[HL1]   P. Hubert, S. Lelièvre, *Prime arithmetic Teichmüller discs in $\mathcal{H}(2)$.* Israel J. Math. **151** (2006), 281-321.

[HL2]   P. Hubert, S. Lelièvre, *Noncongruence subgroups in $\mathcal{H}(2)$.* Int. Math. Res. Notes 2005, no. 1, 47–64.

[Hum]   G. Humbert, *Sur les fonctions abéliennes singulières. I.* J. de Math. (ser. 5) **5** (1899), 233–350 = Œuvres, Gauthier-Villars et Cie., Paris, 1929, pp. 297–401.

[Hu]   A. Hurwitz, *Über Riemann'sche Flächen mit gegebenen Verzweigungspunkten.* Math. Ann. **39** (1891), 1–61.

[Hup]   B. Huppert, *Endliche Gruppen I.* Springer-Verlag, Berlin, 1967.

[I]   Y. Ihara, *On unramified extensions of function fields over finite fields.* In: Adv. Stud. Pure Math. **2** (1983), 89-97.

[Ka1] E. Kani, *Bounds on the number of non-rational subfields of a function field.* Invent. math. **85** (1986), 185-198.

[Ka2] E. Kani, *Elliptic curves on abelian surfaces.* Manuscr. Math. **84** (1994), 199-223.

[Ka3] E. Kani, *The number of curves of genus 2 with elliptic differentials.* J. reine angew. Math. **485** (1997), 93-121.

[Ka4] E. Kani, *The existence of curves of genus 2 with elliptic differentials.* J. Number Theory **64** (1997), 130-161.

[Ka5] E. Kani, *The Hurwitz space of genus 2 covers of an elliptic curve.* Collect. Math. **54** (2003), 1–51.

[Ka6] E. Kani, *The number of genus 2 covers of an elliptic curve.* Manusc. math. **121** (2006), 51–80.

[Ka7] E. Kani, *Mazur's question and modular diagonal quotient surfaces.* In preparation.

[Ka8] E. Kani, *Jacobians isomorphic to a product of two elliptic curves.* Preprint, 38pp.

[KS] E. Kani, W. Schanz *Modular diagonal quotient surfaces.* Math. Z. **227** (1998), 337–366.

[Ki] I. Kiming, *On certain problems in the analytical arithmetic of quadratic forms arising from the theory of curves of genus 2 with elliptic differentials.* Manusc. math. **87** (1995), 101-129.

[KM] N. Katz, B. Mazur, *Arithmetic Moduli of Elliptic Curves.* Princeton U Press, Princeton, 1985.

[Kl] S. Kleiman, *Relative duality for quasi-coherent sheaves.* Comp. Math. 41 (1980), 39–60.

[Kr] A. Krazer, *Lehrbuch der Thetafunktionen.* Leipzig, 1903; Chelsea Reprint, New York, 1970.

[Ku] E. Kunz, *Kähler Differentials.* Vieweg, Wiesbaden, 1986.

[La] S. Lang, *Fundamentals of Diophantine Geometry.* Springer Verlag, New York, 1983.

[Lan] H. Lange, *Normenendomorphismen abelscher Varietäten.* J. reine angew. Math. 290 (1977), 203–213.

[LK] S. Lønsted, S. Kleiman, *Basics on families of hyperelliptic curves.* Compositio Math. **38** (1979), 83–111.

[Ma] B. Mazur, *Perturbations, deformations, and variations (and "near-misses") in geometry, physics and number theory.* Bull. AMS **41** (2004), 307–336.

[Mc1] C. McMullen, *Teichmüller curves in genus 2: discriminant and spin.* Math. Ann. **333** (2005), 87–130.

[Mc2] C. McMullen, *Dynamics of $SL_2(\mathbb{R})$ over moduli space in genus two.* Ann. Math. **165** (2007), 397–456.

[Mi] J.S. Milne, Jacobian Varieties. In: *Arithmetic Geometry.* (G. Cornell, J. Siverman, eds.), Springer-Verlag, New York, 1986; pp. 165–212.

[Ta] G. Tamme, *Teilkörper höheren Geschlechts eines algebraischen Funktionenkörpers.* Arch. Math. **23** (1972), 257–259.

[vdG] G. van der Geer, *Hilbert Modular Surfaces.* Springer-Verlag, Berlin, 1988.

[V1] H. Völklein, *Groups as Galois Groups – an Introduction.* Cambr. Studies in Adv. Math. 53, Cambridge Univ. Press, 1996.

[V2] H. Völklein, Private communication.

[We] S. Wewers, *Construction of Hurwitz spaces.* IEM Preprint **21** (1998), U Essen.

[Wi] H. Wielandt, *Finite permutation groups, 2nd ed.*; New York - London, 1968.

INSTITUTE FOR EXPERIMENTAL MATHEMATICS, UNIVERSITY OF DUISBURG-ESSEN, ELLERN-STRASSE 29, D 45326 ESSEN, GERMANY
E-mail address: frey@iem.uni-due.de

DEPARTMENT OF MATHEMATICS AND STATISTICS, JEFFERY HALL, 99 UNIVERSITY AVENUE, QUEEN'S UNIVERSITY, KINGSTON, ONTARIO, K7L 3N6, CANADA
E-mail address: Kani@mast.queensu.ca

Contemporary Mathematics
Volume **487**, 2009

A note on the Giulietti-Korchmaros maximal curve

Arnaldo Garcia

ABSTRACT. We present a very simple proof of the maximality over $\mathbb{F}_{q^6}$ of the curve introduced by Giulietti and Korchmaros in [GK].

1. Introduction

Let $\mathcal{C}$ be a curve (projective, nonsingular and geometrically irreducible) defined over a finite field k, and let $g(\mathcal{C})$ denote its genus.

We have the following bound on the cardinality of the set $\mathcal{C}(k)$ of k-rational points:

$$(1.1) \qquad \# \mathcal{C}(k) \leq 1 + \# k + 2\sqrt{\# k} \cdot g(\mathcal{C}).$$

The bound above is the so-called Hasse-Weil upper bound. If the cardinality of the finite field is a square, then we say that the curve $\mathcal{C}$ is *maximal* if equality holds in Eq. (1.1).

Suppose $k = \mathbb{F}_{q^2}$ and $\mathcal{C}$ is maximal. From [Ih] we know that

$$(1.2) \qquad g(\mathcal{C}) \leq q(q-1)/2.$$

¿From [RS] we have that the Hermitian curve is the unique maximal curve over $\mathbb{F}_{q^2}$ with genus $g = q(q-1)/2$. The *Hermitian curve* over $\mathbb{F}_{q^2}$ can be given by:

$$(1.3) \qquad X^q + X = Y^{q+1}.$$

It is well-known that subcovers of maximal curves are also maximal, and we have then a natural question:

Question: *Is any maximal curve over $\mathbb{F}_{q^2}$ a subcover of the Hermitian curve ?*

Recently Giulietti and Korchmaros introduced a maximal curve over $\mathbb{F}_{q^6}$ which is not a subcover of the corresponding Hermitian curve for $q \neq 2$ (the Hermitian curve is here given by $X^{q^3} + X = Y^{q^3+1}$). Their curve $\mathcal{C}$ over $\mathbb{F}_{q^6}$ is given by (see

---

1991 *Mathematics Subject Classification.* Primary 11G20, 11D45, 14H50; Secondary 11G20, 11D45, 14H50.

Key words and phrases. Rational points, finite fields, Hasse-Weil upper bound, maximal curves.

The author was partially supported by CNPq-Brazil (470163/06-2 and 307569/06-3).

[GK] and [GGS]):

$$(1.4) \qquad \begin{cases} X^q + X = Y^{q+1} \\ Y^{q^2} - Y = Z^N \text{ with } N = \dfrac{q^3+1}{q+1} \end{cases}$$

The proof in [GK] that $\mathcal{C}$ given by Eq. (1.4) is maximal over $\mathbb{F}_{q^6}$ is based on the fact that the curve $\mathcal{C}$ lies on a Hermitian surface. In [GGS] we have proved in an elementary way a generalization of this result of [GK]; i.e., we have proved that for an odd integer $n \geq 3$, the curve $\mathcal{C}$ given by Eq. (1.4) with $N := (q^n + 1)/(q+1)$ is maximal over the field $\mathbb{F}_{q^{2n}}$. The proof in [GGS] is based on the fact that the curve χ over $\mathbb{F}_{q^{2n}}$ given by

$$(1.5) \qquad Y^{q^2} - Y = Z^N \quad \text{with} \quad N = \frac{q^n+1}{q+1}$$

is a maximal curve over $\mathbb{F}_{q^{2n}}$ (see [GS] and [ABQ]).

Here we give an even simpler proof of the maximality of the curve $\mathcal{C}$ given by Eq. (1.4). This new proof is based on the fact that the Hermitian curve given by Eq. (1.3) is also maximal over $\mathbb{F}_{q^6}$. Since the curve χ over $\mathbb{F}_{q^6}$ given by Eq. (1.5) with $n = 3$ is a subcover of the curve $\mathcal{C}$, we get as a corollary a simpler proof for the maximality of χ (see [GS]).

2. The maximality of the curve

The curve $\mathcal{C}$ over $\mathbb{F}_{q^6}$ given by Eq. (1.4) can also be described by the plane equation:

$$(2.1) \qquad Z^{q^3+1} = \left(\frac{X^{q^2} - X}{X^q + X} \right)^{q+1} \cdot (X^q + X).$$

¿From the theory of Kummer extensions (see Proposition III.7.3 of [S]), it follows easily from Equation (2.1) that the genus satisfies

$$(2.2) \qquad 2\,g(\mathcal{C}) = q^5 - 2q^3 + q^2.$$

To prove the maximality of $\mathcal{C}$ over $\mathbb{F}_{q^6}$ we have to show that the following equality holds:

$$(2.3) \qquad \#\mathcal{C}(\mathbb{F}_{q^6}) = q^8 - q^6 + q^5 + 1.$$

There are $q+1$ points on $\mathcal{C}$ with $X = \infty$ or $X^q + X = 0$; there are $(q^2 - q)(q+1) = q^3 - q$ points on $\mathcal{C}$ with $X \in \mathbb{F}_{q^2}$ and $X^q + X \neq 0$; all those $q^3 + 1 = (q+1) + (q^3 - q)$ points can be shown to be $\mathbb{F}_{q^6}$-rational points on the curve $\mathcal{C}$. Subtracting them from Eq. (2.3), we have to prove that there are exactly

$$q^8 - q^6 + q^5 - q^3 = (q^3 + 1) \cdot (q^5 - q^3)$$

rational points on $\mathcal{C}$ with $Z \in \mathbb{F}_{q^6}$ and $Z \neq 0$. Using that $q^3 + 1$ is the norm exponent in the extension $\mathbb{F}_{q^6}/\mathbb{F}_{q^3}$, we then have to show that

$$(2.4) \qquad \# \left\{ X \in \mathbb{F}_{q^6} \,\middle|\, \left(\frac{X^{q^2} - X}{X^q + X} \right)^{q+1} \cdot (X^q + X) \in \mathbb{F}_{q^3}^* \right\} = q^5 - q^3.$$

Clearly we have that (with $N := (q^3 + 1)/(q + 1)$):

$$(2.5) \qquad \left(\frac{X^{q^2} - X}{X^q + X}\right)^{q+1} \cdot (X^q + X) = X^{q^3} + X - (X^q + X)^N.$$

Since $X^{q^3} + X$ is the trace from $\mathbb{F}_{q^6}$ to $\mathbb{F}_{q^3}$, from Eq. (2.5) we have to show that

$$(2.6) \qquad \# \left\{ X \in \mathbb{F}_{q^6} \backslash \mathbb{F}_{q^2} \,\middle|\, (X^q + X)^N \in \mathbb{F}_{q^3} \right\} = q^5 - q^3.$$

To prove this result in Eq. (2.6) we use that the Hermitian curve $\mathcal{H}$ given by

$$X^q + X = Y^{q+1}$$

is a maximal curve over $\mathbb{F}_{q^6}$ with genus $g = q(q - 1)/2$. Hence we know that

$$(2.7) \qquad \# \mathcal{H}(\mathbb{F}_{q^6}) = q^6 + q^5 - q^4 + 1.$$

As before there are $(q^3 + 1)$ rational points on $\mathcal{H}$ with $X \in \mathbb{F}_{q^2}$ or $X = \infty$. Subtracting them from Eq. (2.7) we get that there are exactly

$$q^6 + q^5 - q^4 - q^3 = (q + 1) \cdot (q^5 - q^3)$$

$\mathbb{F}_{q^6}$-rational points on $\mathcal{H}$ with $X \in \mathbb{F}_{q^6} \backslash \mathbb{F}_{q^2}$. We then conclude that it holds:

$$(2.8) \qquad \# \left\{ X \in \mathbb{F}_{q^6} \backslash \mathbb{F}_{q^2} \,\middle|\, (X^q + X) \text{ is a } (q + 1)\text{-power in } \mathbb{F}_{q^6} \right\} = q^5 - q^3.$$

The proof is now complete since Eq. (2.6) follows now easily from Eq. (2.8). We note again that

$$N \cdot (q + 1) = q^3 + 1$$

is the norm exponent in the extension $\mathbb{F}_{q^6}/\mathbb{F}_{q^3}$. We have then proved:

THEOREM 2.1. *The curve $\mathcal{C}$ given by Eq. (1.4) is a maximal curve over $\mathbb{F}_{q^6}$.*

Since the curve χ given by Eq. (1.5) with $n = 3$ is a subcover of the curve $\mathcal{C}$ we get also:

COROLLARY 2.2. *The curve χ given by Eq. (1.5) with $n = 3$ is a maximal curve over $\mathbb{F}_{q^6}$.*

References

[ABQ] M. Abdon, J. Bezerra and L. Quoos, *Further examples of maximal curves*, preprint.

[GGS] A. Garcia, C. Guneri and H. Stichtenoth, *A generalization of the Giulietti-Korchmros maximal curve*, preprint.

[GS] A. Garcia and H. Stichtenoth, *A maximal curve which is not a Galois subcover of the Hermitian curve*, Bull. Braz. Math. Soc. **37** (2006), 139-152.

[GK] M. Giulietti and G. Korchmaros, *A new family of maximal curves over a finite field*, preprint.

[Ih] Y. Ihara, *Some remarks on the number of rational points of algebraic curves over finite fields*, J. Fac. Sci. Univ. Tokyo **28** (1981), 721-724.

[RS] H.G. Ruck and H. Stichtenoth, *A characterization of Hermitian Function Fields over Finite Fields*, J. Reine Angew. Math. **457** (1994), 185-188.

[S] H.Stichtenoth, *Algebraic Function Fields and Codes*, Springer-Verlag, Berlin, 1993.

IMPA- ESTRADA DONA CASTORINA 110, 22460-320, RIO DE JANEIRO, BRAZIL
Current address: IMPA- Estrada Dona Castorina 110, 22460-320, Rio de Janeiro, Brazil
E-mail address: garcia@impa.br

Contemporary Mathematics
Volume **487**, 2009

Subclose Families, Threshold Graphs, and the Weight Hierarchy of Grassmann and Schubert Codes

Sudhir R. Ghorpade, Arunkumar R. Patil, and Harish K. Pillai

Dedicated to Gilles Lachaud on his sixtieth birthday

ABSTRACT. We discuss the problem of determining the complete weight hierarchy of linear error correcting codes associated to Grassmann varieties and, more generally, to Schubert varieties in Grassmannians. In geometric terms, this corresponds to the determination of the maximum number of $\mathbb{F}_q$-rational points on sections of Schubert varieties (with nondegenerate Plücker embedding) by linear subvarieties of a fixed (co)dimension. The problem is partially solved in the case of Grassmann codes, and one of the solutions uses the combinatorial notion of a close family. We propose a generalization of this to what is called a subclose family. A number of properties of subclose families are proved, and its connection with the notion of threshold graphs and graphs with maximum sum of squares of vertex degrees is outlined.

1. Introduction

It has been almost a decade since the first named author and Gilles Lachaud wrote [5] where alternative proofs of Nogin's results on higher weights of Grassmann codes [14] were given and Schubert codes were introduced. Originally, much of [5] was conceived as a side remark in [6]. But in retrospect, it appears to have been a good idea to write [5] as an independent article and use the opportunity to propose therein a conjecture concerning the minimum distance of Schubert codes. This conjecture has been of some interest, and after being proved, in the affirmative, in a number of special cases (cf. [1, 17, 7, 9]), the general case appears to have been settled very recently by Xiang [19]. The time sees ripe, therefore, to up the ante and think about more general questions. It is with this in view, that we discuss in this paper the problem of determining the complete weight hierarchy of Schubert codes and, in particular, the Grassmann codes. In fact, the case of Grassmann codes and the determination of higher weights in the cases not covered by the result in [14] and [5] has already been considered in some recent work (cf. [10, 11, 8]). What is proposed here is basically a plausible approach to tackle the general case.

2000 *Mathematics Subject Classification.* 94B05, 05C07, 05C35, 14M15.

Key words and phrases. Linear code, higher weight, Grassmann variety, Grassmann code, Schubert variety, Schubert code, threshold graph, optimal graph.

This paper is organized as follows. In Section 2 below we recall the combinatorial notion of a close family, and introduce a more general notion of a subclose family. A number of elementary properties of subclose families are proved here, including a nice duality that prevails among these. Basic notions concerning linear error correcting codes, such as the minimum distance and more generally, the higher weights are reviewed in Section 3. Further, we state here a general conjecture that relates the higher weights of Grassmann codes and Schubert codes with subclose families. Finally, in Section 4, we recall threshold graphs and optimal graphs and then show that, in a special case, subclose families are closely related to these well-studied notions in graph theory. As an application we obtain explicit bounds on the sum of squares of degrees of a simple graph in terms of the number of vertices and edges, which seem to ameliorate and complement some of the known results on this topic that has been of some interest in graph theory (cf. [**3, 2, 15**]).

2. Close Families and Subclose Families

Fix integers ℓ, m such that $1 \leq \ell \leq m$. Set

$$k := \binom{m}{\ell} \quad \text{and} \quad \mu := \max\{\ell, m - \ell\} + 1.$$

Let $[m]$ denote the set $\{1, \ldots, m\}$ of first m positive integers. Given any nonnegative integer j, let $I_j[m]$ denote the set of all subsets of $[m]$ of cardinality j.

Let $\Lambda \subseteq I_\ell[m]$. Following [**6**], we call Λ a *close family* if $|A \cap B| = \ell - 1$ for all $A, B \in \Lambda$. Suppose $|\Lambda| = r$. Then Λ is said to be of *Type I* if there exists $S \in I_{\ell-1}[m]$ and $T \subseteq [m] \setminus S$ with $|T| = r$ such that

$$\Lambda = \{S \cup \{t\} : t \in T\},$$

whereas Λ is said to be of *Type II* if there exists $S \in I_{\ell-r+1}[m]$ and $T \subseteq [m] \setminus S$ with $|T| = r$ such that

$$\Lambda = \{S \cup T \setminus \{t\} : t \in T\}.$$

Basic results about close families are as follows.

PROPOSITION 2.1 (Structure Theorem for Close Families). *Let* $\Lambda \subseteq I_\ell[m]$. *Then* Λ *is close if and only if* Λ *is either of Type I or of Type II.*

This is proved in [**6**, Thm. 4.2]. An immediate consequence is the following.

COROLLARY 2.2. *Let* r *be a nonnegative integer. A close family in* $I_\ell[m]$ *of cardinality* r *exists if and only if* $r \leq \mu$. *In greater details, a close family of Type I in* $I_\ell[m]$ *exists if and only if* $r \leq m - \ell + 1$, *whereas a close family of Type II in* $I_\ell[m]$ *exists if and only if* $r \leq \ell + 1$.

We use this opportunity to state the following elementary result which complements Proposition 2.1. This is not stated explicitly in [**5, 6**], but a related result is proved in [**8**] where we obtain an algebraic counterpart of Proposition 2.1 in the setting of exterior algebras and the Hodge star operator.

PROPOSITION 2.3 (Duality). *Given* $\Lambda \subseteq I_\ell[m]$, *let* $\Lambda^* := \{[m] \setminus A : A \in \Lambda\} \subseteq I_{m-\ell}[m]$. *Then* Λ *is close in* $I_\ell[m]$ *of type I if and only if* Λ^* *is close in* $I_{m-\ell}[m]$ *of type II.*

Proof. Given $S \in I_{\ell-1}[m]$ and $T \subseteq [m] \setminus S$ with $|T| = r$, observe that

$$[m] \setminus (S \cup \{t\}) = ([m] \setminus (S \cup T)) \cup T \setminus \{t\}$$

for every $t \in T$. $\qquad\square$

As explained in [5], Corollary 2.2 essentially accounts for the barrier on r for which the higher weights d_r of Grassmann codes $C(\ell, m)$ are hitherto known (see, e.g., [14, 5]). Recently some attempts have been made to break this barrier (cf. [10, 11, 8]) but the complete weight hierarchy $\{d_r : 1 \le r \le k\}$ is still not known. We will comment more on this in Section 3. For the time being, we introduce a combinatorial generalization of close families which may play some role in the determination of higher weights.

Given a subset $\Lambda = \{A_1, \cdots, A_r\}$ of $I_\ell[m]$, we define

$$K_\Lambda = \sum_{i<j} |A_i \cap A_j|.$$

Further, given any nonnegative integer $r \le k$, we define

$$K_r(\ell, m) := \max\{K_\Lambda : \Lambda \subseteq I_\ell[m] \text{ and } |\Lambda| = r\}.$$

We call Λ a *subclose family* if $K_\Lambda = K_r(\ell, m)$ where $r = |\Lambda|$. It is clear that for each nonnegative integer $r \le k$, there exists a subclose family of cardinality r.

Proposition 2.4. *Given $\Lambda \subseteq I_\ell[m]$, we have*

$$K_\Lambda \le (\ell - 1)\binom{|\Lambda|}{2}.$$

Moreover, equality holds if and only if Λ is a close family. Consequently, for any nonnegative integer $r \le k$, we have

$$K_r(\ell, m) \le (\ell - 1)\binom{r}{2}.$$

Moreover, equality holds if and only if $r \le \mu$.

Proof. The last assertion follows from Corollary 2.2. The remaining assertions are obvious. $\qquad\square$

It is an interesting question to determine $K_r(\ell, m)$ for any r. The first few values are given by the above result. We shall now determine some more. To this end, let us first observe the following analogue of Proposition 2.3.

Proposition 2.5 (First Duality Theorem). *Given $\Lambda \subseteq I_\ell[m]$, consider the family of complements of sets in Λ, viz., $\Lambda^* := \{[m] \setminus A : A \in \Lambda\} \subseteq I_{m-\ell}[m]$. Then Λ is a subclose family in $I_\ell[m]$ if and only if Λ^* is a subclose family in $I_{m-\ell}[m]$. Moreover,*

$$K_r(\ell, m) = \binom{r}{2}(2\ell - m) + K_r(m - \ell, m) \quad \text{for } 0 \le r \le k.$$

Proof: Write A^c for $[m] \setminus A$ for $A \in I_\ell[m]$. Then for any $A, B \in I_\ell[m]$, we have

$$|A \cap B| = m - |A^c \cup B^c| = m - (m - \ell) - (m - \ell) + |A^c \cap B^c| = (2\ell - m) + |A^c \cap B^c|.$$

Thus if $r := |\Lambda|$ and we write $\Lambda = \{A_1, \ldots, A_r\}$, then

$$K_\Lambda = \sum_{1 \le i < j \le r} (2\ell - m) + |A_i^c \cap A_j^c| = \binom{r}{2}(2\ell - m) + K_{\Lambda^*}.$$

Now as Λ varies over families in $I_\ell[m]$ of cardinality r, the dual Λ^* varies over families in $I_{m-\ell}[m]$ of cardinality r. It follows that Λ is a subclose family in $I_\ell[m]$ if and only if Λ^* is a subclose family in $I_{m-\ell}[m]$. Moreover,

$$K_r(\ell, m) = \binom{r}{2}(2\ell - m) + K_r(m - \ell, m) \quad \text{for } 0 \le r \le k. \qquad \square$$

Recall that given any $a, b \in \mathbb{Z}$, the binomial coefficient $\binom{a}{b}$ is defined by

$$\binom{a}{b} := \begin{cases} \dfrac{a(a-1)\cdots(a-b+1)}{b!} & \text{if } b \ge 0, \\ 0 & \text{if } b < 0. \end{cases}$$

With this in view, we may permit a and b to take negative values. We record some elementary properties of binomial coefficients, which will be useful in the sequel.

LEMMA 2.6. *Given any integers a, b, c, d, e, we have the following.*

(i) $\binom{a}{b} = \binom{a}{a-b}$ *if and only if either $a \ge 0$ or $a < b < 0$*

(ii) $\binom{a}{b} = 0$ *if and only if either $b < 0$ or $b > a \ge 0$*

(iii) $\binom{a}{b}\binom{b}{c} = \binom{a}{c}\binom{a-c}{b-c}$.

(iv) $\binom{a+b}{c-e} = \sum_{j=e}^{c} \binom{a+d}{c-j}\binom{b-d}{j-e}$.

(v) *If $a \ge 0$, then* $b\binom{a}{b} = a\binom{a-1}{b-1} = a\binom{a-1}{a-b}$.

PROOF. Both (i) and (ii) are straightforward. Proofs of (iii) and (iv) are also elementary; see, for example, Lemma 3.2 and Corollary 3.4 of [**4**]. Finally, (v) is readily verified when $a \ge 1$ and $b \ge 1$; the case when $a = 0$ or $b \le 0$ follows from (i) and (ii). $\qquad \square$

The value of $K_r(\ell, m)$ for the maximum permissible parameter r is determined below.

PROPOSITION 2.7.

$$K_k(\ell, m) = m\binom{\nu}{2}, \quad \text{where} \quad \nu := \binom{m-1}{\ell-1}.$$

PROOF. Observe that

$$\frac{m}{\ell}\nu = \binom{m}{\ell} = k, \quad \text{that is,} \quad \nu = \frac{\ell k}{m}.$$

Write $I_\ell[m] = \{A_1, \ldots, A_k\}$. Then

$$K_k(\ell, m) = \sum_{1 \le i < j \le k} |A_i \cap A_j| = \frac{1}{2}\left(\sum_{i,j} |A_i \cap A_j| - \sum_{1 \le i = j \le k} \ell\right),$$

and consequently,

$$(2.1) \qquad K_k(\ell, m) = \frac{1}{2}\left[U - k\ell\right], \quad \text{where} \quad U := \sum_{A, B \in I_\ell[m]} |A \cap B|.$$

Thus it suffices to determine U, which is more symmetric than $K_k(\ell, m)$. To find U, note that for any $A, B \in I_\ell[m]$, the intersection $A \cap B$ is a subset E, say, of $[m]$ of cardinality $i \le \ell$. Now,

$$
\begin{aligned}
U &= \sum_{\substack{E \subseteq [m] \\ |E| \le \ell}} \sum_{\substack{A,B \in I_\ell[m] \\ A \cap B = E}} |E| \\
&= \sum_{i=0}^{\ell} \sum_{\substack{E \subseteq [m] \\ |E| = i}} i |\{(A, B) \in I_\ell[m] \times I_\ell[m] : A \cap B = E\}| \\
&= \sum_{i=0}^{\ell} \binom{m}{i} \left[i \binom{m-i}{\ell-i} \binom{m-\ell}{\ell-i} \right] \\
&= \sum_{i=0}^{\ell} i \binom{m}{i} \left[\binom{m-i}{m-\ell} \binom{m-\ell}{\ell-i} \right] \qquad \text{[by Lemma 2.6 (i)]} \\
&= \sum_{i=1}^{\ell} m \left[\binom{m-1}{m-i} \binom{m-i}{m-\ell} \right] \binom{m-\ell}{\ell-i} \qquad \text{[by Lemma 2.6 (v)]} \\
&= m \binom{m-1}{m-\ell} \sum_{i=1}^{\ell} \binom{\ell-1}{\ell-i} \binom{m-\ell}{\ell-i} \qquad \text{[by Lemma 2.6 (iii)]} \\
&= m \binom{m-1}{\ell-1} \sum_{i=1}^{\ell} \binom{\ell-1}{i-1} \binom{m-\ell}{\ell-i} \qquad \text{[by Lemma 2.6 (i)]} \\
&= m \binom{m-1}{\ell-1} \binom{m-1}{\ell-1} \qquad \text{[by Lemma 2.6 (iv)]} \\
&= m\nu^2.
\end{aligned}
$$

Therefore, equation (2.1) becomes

$$
K_k(\ell, m) = \frac{1}{2} [U - k\ell] = \frac{1}{2} \left[m\nu^2 - m\nu \right] = m \binom{\nu}{2},
$$

as desired.   □

The above result will be helpful to establish yet another version of duality among subclose families. But first we need a preliminary result whose proof is similar in spirit to the proof above.

LEMMA 2.8. *Given any* $A \in I_\ell[m]$, *we have*

$$
\sum_{\substack{B \in I_\ell[m] \\ B \ne A}} |A \cap B| = \ell(\nu - 1), \qquad \text{where} \qquad \nu := \binom{m-1}{\ell-1}.
$$

PROOF. As in the proof of Proposition 2.7, we have

$$
\sum_{\substack{B \in I_\ell[m] \\ B \neq A}} |A \cap B| = \sum_{i=0}^{\ell-1} \sum_{\substack{E \subseteq A \\ |E|=i}} \sum_{\substack{B \in I_\ell[m] \\ B \cap A = E}} |E|
$$

$$
= \sum_{i=0}^{\ell-1} \binom{\ell}{i} i \binom{m-\ell}{\ell-i}
$$

$$
= \ell \sum_{i=1}^{\ell-1} \binom{\ell-1}{i-1} \binom{m-\ell}{\ell-i}
$$

$$
= \ell \left[-1 + \sum_{i=1}^{\ell} \binom{\ell-1}{i-1} \binom{m-\ell}{\ell-i} \right]
$$

$$
= \ell \left[\binom{m-1}{\ell-1} - 1 \right],
$$

where the last equality follows from part (iv) of Lemma 2.6.   □

PROPOSITION 2.9 (Second Duality Theorem). *Given $\Lambda \subseteq I_\ell[m]$, consider the complement $\Lambda^c := I_\ell[m] \setminus \Lambda$. Then Λ is a subclose family in $I_\ell[m]$ if and only if Λ^c is a subclose family in $I_\ell[m]$. Moreover, if we let $r := |\Lambda|$ and $\nu := \binom{m-1}{\ell-1}$, then*

$$
(2.2) \qquad K_{\Lambda^c} = m \binom{\nu}{2} - r\ell(\nu - 1) + K_\Lambda.
$$

Consequently,

$$
(2.3) \qquad K_{k-r}(\ell, m) = m \binom{\nu}{2} - r\ell(\nu - 1) + K_r(\ell, m) \quad \text{for } 0 \leq r \leq k.
$$

PROOF. Let $\Lambda \subseteq I_\ell[m]$. Write $\Lambda = \{A_1, \ldots, A_r\}$ and $\Lambda^c = \{B_1, \ldots, B_{k-r}\}$. Then $I_\ell[m] = \{A_1, \ldots, A_r, B_1, \ldots, B_{k-r}\}$, and we clearly have

$$
K_k(\ell, m) = K_{I_\ell[m]} = \sum_{1 \leq i < j \leq r} |A_i \cap A_j| + \sum_{1 \leq i < j \leq k-r} |B_i \cap B_j| + \sum_{\substack{1 \leq i \leq r \\ 1 \leq j \leq k-r}} |A_i \cap B_j|.
$$

Thus, in view of Proposition 2.7, we see that

$$
m \binom{\nu}{2} = K_k(\ell, m) = K_\Lambda + K_{\Lambda^c} + \sum_{A \in \Lambda} \sum_{B \in \Lambda^c} |A \cap B|.
$$

Further, in view of Lemma 2.8, we can write

$$
\sum_{A \in \Lambda} \sum_{B \in \Lambda^c} |A \cap B| = \sum_{A \in \Lambda} \left(\sum_{\substack{B \in I_\ell[m] \\ B \neq A}} |A \cap B| - \sum_{\substack{B \in \Lambda \\ B \neq A}} |A \cap B| \right)
$$

$$
= \sum_{A \in \Lambda} \ell(\nu - 1) - \sum_{A \in \Lambda} \sum_{\substack{B \in \Lambda \\ B \neq A}} |A \cap B|
$$

$$
= r\ell(\nu - 1) - 2K_\Lambda.
$$

It follows that

$$K_{\Lambda^c} = m\binom{\nu}{2} - r\ell(\nu - 1) + K_\Lambda.$$

This implies that $K_{\Lambda^c} \leq m\binom{\nu}{2} - r\ell(\nu-1) + K_r(\ell, m)$, and the equality holds if and only if Λ is subclose. Consequently,

$$K_{k-r}(\ell, m) = m\binom{\nu}{2} - r\ell(\nu - 1) + K_r(\ell, m),$$

and moreover, Λ is subclose if and only if Λ^c is subclose.   □

COROLLARY 2.10. *If $s \in \mathbb{Z}$ is such that $k - \mu \leq s \leq k$, then*

$$K_s(\ell, m) = m\binom{\nu}{2} - \ell(\nu - 1)(k - s) + (\ell - 1)\binom{k - s}{2}, \quad \text{where} \quad \nu := \binom{m - 1}{\ell - 1}.$$

In particular, if $m \geq 4$, then

$$K_s(2, m) = m\binom{m - 1}{2} - 2(k - s)(m - 2) + \binom{k - s}{2}, \quad \text{for} \quad \binom{m - 1}{2} \leq s \leq \binom{m}{2}.$$

PROOF. Given $s \in \mathbb{Z}$ with $k - \mu \leq s \leq k$, observe that $r := k - s$ satisfies $0 \leq r \leq \mu$. Now use (2.3) together with Proposition 2.4 to obtain the first equality. The second equality follows from the first by noting that if $\ell = 2$ and $m \geq 4$, then $\mu = m - 1 = \nu$ and $k - \mu = \binom{m-1}{2}$.   □

EXAMPLE 2.11. Using the above results, one can readily compile a table of values of $K_r(\ell, m)$ for $\ell = 2$ and for small values of m. For example, we have

r	1	2	3	4	5	6	7	8	9	10
$K_r(2,5)$	0	1	3	6	8	12	15	19	24	30

and

r	1	2	3	4	5	6	7	8	9	10	11	12	13	14	15
$K_r(2,6)$	0	1	3	6	10	12	15	19	24	30	34	39	45	52	60

where it may be noted that the barrier μ on the values of r is 4 in the first table and 5 in the second table. That is where the pattern begins to change.

3. Higher Weights of Grassmann Codes and Schubert Codes

We have made it amply clear in the Introduction that the combinatorial considerations in the preceding section were motivated by problems in Coding Theory, more specifically, the determination of the higher weights of linear codes associated to Grassmann and Schubert varieties. In this section, we begin by describing some relevant background, set up some notation, and then state a precise conjecture that relates these higher weights to subclose families.

Fix integers k, n with $1 \leq k \leq n$ and a prime power q. Let C be a linear $[n, k]_q$-code, i.e., let C be a k-dimensional subspace of the n-dimensional vector space $\mathbb{F}_q^n$ over the finite field $\mathbb{F}_q$ with q elements. Given any $x = (x_1, \ldots, x_n)$ in $\mathbb{F}_q^n$, let

$$\text{supp}(x) := \{i : x_i \neq 0\} \quad \text{and} \quad \|x\| := |\text{supp}(x)|$$

denote the *support* and the *(Hamming) norm* of x. More generally, for $D \subseteq \mathbb{F}_q^n$, let

$$\text{supp}(D) := \{i : x_i \neq 0 \text{ for some } x = (x_1, \ldots, x_n) \in D\} \quad \text{and} \quad \|D\| := |\text{supp}(D)|$$

denote the *support* and the *(Hamming) norm* of D. The *minimum distance* or the *Hamming weight* of C is defined by $d(C) := \min\{\|x\| : x \in C \text{ with } x \neq 0\}$. More generally, following [18], for any positive integer r, the rth *higher weight* or the rth *generalized Hamming weight* $d_r = d_r(C)$ of the code C is defined by

$$d_r(C) := \min\left\{\|D\| : D \text{ is a subspace of } C \text{ with } \dim D = r\right\}.$$

Note that $d_1(C) = d(C)$. If C is *nondegenerate*, i.e., if C is not contained in a coordinate hyperplane of $\mathbb{F}_q^n$, then it is easy to see that

$$0 < d_1(C) < d_2(C) < \cdots < d_k(C) = n.$$

See, for example, [16] for a proof as well as a great deal of basic information about higher weights of codes. The set $\{d_r(C) : 1 \leq r \leq k\}$ is often referred to as the *(complete) weight hierarchy* of the code C. It is usually interesting, and difficult, to determine the complete weight hierarchy of a given code.

An equivalent way of describing codes is via the language of projective systems, explained, for example in [16, 14, 5]. A $[n, k]_q$-projective system X is a (multi)set of n points in the projective space $\mathbb{P}^{k-1}$ over $\mathbb{F}_q$. We say X is *nondegenerate* if it is not contained in a hyperplane of $\mathbb{P}^{k-1}$. An $[n, k]_q$-nondegenerate projective system gives rise to a unique (up to isomorphism) nondegenerate $[n, k]_q$-linear code C_X. The minimum distance of C_X corresponds to maximizing the number of points of hyperplane sections of X, while the rth higher weight corresponds to maximizing the number of points of sections of X by codimension r projective linear subspaces. More precisely, for $0 \leq r \leq k$, we have

$$d_r(C_X) = n - \max\left\{|X \cap \Pi| : \Pi \text{ is a projective subspace of codimension } r \text{ in } \mathbb{P}^{k-1}\right\}.$$

Linear codes associated to projective systems given by the $\mathbb{F}_q$-rational points of higher dimensional projective algebraic varieties defined over $\mathbb{F}_q$ have been of much interest lately, and we refer to the recent survey by Little [12] for more on this. We are particularly interested in the case of Grassmann variety $G_{\ell,m}$ and its Schubert subvarieties $\Omega_\alpha = \Omega_\alpha(\ell, m)$ with its nondegenerate Plücker embedding in $\mathbb{P}^{k-1}$ and $\mathbb{P}^{k_\alpha - 1}$, respectively. Here, as in Section 2, ℓ, m are fixed positive integers with $\ell \leq m$ and $k := \binom{m}{\ell}$, while α varies over the natural indexing set for points of $\mathbb{P}^{k-1}$, namely,

$$I(\ell, m) := \{\beta = (\beta_1, \ldots, \beta_\ell) \in \mathbb{Z}^\ell : 1 \leq \beta_1 < \cdots < \beta_\ell \leq m\},$$

and for any $\alpha \in I(\ell, m)$,

$$k_\alpha := |I_\alpha(\ell, m)| \quad \text{where} \quad I_\alpha(\ell, m) := \{\beta \in I(\ell, m) : \beta_i \leq \alpha_i \text{ for all } i = 1, \ldots, \ell\}.$$

We identify $\mathbb{P}^{k_\alpha - 1}$ with $\{p = (p_\beta) \in \mathbb{P}^{k-1} : p_\beta = 0 \text{ for all } \beta \in I(\ell, m) \setminus I_\alpha(\ell, m)\}$ so that $\Omega_\alpha(\ell, m) = G_{\ell,m} \cap \mathbb{P}^{k_\alpha - 1}$. For precise definitions of $G_{\ell,m}$ and Ω_α, and their Plücker embeddings, we refer to [5] and [7] or the references therein. The linear codes corresponding to $G_{\ell,m}$ and $\Omega_\alpha(\ell, m)$ are denoted by $C(\ell, m)$ and $C_\alpha(\ell, m)$ respectively. The length n of $C(\ell, m)$ and n_α of $C_\alpha(\ell, m)$ are respectively given by

$$n = |G_{\ell,m}(\mathbb{F}_q)| = \begin{bmatrix} m \\ \ell \end{bmatrix}_q := \frac{(q^m - 1)(q^m - q) \cdots (q^m - q^{\ell-1})}{(q^\ell - 1)(q^\ell - q) \cdots (q^\ell - q^{\ell-1})} \quad \text{and} \quad n_\alpha = |\Omega_\alpha(\mathbb{F}_q)|.$$

The dimension k of $C(\ell, m)$ and k_α of $C_\alpha(\ell, m)$ are respectively given by

$$k = |I(\ell, m)| = \binom{m}{\ell} \quad \text{and} \quad k_\alpha := |I_\alpha(\ell, m)|.$$

A number of explicit formulas for n_α and k_α are given in [7].

Given any $\Lambda \subseteq I(\ell, m)$, we let Π_Λ denote the intersection of the corresponding Plücker coordinate hyperplanes; more precisely,

$$\Pi_\Lambda := \left\{ p = (p_\beta) \in \mathbb{P}^{k-1} : p_\beta = 0 \text{ for all } \beta \in \Lambda \right\}.$$

Note that Π_Λ is a projective linear subspace of codimension $|\Lambda|$ in $\mathbb{P}^{k-1}$, and also that if $\Lambda \subseteq I_\alpha(\ell, m)$, then $\Pi_\Lambda \cap \mathbb{P}^{k_\alpha - 1}$ is a projective linear subspace of codimension $|\Lambda|$ in $\mathbb{P}^{k_\alpha - 1}$.

There is a natural one-to-one correspondence between the indexing set $I(\ell, m)$ and the set $I_\ell[m]$ defined in the previous section, given simply by

$$\beta = (\beta_1, \ldots, \beta_\ell) \longmapsto \bar\beta = \{\beta_1, \ldots, \beta_\ell\}.$$

With this in view, we shall identify $I(\ell, m)$ with $I_\ell[m]$, and apply the notions and results of Section 2 for $I_\ell[m]$ and its subfamilies to $I(\ell, m)$ and its subfamilies. In particular, we can talk about subclose families in $I(\ell, m)$. We are now ready to propose a plausible fact about the higher weights of $C(\ell, m)$ and $C_\alpha(\ell, m)$.

CONJECTURE 3.1. *Let r be a positive integer. If $r \le k$, then the rth higher weight of the Grassmann code $C(\ell, m)$ is given by*

$$d_r(C(\ell, m)) = \begin{bmatrix} m \\ \ell \end{bmatrix}_q - \max \left\{ |G_{\ell, m}(\mathbb{F}_q) \cap \Pi_\Lambda| : \Lambda \subseteq I(\ell, m) \text{ is subclose and } |\Lambda| = r \right\}.$$

More generally, given any $\alpha \in I(\ell, m)$, if $r \le k_\alpha$, then the rth higher weight of the Schubert code $C_\alpha(\ell, m)$ is given by

$$d_r(C_\alpha(\ell, m)) = n_\alpha - \max \left\{ |\Omega_\alpha(\mathbb{F}_q) \cap \Pi_\Lambda| : \Lambda \subseteq I_\alpha(\ell, m) \text{ is subclose and } |\Lambda| = r \right\}.$$

The evidence we have in favor of this conjecture is as follows.

(1) The conjecture is true in the case of $C(\ell, m)$ for $1 \le r \le \max\{\ell, m-\ell\}+1$. (See [5].)

(2) The conjecture is true in the case of $C(2, m)$ for $r = \max\{2, m-2\}+2$. (See [8].)

(3) The conjecture is true in the case of $C_\alpha(\ell, m)$ for $r = 1$. (See [19].)

(4) The conjecture is true in the case of $C_\alpha(\ell, m)$ where α is a submaximal element of $I_\alpha(\ell, m)$ [so that the corresponding Schubert variety Ω_α is of codimension 1 in $G_{\ell, m}$] for $1 \le r \le \max\{\ell, m - \ell\}$. (See [7].)

It may be remarked that the notion of a subclose family and the above conjecture is similar to, yet distinct from, the notion of a Schubert union introduced in [10] and the corresponding conjecture of Hansen, Johnsen and Ranestad [10, 11] that the higher weights are attained by Schubert unions. It may also be noted that for a given r, there may be more than one subclose family of cardinality r. Thus the above conjecture does not pinpoint to a single such family but simply narrows down the search for such a family.

4. Threshold Graphs, Optimal Graphs and Subclose Families

When $\ell = 2$, the elements of the family $I_\ell[m]$ of ℓ-subsets of $[m] := \{1, \ldots, m\}$ can be viewed as the edges of a graph. It is, therefore, natural to investigate if the combinatorial notions and results in Section 2 have analogues and extensions in the rich and diverse field of graph theory. We will attempt to address these concerns in this section.

Given any $\Lambda \subseteq I_2[m]$, we denote by G_Λ the graph whose vertex set is $[m]$ and the edge set is Λ. Note that this is a simple (undirected) graph. Conversely, any simple graph on $[m]$ is of the form G_Λ for a unique $\Lambda \subseteq I_2[m]$. To say that Λ is close corresponds to saying that any two edges of G_Λ are incident. Thus, Proposition 2.1 corresponds to the following elementary result in graph theory.

PROPOSITION 4.1. *A simple graph in which any two edges are incident is either a star or a triangle.*

The analogue of subclose family is more interesting. Before explaining this, let us recall some notions from graph theory.[1] Let G be a (m, r)-graph, i.e., a graph with m vertices (assumed to be elements of the set $[m]$) and r edges. We denote by $g_i = g_i(G)$ the *degree* of the vertex i, viz., the number of edges emanating from it. The sequence $(g_1, \ldots, g_m)$ is called the *degree sequence* of G. It is well-known and easy to see that

$$(4.1) \qquad \sum_{i=1}^{m} g_i = 2r.$$

A simple graph G is said to be a *threshold graph* if G can be constructed from a one-vertex graph by repeatedly adding an isolated vertex or a universal one (i.e., a vertex adjacent to every other vertex). A simple (m, r)-graph G is said to be (m, r)-*optimal*, or simply, *optimal* if

$$\Sigma(G) := \sum_{i=1}^{m} g_i(G)^2$$

is maximum among all simple (m, r)-graphs. Threshold graphs are a topic of considerable interest in graph theory, and we refer to [13] for more on this. It is easy to see that an optimal graph is a threshold graph (cf. [15, Fact 3]). In [15] it is shown that an optimal graph is one among certain six explicit classes of graphs. However, as the authors of [15] say, the complete characterization of optimal graphs remains an open question. The following explicit bound for $\Sigma(G)$ for a (m, r)-graph G is given by de Caen [3].

$$(4.2) \quad \Sigma(G) \le C(r, m) \quad \text{for } m \ge 2, \quad \text{where} \quad C(r, m) := r \left(\frac{2r}{m-1} + m - 2 \right).$$

A somewhat more general bound has been obtained by Das [2]; however, this bound is not a pure function of m and r, but involves the maximum and the minimum among the vertex degrees $g_1, \ldots, g_m$.

The relation between optimal graphs and subclose families is given below.

PROPOSITION 4.2. *Let Λ be a subset of $I_2[m]$ with $|\Lambda| = r$. Then*

$$K_\Lambda = \frac{1}{2}\Sigma(G_\Lambda) - r \quad \text{or equivalently,} \quad \Sigma(G_\Lambda) = 2K_\Lambda + 2r.$$

Consequently, Λ is a subclose family if and only if G_Λ is an optimal graph. Moreover,

$$\max\{\Sigma(G) : G \text{ is a simple } (m, r)\text{-graph}\} = 2K_r(2, m) + 2r.$$

---

[1]We are using here a notation that is consistent with the notation of Section 2. Inconvenience caused, if any, to graph theorists, who may be more used to letting n be the number of vertices, e the number of edges, and d_i the degree of the vertex i, is regretted.

PROOF. Write $\Lambda = \{A_1, \ldots, A_r\}$. Note that $A_i \cap A_j$ is either empty or singleton for $1 \leq i < j \leq r$. Thus,

$$K_\Lambda = \sum_{i<j} |A_i \cap A_j| = \sum_{i<j} \sum_{v \in A_i \cap A_j} 1 = \sum_{v \in [m]} \sum_{\substack{i<j \\ v \in A_i \cap A_j}} 1 = \sum_{v \in [m]} \binom{g_v}{2}.$$

Further, in view of (4.1), we have

$$K_\Lambda = \frac{1}{2} \sum_{v \in [m]} g_v^2 - r = \frac{1}{2} \Sigma(G_\Lambda) - r.$$

Since G_Λ varies over all simple (m, r)-graphs as Λ varies over subsets of $I_2[m]$ of cardinality r, it follows that Λ is subclose if and only if G_Λ is optimal, and moreover,

$$\max\{\Sigma(G) : G \text{ is a simple } (m, r)\text{-graph}\} = 2K_r(2, m) + 2r,$$

as desired.   □

COROLLARY 4.3. *Assume that $m \geq 4$. Then for any simple (m, r)-graph G, we have*

(4.3) $$\Sigma(G) \leq r(r+1) \quad \text{for } r \leq m - 1,$$

and the equality holds if and only if G is a star with $r + 1$ vertices (and $m - r - 1$ isolated vertices).

PROOF. Since $m \geq 4$, we have $\mu := \max\{2, m - 2\} + 1 = m - 1$. Thus, thanks to Proposition 2.4, we have

$$K_r(2, m) \leq \binom{r}{2} \quad \text{for} \quad r \leq m - 1.$$

Now apply Proposition 4.2 to obtain (4.3). The assertion about the equality follows from Proposition 4.1.   □

Already, the trivial bound given by the Corollary above is superior to de Caen's bound (4.2) in several cases. Indeed, an easy calculation shows that

$$C(r, m) - r(r+1) = \frac{r(m-3)(m-1-r)}{m-1} > 0 \quad \text{for } r < m - 1 \text{ and } m \geq 4.$$

The above result may also be compared with one of the cases where equality holds in a bound for $\Sigma(G)$ obtained by Das [2].

It is well-known that dual graphs $\overline{G}$ of optimal graphs G [defined in such a way that the non-edges of G are the edges of $\overline{G}$] are optimal (cf. [15, Fact 1]). This corresponds precisely to the special case $\ell = 2$ of the Second Duality Theorem (Proposition 2.9). Further, it is easy to see that if $\ell = 2$, then (2.2) corresponds precisely to the following elementary relation for simple (m, r)-graphs G:

$$\Sigma(\overline{G}) = m(m-1)^2 - 4r(m-1) + \Sigma(G).$$

The following bound, dual to the trivial bound given by (4.3), appears to be new.

COROLLARY 4.4. *If G is a simple (m, r)-graph such that $\binom{m-1}{2} \leq r \leq \binom{m}{2}$, then*

$$\Sigma(G) \leq m(m-1)(m-2) + (k-r)(k-r-1) - 4(k-r)(m-2) + 2r, \quad \text{where} \quad k := \binom{m}{2}.$$

PROOF. Follows from Corollary 2.10 and Proposition 4.2.   □

Finally, we remark that the First Duality Theorem (Proposition 2.5) has no analogue in the setting of optimal graphs for the simple reason that it relates optimal graphs to objects that are not graphs, but hypergraphs. Indeed, as far as we know, not much seems to be known about threshold hypergraphs and optimal hypergraphs. Perhaps the notion of a subclose family and the results of Section 2 may be of some help in this direction.

Acknowledgments

We are grateful to Murali Srinivasan for helpful discussions and bringing [15] to our attention.

References

[1] H. Chen, *On the minimum distance of Schubert codes*, IEEE Trans. Inform. Theory **46** (2000), 1535–1538.

[2] K. C. Das, *Maximizing the sum of the squares of the degrees of a graph*, Discrete Math. **285** (2004), 57–66.

[3] D. de Caen, *An upper bound on the sum of the squares of degrees of a graph*, Discrete Math. **185** (1998), 245–248.

[4] S. R. Ghorpade, *Young multitableaux and higher dimensional determinants*, Adv. Math. **121** (1996), 167–195.

[5] S. R. Ghorpade and G. Lachaud, *Higher weights of Grassmann codes*, Coding Theory, Cryptography and Related Areas (Guanajuato, Mexico, 1998), Springer-Verlag, Berlin/Heidelberg, 2000, pp. 122–131.

[6] S. R. Ghorpade and G. Lachaud, *Hyperplane sections of Grassmannians and the number of MDS linear codes*, Finite Fields Appl. **7** (2001), 468–506.

[7] S. R. Ghorpade and M. A. Tsfasman, *Schubert varieties, linear codes and enumerative combinatorics*, Finite Fields Appl. **11** (2005), 684–699.

[8] S. R. Ghorpade, A. R. Patil and H. K. Pillai, *Decomposable subspaces, linear sections of Grassmann varieties, and higher weights of Grassmann codes*, Finite Fields Appl. (2008), doi:10.1016/j.ffa.2008.08.001 [arXiv:0710.5161 (2007)].

[9] L. Guerra and R. Vincenti, *On the linear codes arising from Schubert varieties*, Des. Codes Cryptogr. **33** (2004), 173–180.

[10] J. P. Hansen, T. Johnsen, and K. Ranestad, *Schubert unions in Grassmann varieties*, Finite Fields Appl. **13** (2007), 738–750.

[11] J. P. Hansen, T. Johnsen, and K. Ranestad, *Grassmann codes and Schubert unions*, Arithmetic, Geometry and Coding Theory (AGCT-2005, Luminy), Séminaires et Congrès, Soc. Math. France, Paris, to appear.

[12] J. B. Little, *Algebraic geometry codes from higher dimensional varieties*, arXiv:0802.2349 (2008), 26 pp.

[13] N. V. R. Mahadev and U. N. Peled, *Threshold Graphs and Related Topics*, Ann. Discrete Math. **56**, North Holland, Amsterdam, 1995.

[14] D. Yu. Nogin, *Codes associated to Grassmannians*, Arithmetic, Geometry and Coding Theory (Luminy, 1993), Walter de Gruyter, Berlin/New York, 1996, pp. 145–154.

[15] U. N. Peled, A. Petreschi and A. Sterbini, *(n, e)-Graphs with maximum sum of squares of degrees*, J. Graph Theory **31** (1999), 281–285.

[16] M. A. Tsfasman and S. G. Vlăduţ, *Geometric approach to higher weights*, IEEE Trans. Inform. Theory **41** (1995), 1564–1588.

[17] R. Vincenti, *On some classical varieties and codes*, Rapporto Technico 20/2000, Dip. Mat., Univ. di Perugia, Italy, 2000.

[18] V. K. Wei, *Generalized Hamming weights for linear codes*, IEEE Trans. Inform. Theory **37** (1991), 1412–1418.

[19] Xu Xiang, *On the minimum distance conjecture of Schubert variety codes*, IEEE Trans. Inform. Theory, **54** (2008), 486–488.

DEPARTMENT OF MATHEMATICS, INDIAN INSTITUTE OF TECHNOLOGY BOMBAY,
POWAI, MUMBAI 400076, INDIA.
E-mail address: srg@math.iitb.ac.in
URL: http://www.math.iitb.ac.in/∼srg/

DEPARTMENT OF MATHEMATICS, INDIAN INSTITUTE OF TECHNOLOGY BOMBAY,
POWAI, MUMBAI 400076, INDIA
AND
SHRI GURU GOBIND SHINGHJI INSTITUTE OF ENGINEERING & TECHNOLOGY,
VISHNUPURI, NANDED 431 606, INDIA
E-mail address: arun.iitb@gmail.com

DEPARTMENT OF ELECTRICAL ENGINEERING, INDIAN INSTITUTE OF TECHNOLOGY BOMBAY,
POWAI, MUMBAI 400076, INDIA.
E-mail address: hp@ee.iitb.ac.in

Contemporary Mathematics
Volume **487**, 2009

Characteristic polynomials of automorphisms of hyperelliptic curves

Robert M. Guralnick and Everett W. Howe

ABSTRACT. Let α be an automorphism of a hyperelliptic curve C of genus g and let $\overline{\alpha}$ be the automorphism induced by α on the genus-0 quotient of C by the hyperelliptic involution. Let n be the order of α and let $\overline{n}$ be the order of $\overline{\alpha}$. We show that the characteristic polynomial f of the automorphism α^* of the Jacobian of C is determined by the values of n, $\overline{n}$, and g, unless $n = \overline{n}$, $\overline{n}$ is even, and $(2g+2)/n$ is even, in which case there are two possibilities for f. In every case we give explicit formulas for the possible characteristic polynomials.

1. Introduction

Let α be an automorphism of a genus-g curve C over a field k and let α^* be the corresponding automorphism of the Jacobian of C. Let n be the order of α and let f be the characteristic polynomial of α^*. The values of n and g provide some restrictions on the possible values of f, but in general they do not determine f; for example, a nontrivial involution of a genus-3 curve can have characteristic polynomial equal to $(x-1)^i(x+1)^{6-i}$ for $i \in \{0, 2, 4\}$, and all three possibilities occur.

If C is hyperelliptic, with hyperelliptic involution ι, then the automorphism α gives rise to an automorphism $\overline{\alpha}$ of the genus-0 quotient $C/\langle \iota \rangle$. Let $\overline{n}$ be the order of $\overline{\alpha}$, so that either $n = \overline{n}$ or $n = 2\overline{n}$. The triple $(g, n, \overline{n})$ still does not in general determine f: If C has genus 3 and α and $\overline{\alpha}$ each have order 2, then f can be either $(x-1)^2(x+1)^4$ or $(x-1)^4(x+1)^2$, and both possibilities occur.

The purpose of this note is to show that if C is hyperelliptic, this ambiguity between two possible characteristic polynomials is the worst that can happen; furthermore, the triple $(g, n, \overline{n})$ determines f completely unless $n = \overline{n}$, n is even, and $(2g+2)/n$ is an even integer.

THEOREM 1.1. *Let C be a hyperelliptic curve of genus g over a field k and let α, $\overline{\alpha}$, n, $\overline{n}$, and f be as above.*

2000 *Mathematics Subject Classification.* Primary 14H37; Secondary 14H40.

Key words and phrases. Automorphism, hyperelliptic curve, characteristic polynomial.

The first author was partially supported by NSF grant DMS 0653873.

(1) *If $\bar{n}$ is odd and $n = \bar{n}$, then $2g \equiv 0, -1,$ or $-2 \bmod \bar{n}$, and*

$$
f = \begin{cases}
\dfrac{(x^{\bar{n}} - 1)^{(2g+2)/\bar{n}}}{(x - 1)^2} & \text{if } 2g \equiv -2 \bmod \bar{n}; \\[3mm]
\dfrac{(x^{\bar{n}} - 1)^{(2g+1)/\bar{n}}}{(x - 1)} & \text{if } 2g \equiv -1 \bmod \bar{n}; \\[3mm]
(x^{\bar{n}} - 1)^{2g/\bar{n}} & \text{if } 2g \equiv 0 \bmod \bar{n}.
\end{cases}
$$

(2) *If $\bar{n}$ is odd and $n = 2\bar{n}$, then $2g \equiv 0, -1,$ or $-2 \bmod \bar{n}$, and*

$$
f = \begin{cases}
\dfrac{(x^{\bar{n}} + 1)^{(2g+2)/\bar{n}}}{(x + 1)^2} & \text{if } 2g \equiv -2 \bmod \bar{n}; \\[3mm]
\dfrac{(x^{\bar{n}} + 1)^{(2g+1)/\bar{n}}}{(x + 1)} & \text{if } 2g \equiv -1 \bmod \bar{n}; \\[3mm]
(x^{\bar{n}} + 1)^{2g/\bar{n}} & \text{if } 2g \equiv 0 \bmod \bar{n}.
\end{cases}
$$

(3) *If $\bar{n}$ is even and $n = \bar{n}$, then $2g \equiv -2 \bmod \bar{n}$. Furthermore:*
   (a) *if $(2g + 2)/\bar{n}$ is odd, then*

$$
f = \frac{(x^{\bar{n}} - 1)^{(2g+2)/\bar{n}}}{(x^2 - 1)};
$$

   (b) *if $(2g + 2)/\bar{n}$ is even, then*

$$
f = \frac{(x^{\bar{n}} - 1)^{(2g+2)/\bar{n}}}{(x - 1)^2} \quad \text{or} \quad f = \frac{(x^{\bar{n}} - 1)^{(2g+2)/\bar{n}}}{(x + 1)^2}.
$$

(4) *If $\bar{n}$ is even and $n = 2\bar{n}$, then $2g \equiv 0 \bmod \bar{n}$ and*

$$
f = (x^{\bar{n}} + 1)^{2g/\bar{n}}.
$$

REMARK. Note that in Statements (1) and (2) of the theorem, if $\bar{n} = 1$ then the three expressions in the equality for f are all the same.

REMARK. The ambiguity in Statement (3b) is unavoidable. Suppose α is an automorphism of C for which $n = \bar{n}$, $\bar{n}$ is even, and $(2g + 2)/\bar{n}$ is even. Then α and $\iota\alpha$ give the same values of n and $\bar{n}$, but they have different characteristic polynomials.

One of our motivations for the work in this paper was Proposition 13.1 of [1], which is concerned with automorphisms α of supersingular genus-2 curves C over finite fields of characteristic at least 5. The proposition says in part that if α is such an automorphism, and if n and $\bar{n}$ are as defined above, then the pair $(n, \bar{n})$ appears in the left-hand column of Table 1, and the characteristic polynomial of α^* is as given in the right-hand column of the table. Here we note that Theorem 1.1 shows that the same conclusion holds for automorphisms of arbitrary genus-2 curves over arbitrary fields, with the restrictions on the values of n and $\bar{n}$ coming from the congruence conditions in the theorem.

In Section 2 we prove two lemmas about quotients of hyperelliptic curves by cyclic groups. In Section 3 we use these lemmas to prove Theorem 1.1.

$(n, \overline{n})$	
$(1,1)$	$(x-1)^4$
$(2,1)$	$(x+1)^4$
$(2,2)$	$(x-1)^2(x+1)^2$
$(3,3)$	$(x^2+x+1)^2$
$(4,2)$	$(x^2+1)^2$
$(5,5)$	$x^4+x^3+x^2+x+1$
$(6,3)$	$(x^2-x+1)^2$
$(6,6)$	$(x^2-x+1)(x^2+x+1)$
$(8,4)$	x^4+1
$(10,5)$	$x^4-x^3+x^2-x+1$

TABLE 1. Characteristic polynomials associated to possible values of n and $\overline{n}$ for genus-2 curves [1, Table 4].

Conventions. In this paper, a *curve* will always mean a geometrically irreducible one-dimensional nonsingular scheme over a field k; by the usual equivalence of categories, we could just as well phrase the entire paper in terms of one-dimensional function fields over k. When we speak of the projective line $\mathbf{P}^1$ over a field k, we will usually pick without comment a generator x of its function field, so that we can identify the function field with $k(x)$.

2. Quotients of hyperelliptic curves

Our proof of Theorem 1.1 will depend on two lemmas concerning quotients of hyperelliptic curves, which we state and prove in this section. Throughout this section, C will be a hyperelliptic curve over an algebraically-closed field k, ι will be the hyperelliptic involution on C, and β will be an automorphism of C of order m such that $\iota \notin \langle \beta \rangle$.

Let D be the quotient of C by the group $\langle \beta \rangle$. Since ι is a central element of the automorphism group of C, the automorphism β induces an automorphism $\overline{\beta}$ on the genus-0 curve $\overline{C} := C/\langle \iota \rangle$, and we get a diagram

$$(2.1)$$

$$
\begin{array}{ccc}
C & \xrightarrow{\langle \beta \rangle} & D \\
\downarrow{\scriptstyle 2} & & \downarrow{\scriptstyle 2} \\
\overline{C} & \xrightarrow{\langle \overline{\beta} \rangle} & \overline{D}
\end{array}
$$

where the vertical arrows are quotients by $\langle \iota \rangle$. Let φ be the map from C to $\overline{D}$ and let ψ be the map from $\overline{C}$ to $\overline{D}$. We see that φ and ψ are both Galois covers; the Galois group G of φ is generated by β and ι and is isomorphic to $(\mathbf{Z}/m\mathbf{Z}) \times (\mathbf{Z}/2\mathbf{Z})$, and the Galois group $\overline{G}$ of ψ is generated by $\overline{\beta}$ and is cyclic of order m. Note that $\overline{G}$ is the quotient of G by $\langle \iota \rangle$.

LEMMA 2.1. *Let Q be a point of $\overline{D}$ and let H be the inertia group of Q in the cover φ.*

(1) *If m is odd then H is either the trivial group, the group $\langle \iota \rangle$, the group $\langle \beta \rangle$, or all of G.*

(2) *If m is even and the characteristic of k is not 2, then H is either the trivial group, the group $\langle \iota \rangle$, the group $\langle \beta \rangle$, or the group $\langle \iota\beta \rangle$.*

Before we begin the proof of the lemma we mention some facts about automorphisms of genus-0 curves that we will use repeatedly.

Let X be a genus-0 curve over an algebraically-closed field k, and suppose γ is a finite-order automorphism of X. By choosing an appropriate isomorphism $X \cong \mathbf{P}^1$, we may write the action of γ on the function field of $\mathbf{P}^1$ in one of two forms: either $x \mapsto \xi x$ for a root of unity ξ, or $x \mapsto x + 1$. In the first case the order of γ is not divisible by the characteristic p of k. Furthermore, the quotient map from $\mathbf{P}^1$ to $\mathbf{P}^1$ induced by γ gives a Kummer extension of function fields $k(x) \to k(x)$ that can be written as $x \mapsto x^m$, where m is the order of γ. This map has two ramification points, and each point ramifies totally. When γ can be written $x \mapsto x + 1$ the order of γ is equal to p, and the associated quotient map $\mathbf{P}^1 \to \mathbf{P}^1$ gives an Artin-Schreier extension of function fields $k(x) \to k(x)$ that can be written as $x \mapsto x^p - x$. Only one point of $\mathbf{P}^1$ ramifies in this map, but again the ramification is total.

PROOF OF LEMMA 2.1. Suppose m is odd, so that the Galois group G is cyclic. We know that if a Q ramifies in ψ, then it ramifies completely. Thus, the image of H in $\overline{G}$ is either trivial or all of $\overline{G}$. The only subgroups of G that have these images in $\overline{G}$ are the ones listed in first statement of the lemma.

Suppose m is even and the characteristic p of k is not 2. Since the automorphism $\overline{\beta}$ of $\overline{C}$ has order m, and $m \neq p$, the facts we mentioned before the start of the proof show that that p does not divide m. Thus p does not divide $\#G = 2m$, so all ramification in φ is tame. In particular, the inertia group H is cyclic. The only cyclic subgroups of G whose images in $\overline{G}$ are either trivial or all of $\overline{G}$ are the four groups listed in the second statement. $\square$

LEMMA 2.2. *With notation and assumptions as above, let g be the genus of C, let h be the genus of D, and let e be the number of points of $\overline{D}$ that ramify in both the right and the bottom maps of Diagram (2.1). Then $e \in \{0, 1, 2\}$. If the characteristic of k is not 2, then the relationship between g and h depends on e and on the parity of m as follows:*

	m odd	m even
e = 0	$2h = (2g + 2)/m - 2$	$2h = (2g + 2)/m - 2$
e = 1	$2h = (2g + 1)/m - 1$	$2h = (2g + 2)/m - 1$
e = 2	$2h = 2g/m$	$2h = (2g + 2)/m$

If k has characteristic 2, then m is equal to 2 if it is even, and the relationship between g and h depends on e and on the parity of m as follows:

	m odd	m = 2
e = 0	$2h = (2g + 2)/m - 2$	$2h = g - 1$
e = 1	$2h = (2g + 1)/m - 1$	$2h = g$ if g is even, $2h = g + 1$ if g is odd
e = 2	$2h = 2g/m$	(*not possible*)

PROOF. We know that at most two points ramify in the cover $\psi : \overline{C} \to \overline{D}$, so it follows immediately that e is at most 2.

Let $\mathfrak{d}_C$ and $\mathfrak{d}_D$ denote the differents of the double covers $C \to \overline{C}$ and $D \to \overline{D}$, respectively, and let $\mathfrak{D}_C$ and $\mathfrak{D}_D$ be the discriminants of these covers. Note that we have $\deg \mathfrak{D}_C = \deg \mathfrak{d}_C$ and $\deg \mathfrak{D}_D = \deg \mathfrak{d}_D$. More specifically, if P is a point of $\overline{C}$ at which $\mathfrak{D}_C$ has positive order, then there is a unique point $\mathfrak{p}$ of C over P, and $\mathrm{ord}_{\mathfrak{p}} \mathfrak{d}_C = \mathrm{ord}_P \mathfrak{D}_C$; the analogous statement holds for points of $\overline{D}$. The Riemann-Hurwitz formula [2, Thm. 3.3.5], applied to the double covers $C \to \overline{C}$ and $D \to \overline{D}$, shows that

$$g = -1 + (1/2)\deg \mathfrak{d}_C = -1 + (1/2)\deg \mathfrak{D}_C$$

and

$$h = -1 + (1/2)\deg \mathfrak{d}_D = -1 + (1/2)\deg \mathfrak{D}_D.$$

Therefore, to find the relationship between g and h we need only find the relationship between the degrees of $\mathfrak{D}_C$ and $\mathfrak{D}_D$.

Before we turn to the various cases summarized in the tables in the statement of the lemma, we will sketch out the general method we use to compare the degrees of these two discriminants. Throughout this introductory sketch, we will assume that we are not in the special case where m is even and k has characteristic 2.

Suppose P is a point of $\overline{C}$ that ramifies in the double cover $C \to \overline{C}$. By looking at the lists in Lemma 2.1 of the possible ramification groups for the point $\psi(P)$ in the extension $\varphi : C \to \overline{D}$, we see that $\psi(P)$ must ramify in the double cover $D \to \overline{D}$. In other words, the support of $\mathfrak{D}_C$ is contained in the inverse image under ψ of the support of $\mathfrak{D}_D$.

We divide the support of $\mathfrak{D}_D$ into two sets: Let E be the set of points of $\overline{D}$ that ramify both in $D \to \overline{D}$ and in ψ, and let E' be the set of points of $\overline{D}$ that ramify in $D \to \overline{D}$ but not in ψ. Then we have $e = \#E$, and we set $e' := \#E'$. Let $\mathfrak{D}_C'$ be the part of $\mathfrak{D}_C$ supported on $\psi^{-1}(E')$, and let $\mathfrak{D}_D'$ be the part of $\mathfrak{D}_D$ supported on E'.

Suppose Q is a point of E', and let P be one of the m points in $\psi^{-1}(Q)$. Then locally at P and at Q the extensions $C \to \overline{C}$ and $D \to \overline{D}$ are isomorphic, so the order of $\mathfrak{D}_C$ at P is equal to the order of $\mathfrak{D}_D$ at Q. This shows that $\deg \mathfrak{D}_C' = m \deg \mathfrak{D}_D'$.

All that remains is to find the relationship between the portion of $\mathfrak{D}_C$ supported on $\psi^{-1}(E)$ and the portion of $\mathfrak{D}_D$ supported on E.

Suppose Q is a point of E. For each i let H_i be the i-th ramification group of Q in the double cover $D \to \overline{D}$. By [2, Thm. 3.5.9], the order of $\mathfrak{D}_D$ at Q is equal to $\sum(\#H_i - 1)$, but since each H_i has order 1 or 2, the value of this sum is simply the largest i such that H_i is nontrivial. Let $\mathfrak{q}$ be the point if D lying over Q, and let v be a uniformizer at $\mathfrak{q}$. According to [2, Lem. 3.5.6], the largest value of i such that H_i is nontrivial is the valuation of $v - \iota^* v$ at $\mathfrak{q}$. Thus, $\mathrm{ord}_Q \mathfrak{D}_D = \mathrm{val}_{\mathfrak{q}}(v - \iota^* v)$.

Let P be the unique point of $\overline{C}$ with $\psi(P) = Q$. If P is unramified in the double cover $C \to \overline{C}$ then $\mathfrak{D}_C$ has order 0 at P. If P is ramified, let $\mathfrak{p}$ be the point of C lying over it, and let u be a uniformizer at $\mathfrak{p}$. Arguing as above, we find that $\mathrm{ord}_P \mathfrak{D}_C = \mathrm{val}_{\mathfrak{p}}(u - \iota^* u)$.

With these formulas for $\mathrm{ord}_Q \mathfrak{D}_D$ and $\mathrm{ord}_P \mathfrak{D}_C$ in hand, we turn to the various cases listed in the lemma.

First suppose that the characteristic of k is not 2 and that m is odd. If Q is a point in E, then the inertia group of Q in φ must be G. This shows that the unique point P with $\psi(P) = Q$ is ramified in the double cover $C \to \overline{C}$. Since the characteristic of k is not 2, the point P is tamely ramified. Likewise, Q is tamely ramified in $D \to \overline{D}$. Thus, $\mathfrak{D}_C$ has order 1 at P and $\mathfrak{D}_D$ has order 1 at Q. It follows that

$$\deg \mathfrak{D}_C - e = \deg \mathfrak{D}'_C = m \deg \mathfrak{D}'_D = m(\deg \mathfrak{D}_D - e),$$

which gives $(2g + 2 - e) = m(2h + 2 - e)$, which is what is claimed in the left-hand column of the first table in Lemma 2.2.

Suppose that the characteristic of k is not 2 and that m is even. If Q is a point of E, then the inertia group of Q in the cover φ must be $\langle \iota \beta \rangle$. In this case we see that the unique point P of $\overline{C}$ with $\psi(P) = Q$ does *not* ramify in the double cover $C \to \overline{C}$. This tells us that

$$\deg \mathfrak{D}_C = \deg \mathfrak{D}'_C = m \deg \mathfrak{D}'_D = m(e + \deg \mathfrak{D}_D),$$

which leads to the entries in the right-hand column of the first table in Lemma 2.2.

Now suppose that k has characteristic 2 and that m is odd, and suppose Q is a point of E. Let P be the unique point of $\overline{C}$ with $\psi(P) = Q$. The inertia group of Q in the cover $\varphi : C \to \overline{D}$ must be G, so P is ramified in the double cover $C \to \overline{C}$. As above, let $\mathfrak{p}$ be the point of C lying over P and let $\mathfrak{q}$ be the point of D lying over Q. We would like to compare the order of $\mathfrak{D}_C$ at P to the order of $\mathfrak{D}_D$ at Q. Since these are locally-defined quantities, we may replace the curves in Diagram (2.1) with their completions at $\mathfrak{p}$, $\mathfrak{q}$, P, and Q, respectively. We can then choose a uniformizer u for $\mathfrak{p}$ and a uniformizer v for $\mathfrak{q}$ such that $v = u^m$.

Let $i > 1$ be the valuation at $\mathfrak{p}$ of $u - \iota^* u$. Then we can write

$$\iota^* u = u + cu^i + (\text{higher-order terms}),$$

and raising both sides to the m-th power we find that

$$\iota^* v = v + mcu^{m-1+i} + (\text{higher-order terms}).$$

Since the valuation of $v - \iota^* v$ at $\mathfrak{p}$ is $m - 1 + i$, the valuation j of $v - \iota^* v$ at $\mathfrak{q}$ must be $(m - 1 + i)/m$. In other words, $i = mj - m + 1$. If we let I denote the degree of the portion of $\mathfrak{D}_C$ supported on $\psi^{-1}(E)$, and J the degree of the portion of $\mathfrak{D}_D$ supported on E, then $I = mJ - (m - 1)e$.

We find that

$$\deg \mathfrak{D}_C = \deg \mathfrak{D}'_C + I = m \deg \mathfrak{D}'_D + mJ - (m - 1)e = m \deg \mathfrak{D}_D - (m - 1)e,$$

so that $\deg \mathfrak{D}_C - e = m(\deg \mathfrak{D}_D - e)$, which again leads to $2g + 2 - e = m(2h + 2 - e)$. This gives us the formulas in the left-hand column of the second table in Lemma 2.2.

Finally we consider the case where k has characteristic 2 and m is even. As we noted before the proof of Lemma 2.1, an even-order automorphism of a genus-0 curve in characteristic 2 must have order 2, so $m = 2$. Once again, we define E to be the set of points of $\overline{D}$ that ramify in $\psi : \overline{C} \to \overline{D}$ and in the double cover $D \to \overline{D}$ (so that $e = \#E$), and we define E' to be the set of points that ramify in $D \to \overline{D}$ but not in ψ. As before, we define $\mathfrak{D}'_C$ to be the part of $\mathfrak{D}_C$ supported on $\psi^{-1}(E')$ and $\mathfrak{D}'_D$ to be the part of $\mathfrak{D}_D$ supported on E', and as before, we have $\deg \mathfrak{D}'_C = 2 \deg \mathfrak{D}'_D$.

Since $m = 2$, the map ψ is ramified at a single point, and $e \le 1$. If $e = 0$ then

$$\deg \mathfrak{D}_C = \deg \mathfrak{D}'_C = 2 \deg \mathfrak{D}'_D = 2 \deg \mathfrak{D}_D,$$

and it follows from the Riemann-Hurwitz formula that $2h = g - 1$, as claimed in the second table in Lemma 2.2.

On the other hand, if $e = 1$ we may choose isomorphisms $\overline{C} \cong \mathbf{P}^1$ and $\overline{D} \cong \mathbf{P}^1$ so that ψ corresponds to the function field map $x \mapsto x^2 + x$; then ∞ ramifies in the double cover $D \to \overline{D} \cong \mathbf{P}^1$. The function field $k(D)$ of D is an Artin-Schreier extension of $k(x)$, so it contains an element y not in $k(x)$ such that $y^2 + y$ lies in $k(x)$. The completion of $k(x)$ at ∞ is the ring of Laurent series in $1/x$, and in this completion we can write

$$y^2 + y = a_n x^n + \sum_{i=-\infty}^{n-1} a_i x^i$$

for some integer n and elements a_i of k with $a_n \neq 0$. If $n = 0$ we can replace y with $y + b$ for a constant $b \in k$ with $b^2 + b = a_0$, which has the effect of replacing n with nonzero integer. If n is even and nonzero we may replace y with $y + \sqrt{a_n} x^{n/2}$, which has the effect of replacing n by $n/2$; repeating this reduction, we find that we may assume that n is odd, say $n = 2m - 1$. If n were negative the point ∞ would split in D, contrary to assumption, so n must be positive.

Let $\mathfrak{q}$ be the unique point of D lying over ∞. It is easy to check that $v = y/x^m$ is a uniformizer for $\mathfrak{q}$. As before, the order of $\deg \mathfrak{D}_D$ at ∞ is equal to the valuation of $v - \iota^* v$ at $\mathfrak{q}$. Since $\iota^* v = (y + 1)/x^m$, we find that

$$\mathrm{val}_{\mathfrak{q}}(v - \iota^* v) = \mathrm{val}_{\mathfrak{q}}(1/x^m) = 2 \, \mathrm{val}_{\infty}(1/x^m) = 2m.$$

Now consider the curve C, which is the fiber product of the double cover $D \to \overline{D}$ with $\psi : \overline{C} \to \overline{D}$. Locally at ∞, we obtain C by taking the equality

$$y^2 + y = a_n x^n + \sum_{i=-\infty}^{n-1} a_i x^i$$

and replacing x with $x^2 + x$. This gives us

$$y^2 + y = a_n x^{2n} + a_n x^{2n-1} + (\text{terms in } x^i \text{ with } i < 2n - 1).$$

If $n > 1$, then replacing y with $y + \sqrt{a_n} x^n$ gives us

$$y^2 + y = a_n x^{2n-1} + (\text{terms in } x^i \text{ with } i < 2n - 1),$$

and we find that $\mathrm{ord}_{\infty} \mathfrak{D}_C = 2n = 4m - 2$. But if $n = 1$, the same substitution gives

$$y^2 + y = (a_n + \sqrt{a_n})x + (\text{terms in } x^i \text{ with } i < 1).$$

In this case, $\mathrm{ord}_{\infty} \mathfrak{D}_C = 2 = 4m - 2$ if $a_n \neq \sqrt{a_n}$, and $\mathrm{ord}_{\infty} \mathfrak{D}_C = 0 = 4m - 4$ otherwise.

Applying the Riemann-Hurwitz formula to the double covers $D \to \overline{D}$ and $C \to \overline{C}$ and using the relation $\deg \mathfrak{D}'_C = 2 \deg \mathfrak{D}'_D$, we find that

$$2h = \deg \mathfrak{D}'_D + 2m - 2$$

$$g = \begin{cases} \deg \mathfrak{D}'_D + 2m - 2 & \text{if } \infty \text{ ramifies in } C \to \overline{C} \cong \mathbf{P}^1; \\ \deg \mathfrak{D}'_D + 2m - 3 & \text{if } \infty \text{ is unramified in } C \to \overline{C} \cong \mathbf{P}^1. \end{cases}$$

It follows that

$$2h = \begin{cases} g & \text{if } \infty \text{ ramifies in } C \to \overline{C} \cong \mathbf{P}^1; \\ g + 1 & \text{if } \infty \text{ is unramified in } C \to \overline{C} \cong \mathbf{P}^1. \end{cases}$$

Clearly g is even in the first case and odd in the second, so we get the result given in the second column of the second table of Lemma 2.2. $\qquad \square$

REMARK. One could also prove Lemma 2.2 by using explicit equations and the standard formulas for the genus of a hyperelliptic curve in terms of its defining equation [2, Cor. 3.6.3, Cor. 3.6.9]. For example, suppose that the characteristic p of the base field is not 2, that m is odd and not divisible by p, and that $e = 2$. By choosing appropriate isomorphisms $\overline{C} \cong \mathbf{P}^1 \cong \overline{D}$ and an appropriate model for D, we can assume that ψ is the map $x \mapsto x^m$ and that D is given by $y^2 = xf(x)$ for a separable even-degree polynomial $f(x)$ with $f(0) \neq 0$. Then C has a singular model of the form $y^2 = x^m f(x^m)$ and a nonsingular model of the form $z^2 = xf(x^m)$. In this case one checks that $h = (\deg f)/2$ and $g = (m \deg f)/2$, so $2h = 2g/m$, as claimed by the lemma.

3. Proof of Theorem 1.1.

In this section we prove Theorem 1.1. Let us begin by explaining the basic idea of the proof.

Since the conclusions of Theorem 1.1 are completely geometric, we may assume that k is algebraically closed. The characteristic polynomial f of α^* has degree $2g$; let its complex roots be $\zeta_1, \ldots, \zeta_{2g}$, so that the ζ are all n-th roots of unity. For each divisor d of n let N_d denote the number of the ζ that are primitive d-th roots of unity and let M_d denote the number of the ζ that satisfy $\zeta^d = 1$. Then we have

$$M_d = \sum_{e|d} N_e, \qquad N_d = \sum_{e|d} \mu(d/e) M_e, \quad \text{and} \quad f = \prod_{d|n} \Phi_d^{N_d/\phi(d)},$$

where μ is the Möbius function, ϕ is the Euler ϕ-function, and Φ_d is the d-th cyclotomic polynomial. So to determine f, it is enough to determine the M_d.

For every divisor d of n, let f_d be the characteristic polynomial of the automorphism α^d of C. Then the complex roots of f_d are the d-th powers of the complex roots of f, so M_d is equal to the multiplicity of 1 as a root of f_d. This multiplicity is equal to twice the dimension of the part of the Jacobian on which α^d acts trivially, and this dimension is equal to the genus of the quotient of C by $\langle \alpha^d \rangle$. We see that computing M_d is equivalent to computing the genus of this quotient curve. If the hyperelliptic involution ι lies in $\langle \alpha^d \rangle$ then the genus of the quotient is 0; if not, then the genus is determined by Lemma 2.2. To prove the theorem, all we must do is verify that the values of M_d predicted by the putative characteristic polynomials given in the theorem agree with the values we compute by applying Lemma 2.2.

We consider the four statements of the theorem in turn.

Proof of Statement (1): In this case our assumption is that $n = \bar{n}$ and $\bar{n}$ is odd. For each divisor d of n let D_d be the quotient of C by $\langle \alpha^d \rangle$. Then for each d we have a diagram

(3.1)
$$
\begin{array}{ccccc}
C & \xrightarrow{n/d} & D_d & \xrightarrow{d} & D_1 \\
\downarrow{\scriptstyle 2} & & \downarrow{\scriptstyle 2} & & \downarrow{\scriptstyle 2} \\
\overline{C} & \xrightarrow{n/d} & \overline{D}_d & \xrightarrow{d} & \overline{D}_1.
\end{array}
$$

As in Section 2, we see that the map φ from C to $\overline{D}_1$ is a Galois cover with group $G \cong (\mathbf{Z}/n\mathbf{Z}) \times (\mathbf{Z}/2\mathbf{Z})$. For each d let φ_d be the map from C to $\overline{D}_d$, let E_d be the set of points of $\overline{D}_d$ that ramify going up to D_d and going up $\overline{C}$, and let $e_d = \#E_d$. We will show that e_d is determined by e_1.

Since $\bar{n}$ is odd, Lemma 2.1 tells us that the inertia group of a point of $\overline{D}_1$ in the extension $\varphi : C \to \overline{D}_1$ is either trivial, or $\langle \iota \rangle$, or $\langle \alpha \rangle$, or all of G. A point in E_1 must have ramification group G, and so must lie under a unique point in E_d. Likewise, any point in E_d must lie over a point of $\overline{D}_1$ that has ramification group G, and that therefore lies in E_1. Thus, for every d we have $e_d = e_1$.

If $e_1 = 0$ then Lemma 2.2 shows that $M_d = (2g+2)d/\bar{n} - 2$ for all d. In particular we see that $\bar{n}$ divides $2g + 2$. Also, we check that the polynomial

$$f = \frac{(x^{\bar{n}} - 1)^{(2g+2)/\bar{n}}}{(x-1)^2}$$

produces the correct values of M_d. If $e_1 = 1$ then we have $M_d = (2g+1)d/\bar{n} - 1$, so $\bar{n}$ divides $2g + 1$, and the polynomial

$$f = \frac{(x^{\bar{n}} - 1)^{(2g+1)/\bar{n}}}{(x-1)}$$

gives the correct values of M_d. Finally, if $e_1 = 2$ then $M_d = 2gd/\bar{n}$, so that $\bar{n}$ divides $2g$, and the polynomial

$$f = (x^{\bar{n}} - 1)^{2g/\bar{n}}$$

produces the required values of M_d.

Proof of Statement (2): In this case $n = 2\bar{n}$ and $\bar{n}$ is odd, and we see that $\iota = \alpha^{\bar{n}}$. Let $\alpha_0 = \iota\alpha$, so that α_0 has order $\bar{n}$ and induces an automorphism of order $\bar{n}$ on $\overline{C}$. Then Statement (1) tells us the characteristic polynomial f_0 of α_0^*; furthermore, since $\alpha^* = -\alpha_0^*$, we have $f(x) = f_0(-x)$. This agrees with what is claimed in Statement (2).

Proof of Statement (3): In this case $n = \bar{n}$ and $\bar{n}$ is even, and the analysis is very much like that for Statement (1). For every divisor d of n we let D_d be the quotient of C by $\langle \alpha^d \rangle$, and Diagram (3.1) is again a diagram of Galois extensions, with total Galois group G isomorphic to $(\mathbf{Z}/n\mathbf{Z}) \times (\mathbf{Z}/2\mathbf{Z})$. However, since n is even, G is no longer a cyclic group. As before, we let φ be the map from C to $\overline{D}_1$, we let φ_d be the map from C to $\overline{D}_d$, we let E_d be the set of points of $\overline{D}_d$ that ramify going up to D_d and going up to $\overline{C}$, and we let $e_d = \#E_d$.

Let us first consider the case in which the characteristic of the base field is not equal to 2. Then according to Lemma 2.1, the ramification group of a point Q of $\overline{D}_1$ in the cover φ is either trivial, the group $\langle \iota \rangle$, the group $\langle \alpha \rangle$, or the group $\langle \iota\alpha \rangle$. If a point has one of the first two inertia groups it will not lie in E_d, because it is not ramified in the bottom row of Diagram (3.1). If a point has inertia group $\langle \alpha \rangle$, then it will not lie in E_d because it is not ramified in the extension $D_d \to \overline{D}_d$. But if a point has inertia group $\langle \iota\alpha \rangle$, it will lie in E_d if d is odd, and will not lie in E_d if d is even.

So when the characteristic of the base field is not 2, we see once again that the value of e_d is determined by the value of e_1: We have $e_d = e_1$ if d is odd, and $e_d = 0$ if d is even. If d is odd then n/d is even, so Lemma 2.2 tells us that $M_d = (2g+2)d/n - 2 + e_1$. (Note that since M_1 is twice the genus of D_1, we find that n divides $2g+2$ and that the parity of e_1 is equal to the parity of $(2g+2)/n$.) If d is even then Lemma 2.2 shows that $M_d = (2g+2)d/n - 2$. These values for

M_d are consistent with

$$f = \begin{cases} \dfrac{(x^n - 1)^{(2g+2)/n}}{(x - 1)^2} & \text{if } e_1 = 0; \\[3mm] \dfrac{(x^n - 1)^{(2g+2)/n}}{(x^2 - 1)} & \text{if } e_1 = 1; \\[3mm] \dfrac{(x^n - 1)^{(2g+2)/n}}{(x + 1)^2} & \text{if } e_1 = 2, \end{cases}$$

so these must be the correct values of f.

As we noted above, the parity of $(2g + 2)/n$ is equal to that of e_1. Thus, if $(2g + 2)/n$ is odd then $e_1 = 1$, and we find the value of f given in Statement (3a). If $(2g + 2)/n$ is even then e_1 is either 0 or 2, and we find that f must have one of the two values given in Statement (3b).

Finally, we turn to the case in which the base field has characteristic 2. In this case we must have $n = 2$, so we only have to determine the value of M_1 (since we already know that $M_2 = 2g$). But Lemma 2.2 tells us the possibilities for this value: If g is even then $M_1 = g$, while if g is odd then M_1 is either $g - 1$ or $g + 1$. We check that the values of f given in Statements (3a) and (3b) agree with these values of M_1 and M_2.

Proof of Statement (4): In this case $n = 2\bar{n}$ and $\bar{n}$ is even, and we have $\iota = \alpha^{\bar{n}}$. Taking the quotient of C by $\langle \alpha \rangle$ gives us a Galois extension

$$(3.2) \qquad\qquad C \xrightarrow{2} \overline{C} \xrightarrow{\bar{n}} \overline{D}$$

with group $G \cong \mathbf{Z}/n\mathbf{Z}$, where $\overline{C}$ and $\overline{D}$ are curves of genus 0.

Consider a point Q of $\overline{D}$ that ramifies going up to $\overline{C}$. Then Q must be totally ramified in this extension, so the inertia group of Q in the total extension $C \to \overline{D}$ is a subgroup of G that surjects onto the Galois group of $\overline{C} \to \overline{D}$. The only such subgroup is G itself, so any point of $\overline{D}$ that ramifies going up to $\overline{C}$ must ramify totally in the extension $C \to \overline{D}$.

We see that if d is a divisor of n such that n/d is even, then ι lies in the subgroup $\langle \alpha^d \rangle$, the genus of the quotient of C by this subgroup is 0, and $M_d = 0$. If d is a divisor of n such that n/d is odd, let e_d be the number e associated to α^d as in Lemma 2.2; then e_d is equal to the number of points of $\overline{D}$ that ramify in the degree-$\bar{n}$ extension $\overline{C} \to \overline{D}$, and this value is either 1 or 2, depending on whether or not $\bar{n}$ is divisible by the characteristic of the base field.

Suppose the characteristic of the base field is not equal to 2. Then, since $\bar{n}$ is even, the degree-$\bar{n}$ map $\overline{C} \to \overline{D}$ of genus-0 curves does not give an Artin-Schreier extension of function fields; rather, it gives a Kummer extension, and it follows that there are two points of $\overline{D}$ that totally ramify going up to C. Any other points of $\overline{D}$ that ramify going up to C must have ramification groups of order 2. If there are r of these points, then the Riemann-Hurwitz formula for Galois extensions tells us that

$$2g - 2 = n\left(-2 + 2(1 - 1/n) + r(1 - 1/2)\right) = -2 + r\bar{n},$$

and it follows that $\bar{n}$ divides $2g$.

Also, since we have two points of $\overline{D}$ that ramify totally in $C \to \overline{D}$, we see that $e_d = 2$ whenever n/d is odd, and it follows from Lemma 2.2 that $M_d = 2gd/n$ when

n/d is odd. Combining this with the observation that $M_d = 0$ when n/d is even, we see that the polynomial $f = (x^{\overline{n}} + 1)^{2g/\overline{n}}$ gives the correct values of M_d.

Now suppose that the characteristic of the base field is equal to 2. Then $\overline{n}$ must be equal to 2, and α has order 4. The diagram (3.2) shows that the quotient of C by $\langle \alpha \rangle$ has genus 0, the quotient of C by $\langle \alpha^2 \rangle$ has genus 0, and the quotient of C by $\langle \alpha^4 \rangle$ has genus g. Thus $M_1 = M_2 = 0$ and $M_4 = 2g$. The polynomial that gives rise to these values of M_d is $(x^2 + 1)^g$, which is the polynomial given in Statement (4). $\qquad\qquad\qquad\qquad\qquad\qquad\qquad\qquad\qquad\qquad\qquad\square$

References

[1] Everett W. Howe, Enric Nart, and Christophe Ritzenthaler, *Jacobians in isogeny classes of abelian surfaces over finite fields*, Ann. Inst. Fourier (Grenoble), to appear, available at arXiv: math/0607515v3[math.NT].

[2] David M. Goldschmidt, *Algebraic functions and projective curves*, Graduate Texts in Mathematics, vol. 215, Springer-Verlag, New York, 2003.

DEPARTMENT OF MATHEMATICS, UNIVERSITY OF SOUTHERN CALIFORNIA, LOS ANGELES, CA 90089-2532, USA.

E-mail address: guralnic@usc.edu

URL: http://www-rcf.usc.edu/~guralnic/

CENTER FOR COMMUNICATIONS RESEARCH, 4320 WESTERRA COURT, SAN DIEGO, CA 92121-1967, USA.

E-mail address: however@alumni.caltech.edu

URL: http://alumni.caltech.edu/~however/

Contemporary Mathematics
Volume **487**, 2009

Breaking the Akiyama-Goto cryptosystem

Petar Ivanov and José Felipe Voloch

ABSTRACT. Akiyama and Goto have proposed a cryptosystem based on rational points on curves over function fields (stated in the equivalent form of sections of fibrations on surfaces). It is easy to construct a curve passing through a few given points, but finding the points, given only the curve, is hard. We show how to break their original cryptosystem by using algebraic points instead of rational points and discuss possibilities for changing their original system to create a secure one.

1. Introduction

A basic ingredient of a public-key cryptosystem is a mathematical procedure that is computationally easy to do but hard to undo. The classical example being the fact that multiplying two integers is easy but factoring an integer is hard. This paper explores the following procedure. Let R be a ring and $a, b \in R$. Then it is easy to find polynomials $X(x, y) \in R[x, y]$ with $X(a, b) = 0$. On the other hand, it may be hard, given just X, to find the corresponding a, b. The case $R = \mathbb{Z}$ of course has been given a lot of attention but, for general X the best known method is not much better than brute-force search. Some improvement can be obtained by sieving and lattice reduction techniques but from the perspective of computational complexity, this does not improve on brute-force search. Much of the discussion of the case $R = \mathbb{Z}$ apply to $R = \mathbb{F}_p[t]$ as well. For a discussion of this problem see [4].

A cryptosystem that uses the above problem of "finding points on curves" was proposed by Akiyama and Goto [1] who also proposed a variant in [2]. As we will show in this paper, both variants are insecure. However, the system is not broken by solving the problem of "finding points" and there remains the possibility that a secure cryptosystem can be built around this problem. We briefly discuss this possibility in this paper too. The main motivation of Akiyama and Goto was that, as far it is known, the problem of "finding points on curves" cannot be solved more efficiently by a quantum computer, unlike say, the factoring problem.

2000 *Mathematics Subject Classification.* Primary 14G50, 94A60; Secondary 11G30.
Key words and phrases. Public-key Cryptosystem, Algebraic Curve, Rational Point.

2. The cryptosystem of Akiyama and Goto

In this section we present the cryptosystem described by Akiyama and Goto in [1]. Let p be a prime number, $R = \mathbb{F}_p[t]$ be the polynomial ring over the prime field $\mathbb{F}_p$ and $K = \mathbb{F}_p(t)$ be the field of rational functions over $\mathbb{F}_p$. K is the field of fractions of R. Pick a polynomial in two variables $X(x, y) \in R[x, y]$, together with two points $U = (u_x, u_y) \in R^2$ and $V = (v_x, v_y) \in R^2$, such that $X(U) = X(V) = 0$. In other words, we take an algebraic curve over K together with 2 rational points on the curve. It is easy to find two points and a curve passing through them and we will show how below. On the other hand, if the curve is given (in terms of the polynomial $X(x, y)$), it is a hard mathematical problem to find rational points on it. This fact can be used to build the cryptosystem described in this section.

2.1. Keys and key generation. The secret key consists of the two points $U = (u_x(t), u_y(t))$ and $V = (v_x(t), v_y(t))$, such that either $\deg u_x \neq \deg v_x$ or $\deg u_y \neq \deg v_y$.

The public key consists of four things: the prime number p, the equation $X(x, y) = 0$, defining a curve through U and V, an integer l, which will serve as a lower bound for the degree of a monic irreducible polynomial $f \in R$ and an integer d, satisfying

$$(2.1) \qquad d \geq \max\{\deg u_x, \deg u_y, \deg v_x, \deg v_y\}.$$

Let us write the equation of the curve as

$$X(x, y) = \sum_{i,j} c_{ij} x^i y^j = 0$$

and try to obtain the coefficients $c_{ij} \in R$ in such a way that U and V satisfy it. This means:

$$(2.2) \qquad \sum_{i,j} c_{ij} u_x^i u_y^j = \sum_{i,j} c_{ij} v_x^i v_y^j = 0.$$

If we subtract the second sum from the first we get

$$\sum_{(i,j) \neq (0,0)} c_{ij}(u_x^i u_y^j - v_x^i v_y^j) = 0,$$

which can be written as

$$(2.3) \qquad c_{10}(u_x - v_x) = - \sum_{(i,j) \neq (0,0),(1,0)} c_{ij}(u_x^i u_y^i - v_x^i v_y^i).$$

Now suppose that $(u_x - v_x)|(u_y - v_y)$. Then the right hand side of (2.3) is also divisible by $u_x - v_x$, because $u_x^i u_y^j - v_x^i v_y^j = (u_x^i - v_x^i)u_y^j + v_x^i(u_y^j - v_y^j)$. This suggest the following algorithm for choosing X:

(1) For each pair of indices $(i, j) \neq (0, 0), (1, 0)$ pick a random element $c_{ij} \in R$.
(2) Randomly choose elements $\lambda_x, \lambda_y, v_x, v_y$ in R, such that $\lambda_x | \lambda_y$.
(3) Compute $u_y = \lambda_y + v_y$ and $u_x = \lambda_x + v_x$.
(4) Compute c_{10} from (2.3).
(5) Compute c_{00} from (2.2) as $-c_{00} = \sum_{(i,j) \neq (0,0)} c_{ij} u_x^i v_y^j$.

2.2. Encryption and decryption. Let $m \in R$ be the secret message that we want to encrypt and assume $\deg m < l$. When $p = 2$ we can encode any sequence of k bits as a polynomial of degree k in $\mathbb{F}_2[t]$. The assumptions mean that we need to encrypt the secret message by dividing it into blocks of at most l bits and encrypting each of them individually. The encryption goes like this:

(1) Choose a random polynomial $s(x, y) \in R[x, y]$, which satisfies the following condition

$$(2.4) \qquad (\deg_x s + \deg_y s)d + \deg_t s < l.$$

(2) Choose another random polynomial $r(x, y) \in R[x, y]$ and a monic irreducible polynomial $f \in R$, such that

$$(2.5) \qquad \deg_t f > l.$$

(3) Compute the cypher polynomial $F(x, y) \in R[x, y]$ according to the formula

$$(2.6) \qquad F = m + fs + Xr.$$

Now a person knowing the secret key, namely the points U and V, can easily decypher the encrypted polynomial F in the following way:

(1) Evaluate F at U and V to get polynomials h_1, $h_2 \in R$.

$$h_1 = F(u_x, u_y) = m + fs(u_x, u_y)$$

$$h_2 = F(v_x, v_y) = m + fs(v_x, v_y).$$

(2) Factor $h_1 - h_2$ and find f as the factor with largest degree.
(3) Compute m as the remainder of h_1 when divided by f.

Note that indeed $f(t)$ is the highest degree factor of $h_1(t) - h_2(t)$. We have $h_1(t) - h_2(t) = (s(u_x, u_y) - s(v_x, v_y))f$ and because of the condition (2.5), we only have to show that $\deg(s(u_x, u_y) - s(v_x, v_y)) \leq l$. Suppose that the degree of one of the two polynomials, say $s(u_x, u_y)$, is greater than l. This can only happen if there exist a monomial in $s(x, y) \in R$, say $s_0(x, y) = gx^\alpha y^\beta$, $g \in R$, such that $\deg s_0(u_x, u_y) > l$. If this is the case, then use $\alpha \leq \deg_x s$, $\beta \leq \deg_y s$, (2.1) and (2.4) to get:

$$l < \deg(gu_x^\alpha u_y^\beta) \leq \deg g + (\deg u_x)\alpha + (\deg u_y)\beta \leq \deg_t s + d(\deg_x s + \deg_y s) < l,$$

which is a contradiction. Thus $f(t)$ is indeed the irreducible factor of $h_1(t) - h_2(t)$ with highest degree.

The most time consuming part of this decryption algorithm is to factor the polynomial in step 2. In our case this can be done efficiently using the algorithm of Cantor-Zassenhaus [3]. This is the one of the fastest methods for factoring polynomials over finite fields.

In the description so far we have omitted all the details in choosing the parameters, which concern the security of the cryptosystem and we have listed only the minimal conditions, which have to be imposed to make the decryption possible. However, the algorithm suggested does not always produce a valid decryption. It fails precisely if it happens that $h_1 - h_2 = 0$, i.e. if $s(u_x, u_y) = s(v_x, v_y)$. The probability of this happening is negligible with respect to the degree of s, as discussed in [1], where some values for the parameters are suggested. We follow those suggestions for our experiments, described below.

3. Breaking the cryptosystem

We describe an attack which efficiently breaks the protocol just described. However, although it reveals the secret message efficiently, it says nothing about the secret key, as we will see. Finding two or even one rational point on the curve $X(x, y) = 0$ is a hard problem, which it turns out one does not need to solve in order to reveal the secret message m. The idea is to work in an extension of R, in which we can find points on the curve and then use these points to evaluate the cypher polynomial.

Let $S = R[y]/(X(x, 0))$ and let $\alpha = \pi(x)$ be the image of x in S under the natural projection

$$\pi : R[x] \to R[x]/(X(x, 0)).$$

The point $(\alpha, 0)$ is on the curve $X(x, y) = 0$, because by construction $X(\alpha, 0) = \pi(X(x, 0)) = 0$. We evaluate the cypher polynomial F at $(\alpha, 0)$ to get

(3.1) $$F(\alpha, 0) = m + fs(\alpha, 0) + X(\alpha, 0)r(\alpha, 0) = m + fs(\alpha, 0).$$

We now want to go back to our original ring by applying the trace operator. To be precise, recall that we denoted $K = \mathbb{F}_p(t)$ and let $L = S \otimes_R K$. We have the trace operator $Tr : L \to K$, which satisfies $Tr|_K = [L : K]id$.

Now choose an element $0 \neq \beta \in S$ with $Tr(\beta) = 0$. If $\gamma \in S$, but $\gamma \notin R$, then $\beta = \gamma - \frac{Tr(\gamma)}{n}$, where $n = \deg_x X(x, 0)$, is such a choice, provided $(p, n) = 1$ which we assume for simplicity. Indeed,

$$Tr(\gamma - \frac{Tr(\gamma)}{n}) = Tr(\gamma) - Tr(\frac{1}{n})Tr(\gamma) = Tr(\gamma) - Tr(\gamma) = 0,$$

because $Tr(\frac{1}{n}) = \frac{1}{n}[L : K] = \frac{1}{n}\deg(X(0, y)) = 1$.

Now using (3.1) we get

$$Tr(\beta F(\alpha, 0)) = mTr(\beta) + fTr(\beta s(\alpha, 0)) = fTr(\beta s(\alpha, 0)).$$

In other words, for any choice of β, the adversary can compute $p_\beta = Tr(\beta F(\alpha, 0))$, which is a polynomial in t divisible by f. But f is monic irreducible polynomial of large degree, which allows the adversary to find it, in case $p_\beta \neq 0$. For example he could compute $p_\beta \neq 0$ for several different choices of β, take the greatest common divisor of them, and extract f as the irreducible polynomial of largest degree, which divides the greatest common divisor. The most time consuming computation in this process is the factorization needed to obtain the largest degree irreducible divisor, which is a computation used also in the decryption, as we already saw. We showed how to obtain candidate values for β starting with any $\gamma \in S \setminus R$. A simple choice for γ is to take the powers of α and in our experiments we never needed to try more than 10 different values for γ.

Now when f is known to the adversary, he can easily obtain m in the following way. Apply Tr operator to equation (3.1) to get

(3.2) $$Tr(F(\alpha, 0)) = Tr(m + fs(\alpha, 0)) = nm + nfTr(s(\alpha, 0)).$$

Now nm is the remainder of $F(\alpha, 0)$ when divided by f, because of the conditions $\deg m < l < \deg f$.

The computational steps required for the attack are similar to those required for the decryption, except for the computation of traces. The traces of $\alpha^i, i = 1, \ldots, n - 1$ are obtained from the coefficients of $X(x, 0)$ by Newton's identities and, from those values, the trace of any element of S can be easily obtained by linearity. We implemented the encryption, decryption and attack steps in Pari/GP (code available at http://www.ma.utexas.edu/users/voloch/GP/asc.gp) In the key generation part of the testing script we use the following choice of parameters (which are in the range suggested by [1]): $w = \deg X$ is a random number between 5 and 8, d is chosen to be 50, the coefficients of $X(x, y) \in R[x, y]$, which are polynomials of t are chosen randomly in such a way that their degree is less than or equal to dw. Finally l is chosen to be a random number between $(2w + 4)d$ and $(2w + 4)d + 100$. Finally we took for p all primes up to 31 and also $103, 503, 997$ and 7919 to see how it scaled with p. We performed ten trials for each prime. Our experiment suggests that the time it takes to break the system using our attack takes no longer than six times it takes to decrypt the corresponding message using the secret key.

4. Variants and other attacks

In addition to the attack described above, the Akiyama-Goto cryptosystem is subject to a different attack due to Uchiyama and Tokunaga, [6], which is not as efficient as ours. Both this attack and ours are discussed in [2], where the authors also discuss a new variant of their cryptosystem immune to those attacks. Briefly, this new variant uses the same shape of encryption $F = m + fs + Xr$, except that now $m, f \in \mathbb{F}_p[t, x, y]$. To enable decryption, they need to send two encryptions $F_i = m + fs_i + Xr_i, i = 1, 2$. This new cryptosystem is subject to the following attack:

Let $g = fs_2 - fs_1$. We use a substitution attack as in 6.1.1 of [2]. We have $F_2 - F_1 = g + X(r_2 - r_1)$ We begin by substituting points with coordinates in a finite field satisfying the equation $X = 0$. These points can be easily found by arbitrarily choosing two of the coordinates and solving for the third. We need as many points as there are coefficients in g. As we know $F_2 - F_1$, we can compute its value at these points and therefore we get a set of linear equations for the coefficients of g. So we first find g by solving this system and then we use a multivariate polynomial factoring algorithm (as in, for example, [5]) to find f from g. Once f is found, then we can find m by plugging points satisfying $X = 0$ on $F_i = m + fs_i + Xr_i$ which now are linear in m and s_i.

Another idea of how make the system secure against the attack described in section 3 is the following: Go back to the original system and make the cyphertext be of the form $m + fs + Xr + (x^{p^n} - x)b + (y^{p^n} - y)c$ where b and c are random polynomials in t, x, y. If n is sufficiently large, when we plug in the points U, V we can recover the value of $m + fs$ as the remainder of division by $t^{p^n} - t$, then proceed as before. This is secure against the attack described in section 3 since the value of $x^{p^n} - x$ for $x = \alpha$ as in section 3 will not be divisible by $t^{p^n} - t$, ensuring that only rational points can be used for the decryption. On the other hand, n cannot be made too large or the system might be subject to a substitution attack. It is not clear whether this new system is efficient or secure and it merits further study.

References

1. Akiyama, K., Goto, A.: A Public-key Cryptosystem using Algebraic Surfaces (Extended Abstract), PQCrypto Workshop Record, 2006
2. Akiyama, K., Goto, A.: An improvement of the algebraic surface public-key cryptosystem, Proceedings of SCIS 2008.
3. Cantor, D., Zassenhaus, H.: A New Algorithm for Factoring Polynomials Over Finite Fields, Mathematics of Computation, 36:587-592, 1981.

4. Elkies, N.D.: Rational points near curves and small nonzero $|x^3 - y^2|$ via lattice reduction, Algorithmic number theory (Leiden, 2000), LNCS 1838 (2000), 33–63.
5. Gathen, J. v. z., Kaltofen E., Factorization of multivariate polynomials over finite fields, Math. Comp. 45 (1985), no. 171, 251–261.
6. Uchiyama, S., Tokunaga H., On the Security of the Algebraic Surface Public-key Cryptosystems (Japanese), 2C1-2, SCIS2007 (2007).

DEPT. OF MATHEMATICS, UNIV. OF TEXAS, AUSTIN, TX 78712-0257
E-mail address: pivanov@math.utexas.edu

DEPT. OF MATHEMATICS, UNIV. OF TEXAS, AUSTIN, TX 78712-0257
E-mail address: voloch@math.utexas.edu

Contemporary Mathematics
Volume **487**, 2009

Hyperelliptic Curves, L-polynomials, and Random Matrices

Kiran S. Kedlaya and Andrew V. Sutherland

ABSTRACT. We analyze the distribution of unitarized L-polynomials $\bar{L}_p(T)$ (as p varies) obtained from a hyperelliptic curve of genus $g \leq 3$ defined over $\mathbb{Q}$. In the generic case, we find experimental agreement with a predicted correspondence (based on the Katz-Sarnak random matrix model) between the distributions of $\bar{L}_p(T)$ and of characteristic polynomials of random matrices in the compact Lie group $USp(2g)$. We then formulate an analogue of the Sato-Tate conjecture for curves of genus 2, in which the generic distribution is augmented by 22 exceptional distributions, each corresponding to a compact subgroup of $USp(4)$. In every case, we exhibit a curve closely matching the proposed distribution, and can find no curves unaccounted for by our classification.

1. Introduction

For C a smooth projective curve of genus g defined over $\mathbb{Q}$ and each prime p where C has good reduction, we consider the polynomial $L_p(T)$, the numerator of the zeta function $Z(C/\mathbb{F}_p; T)$. This polynomial is intimately related to many arithmetic properties of the curve, appearing in the Euler product of the L-series

$$L(C, s) = \prod_p L_p(p^{-s})^{-1},$$

the characteristic polynomial of the Frobenius endomorphism,

$$\chi_p(T) = T^{2g} L_p(T^{-1}),$$

and the order of the group of $\mathbb{F}_p$-rational points on the Jacobian of C,

$$\#J(C/\mathbb{F}_p) = L_p(1).$$

In genus 1, C is an elliptic curve, and $L_p(T) = pT^2 - a_pT + 1$ is determined by the trace of Frobenius, a_p. The distribution of a_p as p varies has been and remains a subject of considerable interest, forming the basis of several well known conjectures, including those of Lang-Trotter [**36**] and Sato-Tate [**52**]. Considerable progress has been made on these questions, particularly the latter, much of it quite recently [**4, 5, 6, 21, 38**].

2000 *Mathematics Subject Classification.* Primary 11G40; Secondary 11G30, 05E15.
Key words and phrases. Sato-Tate conjecture, trace of Frobenius, zeta function, Haar measure, moment sequence.
Kedlaya was supported by NSF CAREER grant DMS-0545904 and a Sloan Research Fellowship.

In genus 2, C is a hyperelliptic curve, and the study of $L_p(T)$ may be viewed as a natural generalization of these questions (we also consider hyperelliptic curves of genus 3). Our first objective is to understand the shape of the distribution of $L_p(T)$, which leads us to focus primarily on Sato-Tate type questions.

The random matrix model developed by Katz and Sarnak provides a ready generalization of the Sato-Tate conjecture to higher genera. They show (see Theorems 10.1.18.3 and 10.8.2 in [27]) that over a universal family of hyperelliptic curves of genus g, the distribution of *unitarized* L-polynomials, $\bar{L}_p(T) = L_p(p^{-1/2}T)$, corresponds to the distribution of characteristic polynomials $\chi(T)$ in the compact Lie group $USp(2g)$ (the group of $2g \times 2g$ complex matrices that are both unitary and symplectic). By also considering infinite compact subgroups of $USp(2g)$, we are able to frame a generalization of the Sato-Tate conjecture applicable to any smooth curve defined over $\mathbb{Q}$.[1]

To test this conjecture, we rely on a collection of highly efficient algorithms to compute $L_p(T)$, described by the authors in [28]. The performance of these algorithms has improved dramatically in recent years (due largely to an interest in cryptographic applications [13]). For a hyperelliptic curve, it is now entirely practical to compute $L_p(T)$ for all $p \leq N$ (where the curve has good reduction), with N on the order of 10^8 in genus 2 and more than 10^7 in genus 3. Alternatively, for much smaller N (less than 10^4) one can perform similar computations with 10^{10} curves or more.

We characterize the otherwise overwhelming abundance of data with *moment statistics*. If $\{x_p\}$ is a set of unitarized values derived from $\bar{L}_p(T)$, say $x_p = a_p/\sqrt{p}$, we compute the first several terms of the sequence

$$\text{(1)} \qquad\qquad \text{m}(x_p), \ \text{m}(x_p^2), \ \text{m}(x_p^3), \ \ldots,$$

where $\text{m}(x_p^k)$ denotes the mean of x_p^k over p. Under the conjecture, the moment statistics converge, term by term, to the *moment sequence*

$$\text{(2)} \qquad\qquad \mathbf{E}[X], \ \mathbf{E}[X^2], \ \mathbf{E}[X^3], \ \ldots,$$

where the corresponding random variable X is derived from the characteristic polynomial $\chi(T)$ of a random matrix A, say $X = \text{tr}(A)$. In all the cases of interest to us, the moments $\mathbf{E}[X^n]$ exist and determine the distribution of X. Furthermore, they are integers. The distributions we encounter can typically be distinguished by the first eight terms of (2). It will be convenient to begin our sequences with $\mathbf{E}[X^0] = 1$.

To apply this approach we must determine these moment sequences explicitly. This is an interesting problem in its own right, with applications to representation theory and combinatorics. The particular cases we consider include the elementary and Newton symmetric functions (power sums) of the eigenvalues of a random matrix in $USp(2g)$. We derive explicit formulae for the corresponding moment generating functions. Our results intersect other work in this area, most notably that of Grabiner and Magyar [18], and also Eric Rains [42]. We take a somewhat different approach, relying on the Haar measure to encode the combinatorial structure of the group.

With moment sequences for $USp(2g)$ in hand, we then survey the distributions of $\bar{L}_p(T)$. The case $g = 1$ is easily described, and we do so here. The polynomial

---

[1]With the generous assistance of Nicholas Katz.

$\bar{L}_p(T) = p + a_1 T + 1$ is determined by the coefficient $a_1 = -a_p/\sqrt{p}$, and there are two distributions of a_1 that arise.

For elliptic curves without complex multiplication, the moment statistics of a_1 converge to the corresponding moment sequence in $USp(2)$:

$$1, \ 0, \ 1, \ 0, \ 2, \ 0, \ 5, \ 0, \ 14, \ 0, \ 42, \ \ldots,$$

whose $(2n)$th term is the nth Catalan number. Convergence follows from the Sato-Tate conjecture, which for curves with multiplicative reduction at some prime (almost all curves) is now proven, thanks to the work of Clozel, Harris, Shepherd-Barron, and Taylor [9, 21, 53] (see Mazur [38] for an overview). Testing at least one example of every curve with conductor less than 10^7 (including all the curves in Cremona's tables [11, 50]) revealed no apparent exceptions among curves with purely additive reduction.

For elliptic curves with complex multiplication, the moment statistics of a_1 converge instead to the sequence:

$$1, \ 0, \ 1, \ 0, \ 3, \ 0, \ 10, \ 0, \ 35, \ 0, \ 126, \ \ldots,$$

whose $(2n)$th term is $\binom{2n}{n}/2$ for $n > 0$. This is the moment sequence of a compact subgroup of $USp(2)$, specifically, the normalizer of $SO(2)$ in $SU(2) = USp(2)$. The elements with nonzero traces have uniformly distributed eigenvalue angles, and for elliptic curves with complex multiplication, convergence follows from a theorem of Deuring [12] and known equidistribution results for Hecke characters [35, Ch. XV]. The only other infinite compact subgroup of $USp(2)$ (up to conjugacy) is $SO(2)$, and its moment sequence does not appear to correspond to the moment statistics of any elliptic curve (such a curve would necessarily contradict the Sato-Tate conjecture).

The main purpose of the present work is to undertake a similar study in genus 2. We also lay some groundwork for genus 3, but consider only the case of a typical hyperelliptic curve. Already in genus 2 we find a much richer set of possible distributions.

For the typical hyperelliptic curve in genus 2 (resp. 3), when computed for suitably large N, the moment statistics closely match the corresponding moment sequences in $USp(4)$ (resp. $USp(6)$), as predicted. In genus 2 we can make a stronger statement. In a family of one million curves with randomly chosen coefficients, every single one appeared to have the $\bar{L}_p(T)$ distribution of characteristic polynomials in $USp(4)$. Under reasonable assumptions, we can reject alternative distributions with a high level of statistical confidence.

To find exceptional distributions in genus 2 we must cast our net wider, searching very large families of curves with constrained coefficient values, as well as examples taken from the literature. Such a search yielded over 30,000 nonisomorphic exceptional curves, but among these we find only 22 clearly distinct distributions, each with integer moments (Table 11, p. 29). For each of these distributions we are able to identify a specific compact subgroup H of $USp(4)$ with a matching distribution of characteristic polynomials (Table 13, p. 34). The method we use to construct these subgroups is quite explicit, and a converse statement is very nearly true. Of the subgroups we can construct, only two do not correspond to a distribution we have found; we can rule out one of these and suspect the other can be ruled out also (§ 7.2). We believe we have accounted for all possible L-polynomial distributions of a genus 2 curve.

2. Some Motivating Examples

We work throughout with smooth, projective, geometrically irreducible algebraic curves defined over $\mathbb{Q}$. Recall that for a curve C with good reduction at p, the zeta function $Z(C/\mathbb{F}_p; T)$ is defined by the formal power series

$$Z(C/\mathbb{F}_p; T) = \exp\left(\sum_{k=1}^{\infty} N_k T^k / k\right),$$

where N_k counts the (projective) points on C over $\mathbb{F}_{p^k}$. From the seminal work of Emil Artin [3], we know that $Z(C/\mathbb{F}_p; T)$ is a rational function of the form

$$Z(C/\mathbb{F}_p; T) = \frac{L_p(T)}{(1-T)(1-pT)},$$

where the monic polynomial $L_p(T) \in \mathbb{Z}[T]$ has degree $2g$ (g is the genus of C) and constant coefficient 1.

By the Riemann hypothesis for curves (proven by Weil [56]), the roots of $L_p(T)$ lie on a circle of radius $p^{-1/2}$ about the origin of the complex plane. To study the distribution of $L_p(T)$ as p varies, we use the unitarized polynomial

$$\bar{L}_p(T) = L_p(p^{-1/2}T),$$

which has roots on the unit circle. As $\bar{L}_p(T)$ is a real polynomial of even degree with $\bar{L}_p(0) = 1$, these roots occur in conjugate pairs. We may write

$$\bar{L}_p(T) = T^{2g} + a_1 T^{2g-1} + a_2 T^{2g-2} + \cdots + a_2 T^2 + a_1 T + 1.$$

Since $\bar{L}_p(T)$ has unitary roots, we know that

$$(3) \qquad\qquad\qquad |a_i| \leq \binom{2g}{i},$$

and ask how a_i is distributed within this interval as p varies.

The next three pages show the distribution of a_1 for arbitrarily chosen curves of genus 1, 2, and 3 and various values of N. The coefficient a_1 is the negative sum of the roots of $\bar{L}_p(T)$, and may be written as $a_1 = -a_p/\sqrt{p}$, where a_p is the trace of Frobenius.

Each graph represents a histogram of nearly $\pi(N)$ samples (one for each prime where C has good reduction) placed into approximately $\sqrt{\pi(N)}$ buckets which partition the interval $[-2g, 2g]$ determined by (3). The horizontal axis spans this interval, and the vertical axis has been suitably scaled, with the height of the uniform distribution, $1/(4g)$, indicated by a dotted line.

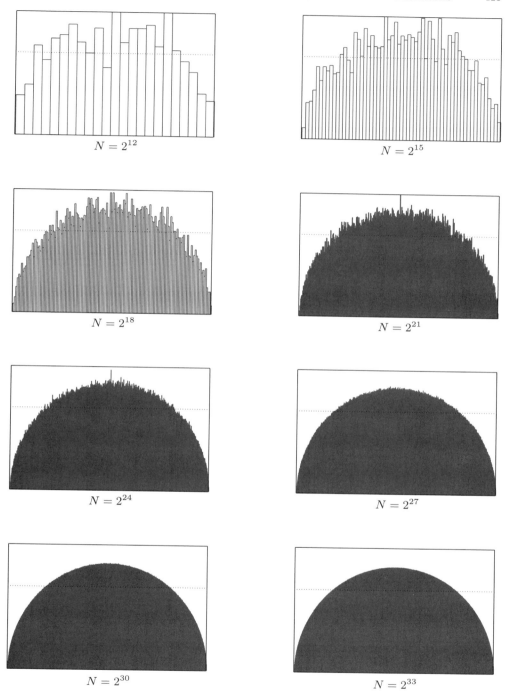

Distribution of $a_1 = -a_p/\sqrt{p}$ for $p \leq N$ with good reduction.

$$y^2 = x^3 + 314159x + 271828.$$

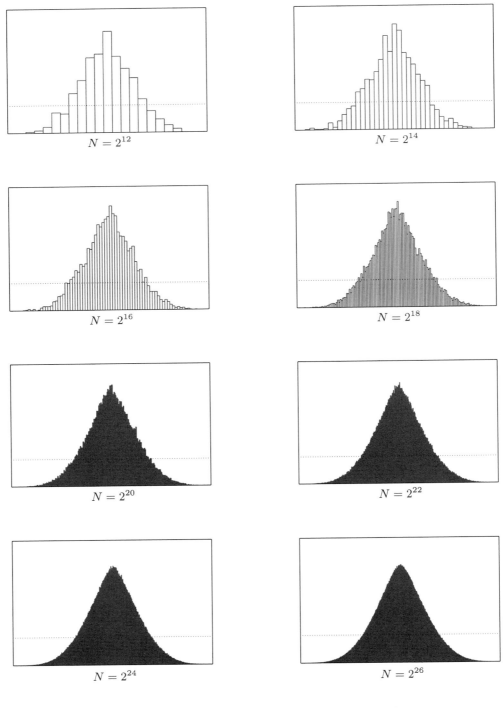

Distribution of $a_1 = -a_p/\sqrt{p}$ for $p \leq N$ with good reduction.

$$y^2 = x^5 + 314159x^3 + 271828x^2 + 1644934x + 57721566.$$

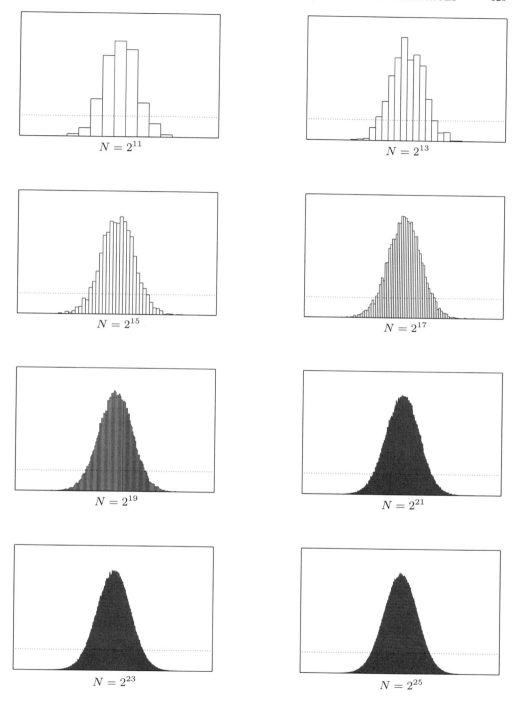

Distribution of $a_1 = -a_p/\sqrt{p}$ for $p \leq N$ with good reduction.
$y^2 = x^7 + 314159x^5 + 271828x^4 + 1644934x^3 + 57721566x^2 + 1618034x + 141021.$

The familiar semicircular shape in genus 1 is the Sato-Tate distribution. The examples in higher genera also appear to converge to distinct distributions. Provided the curve is "typical" (a notion we will define momentarily) this distribution is the same for every curve of a given genus. Even in atypical cases, there is (empirically) a small distinct set of distributions that arise for a given genus. In genus 1, only one exceptional distribution for a_1 is known, exhibited by all curves with complex multiplication:

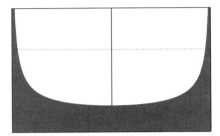

Fig. 1: Distribution of a_1 for $y^2 = x^3 - 15x + 22$.

The central spike has area $1/2$ (asymptotically), arising from the fact that a curve with CM-field $\mathbb{Q}[\sqrt{D}]$ has $a_p = 0$ precisely when D is a not a quadratic residue in $\mathbb{F}_p$ [**12**].

In higher genera, a richer set of exceptional distributions arises. Below is an example for a genus 2 curve whose Jacobian splits as the product of two elliptic curves (one with complex multiplication). Histograms for several other exceptional genus 2 distributions are provided in Appendix II.

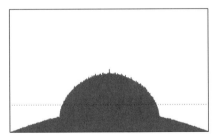

Fig. 2: Distribution of a_1 for $y^2 = x^5 + 20x^4 - 26x^3 + 20x^2 + x$.

3. A Generalized Sato-Tate Conjecture

We wish to give a conjectural basis for these distributions, both in the typical and atypical cases. The formulation presented here follows the model developed by Katz and Sarnak [**27**] and relies heavily on additional detail provided by Nicholas Katz [**26**], whom we gratefully acknowledge. Most of the statements below readily generalize to abelian varieties, but we restrict ourselves to curves defined over $\mathbb{Q}$.

We begin with the Sato-Tate conjecture, which may be stated as follows:

CONJECTURE 1 (Sato-Tate). *For an elliptic curve without complex multiplication, the distribution of the roots $e^{i\theta}$ and $e^{-i\theta}$ of $\bar{L}_p(T)$ for $p \leq N$ converges (as $N \to \infty$) to the distribution given by the measure $\mu = \frac{2}{\pi} \sin^2 \theta d\theta$ over $\theta \in [0, \pi]$.*

As noted earlier, this has been proven for elliptic curves with multiplicative reduction at some prime [21]. Using $a_1 = -2\cos\theta$, one finds that

$$\Pr[a_1 \leq x] = \frac{1}{2\pi} \int_{-2}^{x} \sqrt{4 - t^2} dt,$$

giving the familiar semicircular distribution.

An equivalent formulation of Conjecture 1 is that the distribution of $\bar{L}_p(T)$ corresponds to the distribution of the characteristic polynomial of a random matrix in $USp(2)$. More generally, the Haar measure on $USp(2g)$ provides a natural distribution of unitary symplectic polynomials of degree $2g$: the eigenvalues of a unitary matrix lie on the unit circle and the symplectic condition ensures that they occur in conjugate pairs (giving a polynomial with real coefficients). Conversely, each unitary symplectic polynomial corresponds to a conjugacy class of matrices in $USp(2g)$.

Let the eigenvalues of a random matrix in $USp(2g)$ be $e^{\pm i\theta_1}, \ldots, e^{\pm i\theta_g}$, with $\theta_j \in [0, \pi]$. The joint probability density function on $(\theta_1, \ldots, \theta_g)$ given by the Haar measure on $USp(2g)$ is

$$(4) \qquad \mu(USp(2g)) = \frac{1}{g!} \left(\prod_{j<k} (2\cos\theta_j - 2\cos\theta_k) \right)^2 \prod_j \left(\frac{2}{\pi} \sin^2 \theta_j d\theta_j \right),$$

as shown by Weyl [57, Thm. 7.8B, p. 218] (also see [27, 5.0.4, p. 107]). For $g = 1$, we obtain the Sato-Tate distribution above.

In view of the atypical examples in the previous section, we cannot expect every curve to achieve the distribution given by $\mu(USp(2g))$. One might impose a restriction comparable to that of Sato-Tate by requiring that $\mathrm{End}_{\mathbb{C}}(J(C)) = \mathbb{Z}$, i.e. that the Jacobian have minimal endomorphism ring. While necessary, it is not clear that this restriction is sufficient in general. A stronger condition uses the ℓ-adic representation of $\mathrm{Gal}(\overline{\mathbb{Q}}/\mathbb{Q})$ induced by the Galois action on the Tate module $T_\ell(C)$ (the inverse limit of the ℓ^n-torsion subgroups of $J(C)$). Specifically, we require that the image of the representation

$$\rho_\ell : \mathrm{Gal}(\overline{\mathbb{Q}}/\mathbb{Q}) \to \mathrm{Aut}(T_\ell(C)) \cong GL(2g, \mathbb{Z}_\ell)$$

be Zariski dense in $GSp(2g, \mathbb{Z}_\ell) \subset GL(2g, \mathbb{Z}_\ell)$. We know from results of Serre ([46, Sec. 7, Thm 3] and [44, p. 104]) that if this is achieved for any ℓ, then it holds for all ℓ, and we say such a curve has *large Galois image*.

Each of the conditions below suffices for C to have large Galois image.

(1) C is a genus g curve with g odd, 2, or 6, and $\mathrm{End}_{\mathbb{C}}(J(C)) = \mathbb{Z}$ (Serre [46, 45]). This does not hold in genus 4 (Mumford [40]).
(2) C is a hyperelliptic curve $y^2 = f(x)$ with $f(x)$ of degree $n \geq 5$ and the Galois group of $f(x)$ is isomorphic to S_n or A_n (Zarhin [58]).
(3) C is a genus g curve with good reduction outside of a set of primes S and the mod ℓ reduction of the image of ρ_ℓ is equal to $GSp(2g, \mathbb{Z}/\ell\mathbb{Z})$ for some $\ell \geq c(g, S)$ (Faltings [14], Serre [46, 44]).

Condition 3 was suggested to us by Katz. The constant $c(g, S)$ depends only on g and S. From Faltings [14, Thm. 5], we know that for a given g and S, there are only finitely many nonisomorphic Jacobians of genus g curves with good reduction outside of S. Each such curve has large Galois image if and only if for all $\ell > c$ the mod ℓ image of ρ_ℓ is $GSp(2g, \mathbb{Z}/\ell\mathbb{Z})$, for some constant c. By the results of

Serre cited above, it suffices to find one such ℓ, and taking the maximum of c over the finite set of Jacobians gives the desired $c(g, S)$. At present, effective bounds on $c(g, S)$ are known only in genus 1 [**10, 32**]. Even without effective bounds, Condition 3 gives an easily computable heuristic that is useful in practice.[2]

We now consider the situation for a curve which does not have large Galois image. The simplest case is an elliptic curve with complex multiplication. For such a curve the distribution of a_1 clearly does not match the distribution of traces in $USp(2)$ (see Fig. 1 above). In particular, the density of primes for which $a_1 = -a_p/\sqrt{p} = 0$ is one half. There is, however, a compact subgroup of $USp(2)$ which gives the correct distribution. Consider the subgroup

$$H = \left\{ \begin{pmatrix} \cos\theta & \sin\theta \\ -\sin\theta & \cos\theta \end{pmatrix}, \begin{pmatrix} i\cos\theta & i\sin\theta \\ i\sin\theta & -i\cos\theta \end{pmatrix} : \theta \in [0, 2\pi] \right\}.$$

H contains $SO(2)$ as a subgroup of index 2 (it is in fact the normalizer of $SO(2)$ in $USp(2)$). The elements of H not in $SO(2)$ all have zero traces, giving the desired density of $1/2$. The Haar measure on $SO(2)$ gives uniformly distributed eigenvalue angles, matching the distribution of nonzero traces for elliptic curves with complex multiplication. Note that H is disconnected. This is a common (but not universal) feature among the subgroups we wish to consider.

For an elliptic curve with complex multiplication, the mod ℓ Galois image lies in the normalizer of a Cartan subgroup (see Lang [**34**, Thm 3.2]). For sufficiently large ℓ, one finds that is in fact equal to the normalizer of a Cartan subgroup. There is then a correspondence with the subgroup H above, which is the normalizer of the Cartan subgroup $SO(2)$ in $SU(2) = USp(2)$.

In general, we anticipate a relationship between the ℓ-adic Galois image of a curve C, call it G_ℓ (a subgroup of $GSp(2g, \mathbb{Z}_\ell)$), and a compact subgroup H of $USp(2g)$ whose distribution of characteristic polynomials matches the distribution of unitarized L-polynomials $\bar{L}_p(T)$ of C. One can (conjecturally) describe this relationship, albeit in a nonexplicit manner. Briefly, one takes the Zariski closure of G_ℓ through a series of embeddings:

$$GSp(2g, \mathbb{Z}_\ell) \to GSp(2g, \mathbb{Q}_\ell) \to GSp(2g, \overline{\mathbb{Q}}_\ell) \to GSp(2g, \mathbb{C}).$$

The last step is justified by the fact that $\overline{\mathbb{Q}}_\ell$ and $\mathbb{C}$ are algebraically closed fields containing $\mathbb{Q}$ of equal cardinality, hence isomorphic, and we choose a particular embedding of $\overline{\mathbb{Q}}_\ell$ in $\mathbb{C}$. One then takes the intersection with $Sp(2g)$, obtaining a reductive group over $\mathbb{C}$. After dividing by $\sqrt{p}$, the image of each Frobenius element lies in this intersection, as a diagonalizable matrix with unitary eigenvalues. We now consider a maximal compact subgroup of the intersection, H, lying in $USp(2g)$. Each unitarized Frobenius element is conjugate to some element of H (unique up to conjugacy in H) whose characteristic polynomial is $\bar{L}_p(T)$. When G_ℓ is Zariski dense in $GSp(2g, \mathbb{Z}_\ell)$, we obtain $H = USp(2g)$, but in general, H is some compact subgroup of $USp(2g)$.

There is an analogous construction involving the Mumford-Tate group $\mathrm{MT}(A)$ of an abelian variety A, which contains G_ℓ. The result is the Hodge group $\mathrm{Hg}(A)$, corresponding to our subgroup H above. For simple abelian varieties of low dimension (up to genus 5) the possibilities for $\mathrm{Hg}(A)$ have been classified by Moonen and

---

[2]A fourth condition has recently been proven by Hall [**20**]. In the same paper, Kowalski proves that almost all hyperelliptic curves have large Galois image.

Zarhin [39].[3] For genus 2 curves we consider the classification of H in Section 7, and give what we believe to be a complete list of the possibilities (most of these do not correspond to simple Jacobians). We note here that curves with nonisomorphic Hodge groups may have identical L-polynomial distributions, so the notions are not equivalent.

We can now state the conjecture.

CONJECTURE 2 (Generalized Sato-Tate). *For a curve C of genus g, the distribution of $\bar{L}_p(T)$ converges to the distribution of characteristic polynomials $\chi(T)$ in an infinite compact subgroup $H \subseteq USp(2g)$. Equality holds if and only if C has large Galois image.*

We say that the subgroup H *represents* the L-polynomial distribution of C.

4. Moment Sequences Attached to L-polynomials

Let $P_C(N)$ denote the set of primes $p \leq N$ for which the curve C has good reduction. We may compute $\bar{L}_p(T)$ for all $p \in P_C(N)$, and if x is a quantity derived from $\bar{L}_p(T)$ (e.g., the coefficients a_k or some function of the a_k), we consider the mean value $m(x^n)$ over $p \in P_C(N)$ as an approximation of the nth moment $E[X^n]$ of a corresponding random variable X. Under Conjecture 2, we assume there is a compact subgroup $H \subseteq USp(2g)$ which represents the L-polynomial distribution of C. If X is a real random variable defined as a polynomial of the eigenvalues of an element of H, it clearly has bounded support under the Haar measure on H (the eigenvalues lie on the unit circle). Therefore its moments all exist. Further, one may determine absolute bounds on X depending only on g, which we regard as fixed. Carleman's condition [31, p. 126] then implies that the moment sequence for X uniquely determines its distribution. We summarize this argument with the following proposition.

PROPOSITION 1. *Under Conjecture 2, let H be a compact subgroup of $USp(2g)$ which represents the L-polynomial distribution of a curve C. Let x_p be a real-valued polynomial of the roots of $\bar{L}_p(T)$, and let the random variable X be the corresponding polynomial of the eigenvalues of a random matrix in H. Then the moments of X exist and determine the distribution of X. For all nonnegative integers n, the mean value of x_p^n over $p \in P_C(N)$ converges to $E[X^n]$ as $N \to \infty$.*

We now consider random variables X that are symmetric functions (polynomials) of the eigenvalues, with integer coefficients. In this case, it is easy to see that the moments of X must be integers.

PROPOSITION 2. *Let V be a vector space of finite dimension over $\mathbb{C}$. Let the random variable X be a symmetric polynomial with integer coefficients over the eigenvalues of a random matrix in a compact group $G \subseteq GL(V)$, distributed with Haar measure. Then $E[X^n] \in \mathbb{Z}$ for all nonnegative integers n.*

PROOF. The sum of the eigenvalues, e_1, is the trace of the standard representation V of G, and $E[e_1] = \int_G \operatorname{tr}(A) dA$ counts the multiplicity of the trivial representation in V, an integer. Similarly, the kth symmetric function e_k is the trace of the kth exterior power $\Lambda^k V$, hence $E[e_k]$ must be an integer. The product of the traces of two representations is the trace of their tensor product, hence $E[e_k^n]$

---

[3]We thank David Zywina for bringing this to our attention.

is an integer, as are the moments of any product of elementary symmetric functions. Linearity of expectation implies that $\mathbb{Z}$-linear combinations of products of elementary symmetric functions also have integer moments, and every symmetric polynomial with integer coefficients may be expressed in this way.   □

Proposition 2 is quite useful in practice, particularly when H is unknown. We can "determine" $E[X^n]$ for small n by computing the mean of x_p^n up to a suitable bound N. The value obtained is purely heuristic of course, depending both on Conjecture 2 and an assumption about the rate of convergence. Nevertheless, the consistency of the results so obtained are quite compelling.

For most curves (those with large Galois image) we expect $H = USp(2g)$, and we may compute the distribution of X using the measure μ defined in (4).

4.1. Moment Generating Functions in $USp(2g)$.
Let $\chi(T) = \sum a_k T^k$ be the characteristic polynomial of a random matrix in $USp(2g)$ with eigenvalues $e^{\pm i\theta_1}, \ldots, e^{\pm i\theta_g}$. Up to a sign $(-1)^k$, the a_k are the elementary symmetric functions of the eigenvalues. We also define

$$s_k = \sum_j (e^{ki\theta_j} + e^{-ki\theta_j}) = \sum_j 2\cos k\theta_j,$$

the kth power sums of the eigenvalues (Newton symmetric functions).

We wish to compute the (integer) sequence $M[X] = (1, E[X], E[X^2], \ldots)$, where X is a_k or s_k. We first consider s_k (including $a_1 = -s_1$).[4] We have

$$(5) \qquad M[s_k](n) = E[s_k^n] = \int_V \left(\sum_j 2\cos k\theta_j\right)^n \mu,$$

where $V = [0, \pi]^g$ denotes the volume of integration. If we expand (5) using the formula for μ in (4), we need only consider univariate integrals of the form

$$(6) \qquad C_k^m(n) = \frac{1}{\pi}\int_0^\pi (2\cos k\theta)^n (2\cos\theta)^m (2\sin^2\theta)d\theta.$$

We regard C_k^m as a sequence indexed by $n \in \mathbb{Z}^+$ (in fact, an integer sequence), and define the exponential generating function (egf) of C_k^m by

$$\mathcal{C}_k^m(z) = \sum_{n=0}^\infty C_k^m(n)\frac{z^n}{n!}.$$

Similarly, $\mathcal{M}[s_k]$, the egf of $M[s_k]$, is the moment generating function of s_k. We will compute C_k^m in terms of sequences B_ν, defined by[5]

$$(7) \qquad B_\nu(n) = \binom{n}{\frac{n+\nu}{2}},$$

where the binomial coefficient is zero when $(n + \nu)/2$ is not an integer in the interval $[0, n]$. With this understanding, we allow ν to take arbitrary values, with B_ν identically zero for $\nu \notin \mathbb{Z}$. Letting $\mathcal{B}_\nu(z) = \sum_{n=0}^\infty B_\nu(n)\frac{z^n}{n!}$, we find that

$$(8) \qquad \mathcal{B}_\nu(z) = \mathcal{I}_\nu(2z) = \sum_{n=0}^\infty \frac{z^{2n+\nu}}{n!(n+\nu)!}, \qquad\qquad \text{for } \nu \in \mathbb{Z}.$$

---

[4]The other a_k are addressed in Section 4.4 (for $g \leq 3$).
[5]$B_\nu(n)$ counts paths of length n from 0 to ν on the real line using step set $\{\pm 1\}$.

The function $\mathcal{I}_\nu(z)$ is a hyperbolic Bessel function (of the first kind, of order ν) [**49**, Ch. 49-50]. Hyperbolic Bessel functions are defined for nonintegral values of ν (and are not identically zero), so the condition $\nu \in \mathbb{Z}$ in (8) should be duly noted.

We can now state a concise formula for $\mathcal{M}[s_k]$.

THEOREM 1. *Let s_k denote the kth power sum of the eigenvalues of a random element of $USp(2g)$. The moment generating function of s_k is*

$$(9) \qquad \mathcal{M}[s_k] = \det_{g \times g} \left(\mathcal{C}_k^{i+j-2} \right),$$

where $\mathcal{C}_k^m(z)$ is given by

$$(10) \qquad \mathcal{C}_k^m = \sum_j \binom{m}{j} \left(\mathcal{B}_{(2j-m)/k} - \mathcal{B}_{(2j-m+2)/k} \right),$$

with $\mathcal{B}_\nu$ defined as above.

We postpone the proof until Section 4.2.

The determinantal formula in Theorem 1 contains some redundancy (e.g., when $g = 3$ the term $C_k^1 C_k^2 C_k^3$ appears twice), but we find the simple form of Theorem 1 well suited to both hand and machine computation. For example, when $g = 2$ and $k = 1$ we have

$$\mathcal{M}[s_1] = \mathcal{C}_1^0 \mathcal{C}_1^2 - \mathcal{C}_1^1 \mathcal{C}_1^1 = (\mathcal{B}_0 - \mathcal{B}_2)(\mathcal{B}_0 - \mathcal{B}_4) - (\mathcal{B}_1 - \mathcal{B}_3)^2.$$

From (15) of Section 4.3, we obtain the identity[6]

$$(11) \qquad M[s_1](2n) = c(n)c(n+2) - c(n+1)^2,$$

where $c(n)$ is the nth Catalan number. The odd moments are zero and the even moments form sequence A005700 in the On-line Encyclopedia of Integer Sequences (OEIS) [**47**]. A complete list of the sequences $M[s_k]$ and $g \leq 3$ can be found in Section 4.3.

The sequence A005700 $= (1, 1, 3, 14, 84, 594, \dots)$ is well known. It counts lattice paths of length $2n$ in $\mathbb{Z}^2$ with step set $\{(\pm 1, 0), (0, \pm 1)\}$ that return to the origin and are constrained by $x_1 \geq x_2 \geq 0$. In general, the sequence $M[s_1]$ counts returning lattice paths in $\mathbb{Z}^g$ which remain in the region $x_1 \geq \dots \geq x_g \geq 0$. This follows from a general result of Grabiner and Magyar[7] which relates the decomposition of tensor powers of certain Lie group representations to lattice paths in a chamber of the associated Weyl group [**18**]: as in the proof of Proposition (2), interpret $E[s_1^n]$ as the multiplicity of the trivial representation in $V^{\otimes n}$, where V is the standard representation of $USp(2g)$, then apply Theorem 2 of [**18**].

For the group $USp(2g)$ and $k = 1$, our results intersect those of [**18**], where an equivalent determinantal formula is obtained by counting lattice paths in the Weyl chamber of the corresponding Lie algebra. By contrast, we compute $\mathcal{M}[s_k]$ directly from the measure $\mu(USp(2g))$, using only elementary methods. The Haar measure, via the Weyl integration formula, effectively encodes the relevant combinatorial content. Particularly when $k > 1$, this is simpler than a combinatorial or

---

[6]The similarity of $\mathcal{C}_1^0 \mathcal{C}_1^2 - \mathcal{C}_1^1 \mathcal{C}_1^1$ and $c(n)c(n+2) - c(n+1)^2$ is somewhat misleading, both expressions involve Catalan numbers, but the terms do not correspond.

[7]As explained to us by Arun Ram, this equality may be interpreted in terms of crystal bases, which leads directly to analogues for groups other than $USp(2g)$.

representation-theoretic approach. More generally, Haar measure, and moment sequences in particular, can provide convenient access to the combinatorial structure of compact groups.

Determinantal formulas of the type above arise in many combinatorial questions related to lattice paths and Young tableaux (see [**16**, **8**] for examples). One might ask what the moment sequences for s_k count when $k > 1$. For $g = 2$, some answers may be found in the OEIS (see links in Table 2).

Before proving Theorem 1, we note the following corollary.

COROLLARY 1. *For all $k > 2g$, the random variables s_k are identically distributed with moment generating function $\mathcal{M}[s_k] = (\mathcal{B}_0)^g$.*

PROOF. From (9) of Theorem 1, $\mathcal{M}[s_k]$ is a polynomial expression in $\mathcal{C}_k^m$ with $m \leq 2g - 2$. From (10), it follows that $\mathcal{C}_k^m = \binom{m}{m/2}\mathcal{B}_0 - \binom{m}{m/2-1}\mathcal{B}_0$ for all $k > m+2$, since $\mathcal{B}_\nu$ is zero for nonintegral ν. Thus $\mathcal{M}[s_k]$ is an integer multiple of $\mathcal{B}_0^g$, and $\mathcal{M}[s_k]$ must be equal to $\mathcal{B}_0^g$, since $M[s_k](0) = 1 = B_0(0)$.[8] $\square$

The distribution of s_k given by Corollary 1 corresponds to the trace of a random matrix under a uniform distribution of $\theta_1, \ldots, \theta_g$. This is a special case of a general phenomenon first noticed by Eric Rains, who has proven similar results for all the compact classical groups [**41**, **42**]. The sudden transition to a fixed distribution is quite startling when first encountered; one might naïvely expect the distribution of s_k to gradually converge as $k \to \infty$. There is, however, an elementary explanation (see the proof of Lemma 3 below).

4.2. Proof of Theorem 1. We rearrange the integral for $M[s_k](n) = E[s_k^n]$ to obtain a determinantal expression in C_k^m. Lemma 3 then evaluates $C_k^m(n)$.

PROOF OF THEOREM 1. Let $w_j = 2\cos k\theta_j$ for $1 \leq j \leq g$. Then $s_k = \sum w_j$ and the integral for $M[s_k](n)$ in (5) becomes

$$(12) \qquad M[s_k](n) = \int_V \left(\sum_j w_j\right)^n \mu = \sum_v n_v \int_V \prod_j w_j^{v_j} \mu$$

where $V = [0, \pi]^g$ is the volume of integration, μ is the Haar measure on $USp(2g)$ defined in (4), v ranges over vectors of g nonnegative integers, and $n_v = \binom{n}{v_1, \ldots, v_g}$. Now let $x_j = 2\cos\theta_j$ and $y_j = 2\sin^2\theta_j$ so that (4) becomes

$$\mu = \frac{1}{g!\pi^g}\prod_{i<j}(x_i - x_j)^2 \prod_j (y_j d\theta_j).$$

By Lemma 1, we may write this as

$$(13) \qquad \mu = \frac{1}{g!\pi^g}\sum_\sigma \det_{g \times g}\left(x_{\sigma(j)}^{i+j-2}\right)\prod_j (y_j d\theta_j),$$

where σ ranges over permutations of $\{1, \ldots, g\}$. From the definition of $C_k^m(n)$ in (6) we have $C_k^m(n) = \frac{1}{\pi}\int_0^\pi w_j^n x_j^m y_j d\theta_j$, for any j. Combining (12) and (13),

$$M[s_k](n) = \sum_v n_v \frac{1}{g!}\sum_\sigma \det_{g \times g}\left(C_k^{i+j-2}(v_j)\right) = \sum_v n_v \det_{g \times g}\left(C_k^{i+j-2}(v_j)\right).$$

---

[8]In fact, for $k > 2g$, factoring out $\mathcal{B}_0$ from $\mathcal{M}[s_k]$ leaves the Hankel determinant of the sequence $c(0)$, 0, $c(1)$, 0, $c(2)$, $\ldots$, which is 1 (see [**1**]).

In terms of egfs (by Lemma 2), we then have the desired expression

$$\mathcal{M}[s_k] = \det_{g \times g} \left(\mathcal{C}_k^{i+j-2} \right).$$

Applying Lemma 3 to evaluate the integral $C_k^m(n)$ completes the proof.   □

Lemmas (1) and (2) are elementary facts; lacking a ready reference, we provide short proofs.

LEMMA 1. *If $r_1, \ldots, r_g$ are indeterminates in a commutative ring, then*

$$\prod_{j<k}(r_j - r_k)^2 = \sum_\sigma \det_{g \times g} \left(r_{\sigma(j)}^{i+j-2} \right),$$

where σ ranges over permutations of $\{1, \ldots, g\}$.

PROOF. Recall that $\prod_{j<k}(r_j - r_k)$ is the Vandermonde determinant [**29**, p. 71]. Define $r^\alpha = \prod_j r_j^{\alpha(j)}$ for any function $\alpha : \{1, \ldots, g\} \to \mathbb{Z}^+$, and for a permutation σ let $r_\sigma = (r_{\sigma(1)}, \ldots, r_{\sigma(g)})$. Then $\prod_{j<k}(r_j - r_k)^2$ is given by

$$\left(\det \left(r_i^{j-1} \right) \right)^2 = \left(\sum_\pi \operatorname{sgn}(\pi) r^{\pi-1} \right)^2 = \sum_{\pi,\phi} \operatorname{sgn}(\pi)\operatorname{sgn}(\phi) r^{\pi+\phi-2}$$

$$= \sum_{\pi,\phi} \operatorname{sgn}(\pi\phi) r_\phi^{\pi(\phi^{-1})+\mathrm{id}-2} = \sum_{\pi,\phi} \operatorname{sgn}(\pi\phi^{-1}) r_\phi^{\pi(\phi^{-1})+\mathrm{id}-2}$$

$$= \sum_\sigma \sum_\phi \operatorname{sgn}(\phi) r_\sigma^{\phi+\mathrm{id}-2} = \sum_\sigma \det \left(r_{\sigma(j)}^{i+j-2} \right),$$

and the lemma is proven.   □

LEMMA 2. *For $1 \le i, j \le m$, let $\mathcal{A}_{i,j} \in \mathbb{C}[[z]]$. Then*

$$[n] \det_{m \times m} \left(\mathcal{A}_{i,j} \right) = \sum_{n_1,\ldots,n_m} \binom{n}{n_1, \ldots, n_m} \det_{m \times m} \left([n_j]\mathcal{A}_{i,j} \right),$$

where $[n]\mathcal{F}$ denotes the coefficient of $z^n/n!$ in $\mathcal{F} \in \mathbb{C}[[z]]$.

PROOF. For any $\mathcal{F}_1, \ldots, \mathcal{F}_m \in \mathbb{C}[[z]]$ we have

$$[n] \prod_j \mathcal{F}_j = \sum_{n_1,\ldots,n_m} \binom{n}{n_1, \ldots, n_m} \prod_j [n_j]\mathcal{F}_j.$$

It follows that

$$[n] \det_{m \times m} \left(\mathcal{A}_{i,j} \right) = \sum_\sigma \operatorname{sgn}(\sigma) \sum_i [n] \prod_j \mathcal{A}_{\sigma(i),j}$$

$$= \sum_\sigma \operatorname{sgn}(\sigma) \sum_i \sum_{n_1,\ldots,n_m} \binom{n}{n_1, \ldots, n_m} \prod_j [n_j]\mathcal{A}_{\sigma(i),j}$$

$$= \sum_{n_1,\ldots,n_m} \binom{n}{n_1, \ldots, n_m} \det_{m \times m} \left([n_j]\mathcal{A}_{i,j} \right),$$

as desired.   □

Lemma 3 computes the integral C_k^m.

LEMMA 3. *Define* $C_k^m(n) = \frac{1}{\pi}\int_0^\pi (2\cos k\theta)^n (2\cos\theta)^m (2\sin^2\theta)d\theta$, *for positive integers* k *and nonnegative integers* m *and* n, *and let* $B_\nu(n) = \binom{n}{\frac{n+\nu}{2}}$. *Then*

$$C_k^m(n) = \sum_j \binom{m}{j}\left(B_{(2j-m)/k}(n) - B_{(2j-m+2)/k}(n)\right)$$

for all nonnegative integers n.

PROOF. We may write $C_k^m(n)$ as

$$\frac{1}{\pi}\int_0^\pi \left(e^{ik\theta} + e^{-ik\theta}\right)^n \left(e^{i\theta} + e^{-i\theta}\right)^m \left(1 - \left(e^{2i\theta} + e^{-2i\theta}\right)/2\right)d\theta.$$

In terms of $\delta(t) = \frac{1}{\pi}\int_0^\pi e^{it\theta}d\theta$, one finds that $C_k^m(n)$ is equal to

$$\sum_r\sum_j \binom{n}{r}\binom{m}{j}\delta(k(2r-n) + 2j - m)$$

$$-\frac{1}{2}\sum_r\sum_j \binom{n}{r}\binom{m}{j}\delta(k(2n-r) + 2j - m + 2)$$

$$-\frac{1}{2}\sum_r\sum_j \binom{n}{r}\binom{m}{j}\delta(k(2n-r) + 2j - m - 2).$$

As $C_k^m(n)$ is a real number, we need only consider the real parts of the sums above. For real t, the real part of $\delta(t)$ is nonzero only when $t = 0$, in which case it is 1. Hence, in Iverson notation, $\Re(\delta(t)) = [t = 0]$ holds for all $t \in \mathbb{R}$.[9]

The real parts of the second two sums are equal, since we may replace j with $m - j$ and r with $n - r$ in the last sum and then apply $[t = 0] = [-t = 0]$. Interchanging the order of summation we obtain

$$\sum_j\sum_r \binom{m}{j}\binom{n}{r}\left([k(2r-n) + 2j - m = 0] - [k(2n-r) + 2j - m + 2 = 0]\right).$$

We now note that

$$\sum_r \binom{n}{r}[2r - n = \nu] = \binom{n}{\frac{n+\nu}{2}} = B_\nu(n),$$

for all nonnegative integers n and arbitrary ν. Thus we have

$$C_k^m(n) = \sum_j \binom{m}{j}\left(B_{(m-2j)/k}(n) - B_{(m-2j-2)/k}(n)\right).$$

Applying the identity $B_\nu = B_{-\nu}$ completes the proof.   □

4.3. Explicit Computation of $M[s_k]$ in $USp(2g)$, for $g \leq 3$. For $g \leq 3$, Theorem 1 gives:

  $g{=}1$: $M[s_k] = C_k^0$;
  $g{=}2$: $M[s_k] = C_k^0 C_k^2 - C_k^1 C_k^1$;
  $g{=}3$: $M[s_k] = C_k^0 C_k^2 C_k^4 + 2C_k^1 C_k^2 C_k^3 - C_k^0 C_k^3 C_k^3 - C_k^1 C_k^1 C_k^4 - C_k^2 C_k^2 C_k^2$.

We will compute these moment sequences explicitly.

---

[9]The function $[P]$ is 1 when the boolean predicate P is true and 0 otherwise, see [**30**].

$k\backslash m$	0	1	2	3	4	5	6
1	$\mathcal{C}$	$\mathbf{D}\mathcal{C}$	$\mathbf{D}^2\mathcal{C}$	$\mathbf{D}^3\mathcal{C}$	$\mathbf{D}^4\mathcal{C}$	$\mathbf{D}^5\mathcal{C}$	$\mathbf{D}^6\mathcal{C}$
2	$\mathcal{A}$	0	$\mathcal{C}$	0	$(\mathbf{D}+2)\mathcal{C}$	0	$(\mathbf{D}+2)^2\mathcal{C}$
3	$\mathcal{B}$	$-\mathcal{D}$	$\mathcal{B}$	$-\mathcal{D}$	$\mathcal{B}+\mathcal{C}$	$-\mathcal{D}$	$\mathcal{B}+4\mathcal{C}$
4	$\mathcal{B}$	0	$\mathcal{A}$	0	$2\mathcal{A}$	0	$4\mathcal{A}+\mathcal{C}$
5	$\mathcal{B}$	0	$\mathcal{B}$	$-\mathcal{D}$	$2\mathcal{B}$	$-3\mathcal{D}$	$5\mathcal{B}$
6	$\mathcal{B}$	0	$\mathcal{B}$	0	$\mathcal{A}+\mathcal{B}$	0	$4\mathcal{A}+\mathcal{B}$
7	$\mathcal{B}$	0	$\mathcal{B}$	0	$2\mathcal{B}$	$-\mathcal{D}$	$5\mathcal{B}$
8	$\mathcal{B}$	0	$\mathcal{B}$	0	$2\mathcal{B}$	0	$5\mathcal{B}-\mathcal{D}$
9	$\mathcal{B}$	0	$\mathcal{B}$	0	$2\mathcal{B}$	0	$5\mathcal{B}$

TABLE 1. Exponential Generating Functions $\mathcal{C}_k^m$.

For convenience, define

$$A = B_0 - B_1 \qquad \text{A126930}=(1, -1, 2, -3, 6, -10, 20, -35, 70, -126, \dots),$$
$$B = B_0 \qquad \text{A126869}=(1, 0, 2, 0, 6, 0, 20, 0, 70, 0, 252, \dots),$$
$$C = B_0 - B_2 \qquad \text{A126120}=(1, 0, 1, 0, 2, 0, 5, 0, 14, 0, 42, \dots),$$
$$D = B_1 \qquad \text{A138364}=(0, 1, 0, 3, 0, 10, 0, 35, 0, 126, 0, \dots),$$

and let $\mathcal{A}, \mathcal{B}, \mathcal{C},$ and $\mathcal{D}$ denote the corresponding egfs.

LEMMA 4. *Let $\mathbf{D}$ denote the derivative operator.*

1. $\mathcal{C}_1^m = \mathbf{D}^m\mathcal{C}.$
2. $\mathcal{C}_2^m = (\mathbf{D}+2)^{(m-2)/2}\mathcal{C},$ *for even $m > 0$.*
3. $\mathcal{C}_k^m = 0,$ *for m odd and k even.*
4. $\mathcal{C}_k^m = C(m)\mathcal{B},$ *for $k = m+1 > 1$, or $k > m+2$.*

PROOF. Recall from (10) of Theorem 1 that

$$\mathcal{C}_k^m = \sum_j \binom{m}{j}\left(\mathcal{B}_{(2j-m)/k} - \mathcal{B}_{(2j-m+2)/k}\right).$$

By Pascal's identity, we have

(14) $$\mathbf{D}\mathcal{B}_\nu = \mathcal{B}_{\nu+1} + \mathcal{B}_{\nu-1}.$$

The proofs of statements 1 and 2 are then straightforward inductions on m. Statements 3 and 4 follow immediately from Lemma 3. $\square$

Applying Lemmas 3 and 4, we compute Table 1. From Table 1 and (9) of Theorem 1 we obtain a closed form for $\mathcal{M}[s_k]$ in terms of $\mathcal{A}, \mathcal{B}, \mathcal{C},$ and $\mathcal{D}$. To determine $M[s_k](n)$ we must compute linear combinations of multinomial convolutions of various B_j. This is reasonably efficient for small values of g and n, however we can speed up the process significantly with the following lemma.

LEMMA 5. *Let* $\mathcal{B}_\nu(z) = \sum_{n=0}^\infty B_\nu(n) z^n/n!$, *where* $B_\nu(n) = \binom{n}{\frac{n+\nu}{2}}$. *Then*

$$\mathcal{B}_a(z)\mathcal{B}_b(z) = \sum_{n=0}^\infty B_{a+b}(n) B_{a-b}(n) \frac{z^n}{n!}$$

for all $a, b \in \mathbb{Z}$.

PROOF. The coefficient of $z^n/n!$ on both sides of the equality count lattice paths from $(0,0)$ to (a,b) in $\mathbb{Z}^2$ with step set $\{(\pm 1, 0), (0, \pm 1)\}$. This is immediate for the LHS. A simple bijective proof for the RHS appears in [**19**]. □

In genus 2, Lemma 5 gives us a closed form for $\mathcal{M}[s_k](n)$ in terms of binomial coefficients (or Catalan numbers). For example, from

$$\mathcal{M}[s_1] = \mathcal{C}_1^0 \mathcal{C}_1^2 - \mathcal{C}_1^1 \mathcal{C}_1^1 = (\mathcal{B}_0 - \mathcal{B}_2)(\mathcal{B}_0 - \mathcal{B}_4) - (\mathcal{B}_1 - \mathcal{B}_3)^2,$$

we obtain

$$M[s_1] = B_0^2 - B_4^2 - B_2^2 + B_2 B_6 - B_0 B_2 + 2 B_2 B_4 - B_0 B_6$$
$$= (B_0 - B_2)(2(B_0 - B_4) + B_2 - B_6) - (B_0 - B_4)^2.$$

This may be expressed more compactly as

$$(15) \qquad M[s_1](n) = C(n) C(n+2) - C(n+4).$$

Here $C(2n) = c(n)$ is the nth Catalan number, giving the identity (11) noted earlier. Similar formulas for the other $M[s_k]$ in genus 2 are listed in Table 2. In higher genera we do not obtain a closed form, but computation is considerably faster with Lemma 5; in genus 4 we use $O(n)$ multiplications to compute $M[s_k](n)$, rather than $O(n^3)$.

By Corollary 1, the sequences for $k > 2g$ are all the same, so it suffices to consider $k \le 2g+1$. For even $k \le 2g$ we find that $E[s_k] = -1$, hence we also consider $M[s_k + 1] = M[s_k^+]$, the sequence of central moments. These may be obtained by computing the binomial convolution of $M[s_k]$ with the sequence $(1, 1, 1, \ldots)$. In genus 1 we obtain

$$M[s_2^+] = (1, 0, 1, 1, 3, 6, 15, 36, 91, 232, 603, 1585, \ldots) \qquad \text{A005043},$$

and in genus 2 we find

$$M[s_2^+] = (1, 0, 2, 1, 11, 16, 95, 232, 1085, 3460, 14820 \ldots) \qquad \text{A138351},$$

$$M[s_4^+] = (1, 0, 3, 1, 21, 26, 215, 493, 2821, 9040, 43695, \ldots) \qquad \text{A138354}.$$

4.4. Explicit Computation of $M[a_k]$ in $USp(2g)$ for $g \le 3$. To complete our study of moment sequences, we now consider the coefficients a_k of the characteristic polynomial $\chi(T)$ of a random matrix in $USp(2g)$. We have already addressed $a_1 = -s_1$. For $k > 1$, the Newton identities allow us to express a_k in terms of s_1, ..., s_k, however this does not allow us to easily compute $M[a_k]$ from the sequences $M[s_j]$ (the covariance among the s_j is nonzero). Instead, we note that by writing

$$(16) \qquad \chi(T) = \prod_j \left((T - e^{i\theta_j})(T - e^{-i\theta_j}) \right) = \prod_j (T^2 - 2\cos\theta_j T + 1),$$

g	k	$M[s_k](n) = E[s_k^n]$	OEIS
1	1	$C(n)$	
		1, 0, 1, 0, 2, 0, 5, 0, 14, 0, 42, 0, 132, 0, 429, ...	A126120
	2	$A(n)$	
		1, -1, 2, -3, 6, -10, 20, -35, 70, -126, 252, -462, 924, ...	A126930
	3	$B(n)$	
		1, 0, 2, 0, 6, 0, 20, 0, 70, 0, 252, 0, 924, 0, 3432, ...	A126869
2	1	$C(n)C(n+4) - C(n+2)^2$	
		1, 0, 1, 0, 3, 0, 14, 0, 84, 0, 594, 0, 4719, 0, 40898, ...	A138349
	2	$C(n)D(n+1) - D(n)C(n+1)$	
		1, -1, 3, -6, 20, -50, 175, -490, 1764, -5292, 19404, ...	A138350
	3	$B(n)C(n)$	
		1, 0, 2, 0, 12, 0, 100, 0, 980, 0, 10584, 0, 121968, ...	A000888*
	4	$B(n)^2 - D(n)^2$	
		1, -1, 4, -9, 36, -100, 400, -1225, 4900, -15876, 63504, ...	A018224*
	5	$B(n)^2$	
		1, 0, 4, 0, 36, 0, 400, 0, 4900, 0, 63504, 0, 853776, ...	A002894*

TABLE 2. Moment Sequences $M[s_k]$ for $g \le 2$.

X	$M[X](n) = E[X^n]$	OEIS
s_1	1, 0, 1, 0, 3, 0, 15, 0, 104, 0, 909, 0, 9449, 0, 112398, ...	A138540
s_2	1, -1, 3, -7, 24, -75, 285, -1036, 4242, -16926, 73206, ...	A138541
s_2^+	1, 0, 2, 0, 11, 1, 95, 36, 1099, 982, 15792, 25070,...	A138542
s_3	1, 0, 3, 0, 26, 0, 345, 0, 5754, 0, 110586, 0, 2341548 ...	A138543
s_4	1, -1, 4, -9, 42, -130, 660, -2415, 12810, -51786, 281736, ...	A138544
s_4^+	1, 0, 3, 1, 27, 26, 385, 708, 7231, 20296, 164277, ...	A138545
s_5	1, 0, 4, 0, 42, 0, 660, 0, 12810, 0, 281736, 0, 6727644, ...	A138546
s_6	1, -1, 6, -15, 90, -310, 1860, -7455, 44730, -195426, ...	A138547
s_6^+	1, 0, 5, 1, 63, 46, 1135, 1800, 25431, 66232, 666387, ...	A138548
s_7	1, 0, 6, 0, 90, 0, 1860, 0, 44730, 0, 1172556, ...	A002896

TABLE 3. Moment Sequences $M[s_k]$ and $M[s_{2k}^+]$ for $g = 3$.[†]

*The OEIS sequence differs slightly.
[†] The notation X^+ denotes the random variable $X + 1$.

each a_k may be expressed as a polynomial in $2\cos\theta_1, \ldots, 2\cos\theta_g$. It follows that we may compute $M[a_k]$ in terms of the sequences C_1^m already considered. The following proposition addresses the cases that arise for $g \le 3$.

PROPOSITION 3. *Let $C(n) = B_0(n) - B_2(n)$ as above and let a_k be the coefficient of T^k in $\chi(T)$, the characteristic polynomial of a random matrix in $USp(2g)$. For $g = 2$ we have*

$$(17) \qquad E[(a_2 - 2)^n] = C(n)C(n+2) - C(n+1)^2,$$

and if $g = 3$ then

$$(18) \qquad E[(a_2 - 3)^n] = \sum_{n_1, n_2, n_3} \binom{n}{n_1, n_2, n_3} \det_{3\times 3} C(n - n_j + i + j - 2).$$

Also for $g = 3$ we have

$$(19) \qquad E[a_3^n] = \sum_{n_1, n_2, n_3, m} \binom{n}{n_1, n_2, n_3, m} 2^{n-m} \det_{3\times 3} C(m + n_j + i + j - 2).$$

PROOF. In genus 2, equation (16) gives $a_2 - 2 = (2\cos\theta_1)(2\cos\theta_2)$ and we have

$$(20) \qquad E[(a_2 - 2)^n] = \int_0^\pi \int_0^\pi (2\cos\theta_1)^n (2\cos\theta_2)^n \mu.$$

By Lemma 3, we note that

$$(21) \qquad C(n + m) = C_1^m(n) = \frac{1}{\pi} \int_0^\pi (2\cos\theta)^n (2\cos\theta)^m (2\sin^2\theta) d\theta.$$

Expanding (20) and applying (21) yields (17). In genus 3 we write

$$a_2 - 3 = (2\cos\theta_1)(2\cos\theta_2) + (2\cos\theta_1)(2\cos\theta_3) + (2\cos\theta_2)(2\cos\theta_3),$$

and apply Lemma 1 to write the expanded integral for $E[(a_2-3)^n]$ in determinantal form to obtain (18) (we omit the details). For (19), note that

$$a_3 = (2\cos\theta_1)(2\cos\theta_2)(2\cos\theta_3) + 2(2\cos\theta_1 + 2\cos\theta_2 + 2\cos\theta_3),$$

and proceed similarly.   □

Taking the binomial convolution of $M[a_2 - g]$ with the sequence $(1, g, g^2, \ldots)$ gives the moment sequence for a_2 in genus g. One finds that $E[a_2] = 1$, hence we also consider the sequence of central moments, $M[a_2 - 1]$. Table 4 gives the complete set of moment sequences for a_k in genus $g \le 3$, including $a_1 = -s_1$.

5. Moment Statistics of Hyperelliptic Curves

Having computed moment sequences attached to characteristic polynomials of random matrices in $USp(2g)$, we now consider the corresponding moment statistics of a hyperelliptic curve. Under Conjecture 2, the latter should converge to the former, provided the curve has large Galois image. Tables 5-7 list moment statistics of a hyperelliptic curves known to have large Galois image.

These tables were constructed by computing $\bar{L}_p(T)$ to determine sample values of a_k and s_k for each p (the a_k are the coefficients, the s_k are derived via the Newton identities). Central moment statistics of $X = a_k - E[a_k]$ and $X = s_k - E[s_k]$ were then computed by averaging X^n over all $p \le N$ where the curve has good reduction.

g	X	$M[X](n) = E[X^n]$	OEIS
1	a_1	1, 0, 1, 0, 2, 0, 5, 0, 14, 0, 42, 0, 132, ...	A126120
2	a_1	1, 0, 1, 0, 3, 0, 14, 0, 84, 0, 594, 0, 4719, ...	A138349
	a_2	1, 1, 2, 4, 10, 27, 82, 268, 940, 3476, 13448, ...	A138356
	a_2^-	1, 0, 1, 0, 3, 1, 15, 15, 105, 190, 945, 2410,...	A095922
3	a_1	1, 0, 1, 0, 3, 0, 15, 0, 104, 0, 909, 0, 9449, ...	A138540
	a_2	1, 1, 2, 5, 16, 62, 282, 1459, 8375, 52323 ...	A138549
	a_2^-	1, 0, 1, 1, 5, 16, 75, 366, 2016, 11936, 75678, ...	A138550
	a_3	1, 0, 2, 0, 23, 0, 684, 0, 34760, 0, 2493096, ...	A138551

TABLE 4. Moment Sequences $M[a_k]$ and $M[a_{2k}^-]$ for $g \leq 3$.[†]

[†]The notation X^- denotes the random variable $X - 1$.

The values of $E[a_k]$ and $E[s_k]$ are as determined in the previous two sections, 0 for k odd or $k > 2g$ and ± 1 otherwise.

Using central moments, we find the moment statistic M_1 (the mean of x) very close to zero ($|M_1| < 0.001$), so we list only $M_2, \ldots, M_{10}$ for each x. Beneath each row the corresponding moments for $USp(2g)$ are listed. Note that the value of N is not the same in each table (we are able to use larger N in lower genus), and the number of sample points is approximately $\pi(N) \approx N/\log N$.

Tables 8 and 9 show the progression of the moment statistics in genus 2 and 3 as N increases, giving a rough indication of the rate of convergence and the degree of uncertainty in the higher moments.

The agreement between the moment statistics listed in Tables 5-7 and the moment sequences computed in Sections 4.3-4.4 is consistent with Conjecture 2. Indeed, on the basis of these results we can quite confidently reject certain alternative hypotheses.

As an example, consider the fourth moment of a_1 in genus 2. The value $M_4 = 3.004$ represents the average of 3,957,807 data points ($\approx \pi(2^{26})$). A uniform distribution on a_1 would imply the mean value of a_1^4 is greater than 50. The probability of then observing $M_4 = 3.004$ over a sample of nearly four million data points is astronomically small. A uniform distribution on the eigenvalue angles gives a mean of 36 yielding a similarly improbable event. In fact, let us suppose only that a_1^4 has a distribution with integer mean not equal to 3. We can then bound the standard deviation by 256 (since $a_1^8 \leq 256^2$) and apply a Z-test to obtain an event probability less than one in a trillion.

X	M_2	M_3	M_4	M_5	M_6	M_7	M_8	M_9	M_{10}
a_1	1.000	0.000	2.000	0.000	5.000	0.001	14.000	0.002	42.000
	1	0	2	0	5	0	14	0	42
s_2^+	1.000	1.000	3.000	6.000	15.001	36.003	91.010	232.03	603.11
	1	1	3	6	15	36	91	232	603
s_3	2.000	0.000	6.000	0.000	20.000	0.000	69.998	0.000	252.00
	2	0	6	0	20	0	70	0	252
s_4	2.000	0.000	6.000	0.000	20.000	0.000	70.001	-0.001	252.00
	2	0	6	0	20	0	70	0	252

TABLE 5. Central moment statistics of a_k and s_k in genus 1, $N = 2^{35}$.

$$y^2 = x^3 + 314159x + 271828$$

.

X	M_2	M_3	M_4	M_5	M_6	M_7	M_8	M_9	M_{10}
a_1	1.001	-0.001	3.004	-0.006	14.014	-0.031	84.041	-0.178	594.02
	1	0	3	0	14	0	84	0	594
a_2^-	1.001	0.000	3.003	0.997	15.013	14.964	105.10	190.00	947.38
	1	0	3	1	15	15	105	190	945
s_2^+	2.001	1.002	11.014	16.044	95.247	232.90	1089.3	3476.4	14891
	2	1	11	16	95	232	1085	3460	14820
s_3	2.002	0.001	12.014	-0.001	100.14	-0.147	981.54	-2.850	10603
	2	0	12	0	100	0	980	0	10584
s_4^+	3.004	1.010	21.049	26.150	215.66	500.32	2830.6	9075.6	43836
	3	1	21	26	215	493	2821	9040	43695
s_5	3.996	-0.015	35.958	-0.211	399.62	-3.152	4897.2	-47.602	63492
	4	0	36	0	400	0	4900	0	63504
s_6	3.999	-0.002	35.983	-0.023	399.81	-0.490	4898.0	-9.460	63487
	4	0	36	0	400	0	4900	0	63504

TABLE 6. Central moment statistics of a_k and s_k in genus 2, $N = 2^{26}$.

$$y^2 = x^5 + 314159x^3 + 271828x^2 + 1644934x + 57721566.$$

X	M_2	M_3	M_4	M_5	M_6	M_7	M_8	M_9	M_{10}
a_1	0.999	-0.007	2.995	-0.046	14.968	-0.343	103.76	-2.723	906.67
	1	0	3	0	15	0	104	0	909
a_2^-	0.999	0.996	4.992	15.982	75.049	366.87	2023.7	11990	75992
	1	1	5	16	75	366	2016	11936	75678
a_3	1.996	-0.034	22.940	-0.950	684.23	-22.334	34938	2360.8	2512126
	2	0	23	0	684	0	34760	0	2493096
s_2^+	2.000	0.001	10.998	0.969	94.977	35.182	1099.5	966.05	15812
	2	0	11	1	95	36	1099	982	15792
s_3	2.996	0.008	25.953	0.129	344.64	2.935	5759.4	73.138	111003
	3	0	26	0	345	0	5754	0	110586
s_4^+	3.002	0.980	27.023	25.574	384.80	697.45	7207.4	20004	163235
	3	1	27	26	385	708	7231	20296	164277
s_5	3.995	-0.036	41.906	-0.719	658.28	-16.625	12776	-428.23	281027
	4	0	42	0	660	0	12810	0	281736
s_6^+	5.001	0.968	63.005	45.334	1134.1	1782.0	25376	65650	663829
	5	1	63	46	1135	1800	25431	66232	666387
s_7	6.000	0.002	90.015	0.356	1860.6	13.010	44746	380.75	1172844
	6	0	90	0	1860	0	44730	0	1172556

TABLE 7. Central moment statistics of a_k and s_k in genus 3, $N = 2^{25}$.
$y^2 = x^7 + 314159x^5 + 271828x^4 + 1644934x^3 + 57721566x^2 + 1618034x + 141021$.

N	M_1	M_2	M_3	M_4	M_5	M_6	M_7	M_8
2^{11}	-0.071	1.031	-0.276	3.167	-2.295	15.250	-21.145	97.499
2^{12}	-0.036	1.112	-0.087	3.565	-0.475	17.251	-4.539	105.082
2^{13}	-0.067	1.085	-0.249	3.407	-1.567	16.537	-12.893	103.344
2^{14}	-0.046	1.029	-0.232	3.181	-1.529	15.795	-13.309	104.558
2^{15}	-0.044	1.031	-0.121	3.256	-0.428	16.325	-2.396	107.173
2^{16}	-0.025	1.022	-0.069	3.143	-0.251	15.251	-1.673	96.837
2^{17}	-0.016	1.011	-0.041	3.079	-0.204	14.594	-1.717	88.871
2^{18}	-0.009	1.002	-0.022	3.041	-0.138	14.441	-1.456	88.636
2^{19}	-0.002	1.003	-0.013	3.031	-0.108	14.259	-1.023	86.288
2^{20}	0.001	0.998	0.001	3.003	-0.041	14.126	-0.687	85.815
2^{21}	-0.000	1.003	-0.002	3.016	-0.045	14.088	-0.577	84.746
2^{22}	0.002	1.002	0.009	3.013	0.037	14.058	0.101	84.166
2^{23}	0.001	1.001	0.002	3.006	0.001	13.999	-0.103	83.715
2^{24}	0.000	1.001	0.001	3.002	0.008	13.964	0.036	83.346
2^{25}	0.000	1.000	-0.000	2.995	-0.010	13.950	-0.120	83.500
2^{26}	0.000	1.001	-0.001	3.004	-0.006	14.014	-0.031	84.041

TABLE 8. Convergence of moment statistics for a_1 in genus 2 as N increases.

$$y^2 = x^5 + 314159x^3 + 271828x^2 + 1644934x + 57721566.$$

N	M_1	M_2	M_3	M_4	M_5	M_6	M_7	M_8
2^{11}	0.033	0.942	0.204	2.611	1.313	11.365	9.198	62.068
2^{12}	0.009	0.921	0.092	2.447	0.649	10.149	4.617	52.838
2^{13}	0.015	0.961	0.075	2.676	0.535	11.971	4.345	69.641
2^{14}	-0.011	0.983	-0.060	2.893	-0.245	14.316	0.704	99.690
2^{15}	-0.005	1.011	-0.018	3.134	-0.067	16.286	0.836	116.675
2^{16}	-0.017	1.007	-0.054	3.154	-0.105	16.813	2.952	127.212
2^{17}	-0.006	0.993	-0.024	3.027	-0.041	15.431	1.622	109.717
2^{18}	-0.005	0.996	-0.026	3.006	-0.110	15.196	0.239	106.901
2^{19}	-0.001	0.999	-0.013	2.985	-0.087	14.793	-0.418	101.662
2^{20}	0.000	0.989	-0.007	2.934	-0.072	14.440	-0.759	98.109
2^{21}	0.003	0.997	0.003	2.979	-0.017	14.796	-0.562	101.690
2^{22}	0.002	0.999	0.005	3.003	0.038	15.098	0.446	105.733
2^{24}	0.000	1.001	0.001	3.015	-0.005	15.138	-0.102	105.418
2^{24}	0.000	0.999	-0.004	2.990	-0.043	14.916	-0.397	103.271
2^{25}	0.000	0.999	-0.007	2.995	-0.046	14.968	-0.343	103.755

TABLE 9. Convergence of moment statistics for a_1 in genus 3 as N increases.

$$y^2 = x^7 + 314159x^5 + 271828x^4 + 1644934x^3 + 57721566x^2 + 1618034x + 141021.$$

#	N	%	ΔM_1	ΔM_2	ΔM_3	ΔM_4	ΔM_5	ΔM_6	ΔM_7	ΔM_8
10^6	2^{16}	50	0.008	0.012	0.031	0.072	0.204	0.563	1.685	5.104
		90	0.020	0.029	0.076	0.177	0.497	1.371	4.120	12.397
		99	0.032	0.045	0.120	0.277	0.781	2.156	6.512	19.633
10^4	2^{20}	50	0.002	0.003	0.009	0.020	0.057	0.159	0.470	1.433
		90	0.006	0.008	0.021	0.049	0.138	0.384	1.154	3.485
		99	0.009	0.013	0.033	0.078	0.214	0.604	1.801	5.432
10^2	2^{24}	50	0.001	0.001	0.002	0.005	0.017	0.044	0.138	0.424
		90	0.002	0.002	0.006	0.013	0.035	0.101	0.277	0.933
		99	0.002	0.003	0.008	0.019	0.054	0.165	0.543	1.519

TABLE 10. Moment deviations for families of random genus 2 curves.

After perusing the data in Tables 5-9, several questions come to mind:

(1) What is the rate of convergence as N increases?
(2) Are these results representative of typical curves?
(3) Can we distinguish exceptional distributions that may arise?

Question 1 has been considered in genus 1 (but not, to our knowledge, in higher genera). For an elliptic curve without complex multiplication, the conjectured discrepancy between the observed distribution and the Sato-Tate prediction is $O(N^{-1/2})$ (see Conjecture 1 of [2] or Conjecture 2.2 of [38]). This conjecture implies that the generalized Riemann Hypothesis then holds for the L-series of the curve [2, Theorem 2]. We will not attempt to address this question here, other than noting that the figures listed in Tables 8 and 9 are not inconsistent with a convergence rate of $O(N^{-1/2})$.

5.1. Random Families of Genus 2 Curves. We can say more about Questions 2 and 3, at least in genus 2. To address Question 2 we tested over a million randomly generated curves of the form

$$y^2 = x^5 + f_4 x^4 + f_3 x^3 + f_2 x^2 + f_1 x + f_0,$$

with the integer coefficients $f_0, \ldots, f_5$ obtained from a uniform distribution on the interval $[-2^{63} + 1, 2^{63} - 1]$.

Table 10 describes the distribution of moment statistics for a_1 over three sets of computations: one million curves with $N = 2^{16}$, ten thousand curves with $N = 2^{20}$, and one hundred curves with $N = 2^{24}$. The rows list bounds on the deviation from the moment sequence for a_1 in $USp(2g)$ that apply to $m\%$ of the curves, with m equal to 50, 90, or 99. One sees close agreement with the predicted moment statistics, with ΔM_n decreasing as N increases. The maximum deviation in M_4 observed for any curve was 0.56 with $N = 2^{16}$.

We also looked for exceptional distributions among the outliers, considering the possibility that one or more curves in our random sample might not have large Galois image. From our initial family of one million random curves we selected one thousand curves whose moment statistics showed the greatest deviation from the predicted values. We recomputed the a_1 moment statistics of these curves, with the bound N increased from 2^{16} to 2^{20}. In each and every case, we saw convergence toward the moment sequence for a_1 in $USp(4)$,

(A138349) $M[a_1] = 1,\ 0,\ 1,\ 0,\ 3,\ 0,\ 14,\ 0,\ 84,\ 0,\ 594,\ \ldots,$

as predicted by Conjecture 2 for curves with large Galois image. An additional test of one hundred of the most deviant curves from within this group with $N = 2^{24}$ yielded further convergence, with $\Delta M_n < 1$ for all $n \leq 8$. This is strong evidence that all the curves in our original random sample had large Galois image; every exceptional distribution we have found for a_1 in genus 2 has $\Delta M_8 > 200$.

One may ask whether convergence of the a_1 moment statistics to $M[a_1]$ is enough to guarantee that the L-polynomial distribution of the curve is represented by $USp(2g)$, since we have not examined the distribution of a_2. If we assume the distribution of $\bar{L}_p(T)$ is given by some infinite compact subgroup of $USp(4)$ (as in Conjecture 2), then it suffices to consider a_1. In fact, under this assumption, much more is true: if the fourth moment statistic of a_1 converges to 3, then the distribution of $\bar{L}_p(T)$ converges to the distribution of $\chi(T)$ in $USp(4)$. This remarkable phenomenon is a consequence of *Larsen's alternative* [37, 25].

5.2. Larsen's Alternative.

To apply Larsen's alternative we need to briefly introduce a representation theoretic definition of "moment" which will turn out to be equivalent to the usual statistical moment in the case of interest to us. Here we parallel the presentation in [25, Section 1.1], but assume G to be compact rather than reductive. Let V be a complex vector space of dimension at least two and $G \subset GL(V)$ a compact group. Define

(22) $M_{a,b}(G, V) = \dim_{\mathbb{C}}(V^{\otimes a} \otimes \check{V}^{\otimes b})^G,$

and set $M_{2n}(G, V) = M_{n,n}(G, V)$. Let $\chi(A) = \mathrm{tr}(A)$ denote the character of V as a G-module (the standard representation of G). We then have

$$M_{2n}(G, V) = \int_G \chi(A)^n \bar{\chi}(A)^n dA = \int_G |\chi(A)|^{2n} dA.$$

We now specialize to the case $V = \mathbb{C}^{2g}$ and suppose $G \subset USp(2g)$. Then

$$M_{2n}(G, V) = \int_G (\mathrm{tr}(A))^{2n} dA = E[(\mathrm{tr}(A))^{2n}] = M[a_1](2n),$$

where $a_1 = -\mathrm{tr}(A)$ and $M[a_1](n) = E[a_1^n]$ as usual. We can now state Larsen's alternative as it applies to our present situation.

THEOREM 2 (Larsen's Alternative). *Let V a complex vector space of even dimension greater than 2 and suppose G is a compact subgroup of $USp(V)$. If $M_4(G, V) = 3$, then either G is finite or $G = USp(V)$.*

This is directly analogous to Part 3 of Theorem 1.1.6 in [25], and the proof is the same.

COROLLARY 2. *Let C be a curve of genus $g > 1$. Under Conjecture 2, the distribution of $\bar{L}_p(T)$ converges to the distribution of $\chi(T)$ in $USp(2g)$ if and only if the fourth moment statistic of a_1 converges to 3.*

The corollary provides a wonderfully effective way for us to distinguish curves with exceptional $\bar{L}_p(T)$ distributions.

6. Exceptional $\bar{L}_p(T)$ Distributions in Genus 2

While we were unable to find *any* exceptional $\bar{L}_p(T)$ distributions among random genus 2 curves with large coefficients, if one restricts the size of the coefficients such cases are readily found. We tested every curve of the form $y^2 = f(x)$, with $f(x)$ a monic polynomial of degree 5 with coefficients in the interval $[-64, 64]$, more than 2^{35} curves. As not every hyperelliptic curve of genus 2 can be put in this form, we also included curves with $f(x)$ of degree 6 (not necessarily monic) and coefficients in the interval $[-16, 16]$. With such a large set of curves to test, we necessarily used a much smaller value of N, approximately 2^{12}. We computed the moment statistics of a_1 using parallel point-counting techniques described in [28] to process 32 curves at once.

To identify exceptional curves, we applied a heuristic filter to bound the deviation of the fourth and sixth moment statistics from the $USp(2g)$ values $M[a_1](4) = 3$ and $M[a_1](6) = 14$. By Larsen's alternative, it suffices only to consider the fourth moment, however we found the sixth moment also useful as a distinguishing metric: the smallest sixth moment observed among any of the exceptional distributions was 35, compared to 14 in the typical case. A combination of the two moments proved to be most effective.

Searching for a small subset of exceptional curves in a large family using a statistical test necessarily generates many false positives: nonexceptional curves which happen to deviate significantly from the $USp(2g)$ distribution for $p \leq N$. The filter criteria were tuned to limit this, at the risk of introducing more false negatives (unnoticed exceptional curves). After filtering the entire family with $N \approx 2^{12}$, the remaining curves were filtered again with $N = 2^{16}$ to remove false positives. Finally, we restricted the resulting list to curves with distinct Igusa invariants [24], leaving a set of some 30,000 nonisomorphic curves with (apparently) exceptional distributions.

One additional criterion used to distinguish distributions was the ratio $z(C, N)$ of zero traces, that is, the proportion of primes p for which $a_p = 0$, among $p \leq N$ where C has good reduction. For a typical curve, $z(C, N) \to 0$ as $N \to \infty$, but for many exceptional distributions, $z(c, N)$ converges to a nonzero rational number. In most cases where this arises, one can readily compute

$$(23) \qquad \lim_{N \to \infty} z(C, N) = z(C),$$

using the Hasse-Witt matrix. This is described in detail in [51], and the following is a typical example. One can show that the curve $y^2 = x^6 + 2$ has $a_p = 0$ unless p is of the form $p = 6n + 1$, in which case

$$(24) \qquad a_p \equiv \binom{3n}{n} 2^n (2^n + 1) \mod p.$$

It follows that for $p = 6n + 1$ we have $a_p = 0$ if and only if $2^n \equiv -1 \mod p$. The integer $2^n = 2^{(p-1)/6}$ is necessarily a sixth root of unity mod p, and exactly one of these is congruent to -1. By the Čebotarev density theorem, this occurs for a set of density $1/6$ among primes of the form $p = 6n + 1$. Combine this with the fact that $a_p = 0$ when p is not of this form and we obtain $z(C) = 7/12$.

Under Conjecture 2, one can show that for any curve C (of arbitrary genus), $z(C)$ must exist and is a rational number, however we need not assume this here.[10] For each nonzero $z(C)$ in Table 11, we have identified a specific curve exhibiting the distribution for which one can show of $\lim_{N\to\infty} z(C, N) = z(C)$. Typically, the Hasse-Witt matrix gives a lower bound on $\liminf_{N\to\infty} z(C, N)$, and by computing $G[\ell]$, the mod ℓ image of the Galois representation in $GSp(2g, Z/\ell z)$, one obtains the density of nonzero traces mod ℓ, establishing an upper bound on $\limsup_{N\to\infty} z(c, N)$. It is generally not difficult to find an ℓ for which these two bounds are equal (the Čebotarev density theorem is invoked on both sides of the argument).

After sorting the sequences of moment statistics and considering the values of $z(C)$ among our set of more than 30,000 exceptional curves, we were able to identify only 22 distributions that were clearly distinct (within the precision of our computations). We also tested a wide range of genus 2 curves taken from the literature [**7, 15, 17, 22, 23, 33, 43, 48, 54, 55**], most with coefficient values outside the range of our search family. In every case the a_1 moment statistics appeared to match one of our previously identified distributions. Conversely, several of the distributions found in our search did not arise among the curves we tested from the literature.

Table 11 lists the 23 distinct distributions for a_1 we found for genus two curves, including the typical case, which is listed first. The value of $z(C)$ and the first six moments of a_1 suffice to distinguish every distribution we have found. We also list the eighth moment statistic, which, while not accurate to the nearest integer, is almost certainly within one percent of the "true" value. We list only the moment statistics of a_1. Histograms of the first twelve a_1 distributions can be found in Appendix II. Additional a_1 histograms, along with moment statistics and histograms for a_2 and s_k are available at `http://math.mit.edu/~drew/`.

The third distribution in Table 11 went unnoticed in our initial analysis (we later found several examples that had been misclassified) and is not a curve taken from the literature. We constructed the curve

$$(C) \qquad\qquad y^2 = x^5 + 20x^4 - 26x^3 + 20x^2 + x$$

to have a split Jacobian, isogenous to the product of the elliptic curve

$$(E_1) \qquad\qquad y^2 = x^3 - 11x + 14,$$

which has complex multiplication, and the elliptic curve

$$(E_2) \qquad\qquad y^2 = x^3 + 4x^2 - 4x,$$

which does not. For every p where C has good reduction, the trace of Frobenius is simply the sum of the traces of E_1 and E_2. As E_1 and E_2 are not isogenous (over $\mathbb{C}$), we expect their a_1 distributions to be uncorrelated.[11] It follows that the moment sequence of a_1 for the curve C is simply the binomial convolution of the moment sequences of a_1 for the curves E_1 and E_2. Thus we expect the moment

---

[10]A compact subgroup of $USp(2g)$ has a finite number of connected components. The density of zero trace elements must be zero or one on each component.

[11]The genus 1 traces may be correlated in a way that does not impact $a_1 = -a_p/\sqrt{p}$, e.g., both curves might have the property that p mod 3 determines a_p mod 3.

#	$z(C)$	M_2	M_4	M_6	M_8	$f(x)$
1	0	1	3	14	84	$x^5 + x + 1$
2	0	2	10	70	588*	$x^5 - 2x^4 + x^3 + 2x - 4$
3	0	2	11	90	888*	$x^5 + 20x^4 - 26x^3 + 20x^2 + x$
4	0	2	12	110	1203*	$x^5 + 4x^4 + 3x^3 - x^2 - x$
5	0	4	32	320	3581*	$x^5 + 7x^3 + 32x^2 + 45x + 50$
6	1/6	2	12	100	979*	$x^5 - 5x^3 - 5x^2 - x$
7	1/4	2	12	100	1008*	$x^5 + 2x^4 + 2x^2 - x$
8	1/4	2	12	110	1257*	$x^5 - 4x^4 - 2x^3 - 4x^2 + x$
9	1/2	1	5	35	293*	$x^5 - 2x^4 + 11x^3 + 4x^2 + 4x$
10	1/2	1	6	55	601*	$x^5 - 2x^4 - 3x^3 + 2x^2 + 8x$
11	1/2	2	16	160	1789*	$x^5 + x^3 + x$
12	1/2	2	18	220	3005*	$x^5 - 3x^4 + 19x^3 + 4x^2 + 56x - 12$
13	1/2	4	48	640	8949*	$x^6 + 1$
14	7/12	1	6	50	489*	$x^5 - 4x^4 - 3x^3 - 7x^2 - 2x - 3$
15	7/12	2	18	200	2446*	$x^6 + 2$
16	5/8	1	6	50	502*	$x^5 + x^3 + 2x$
17	5/8	2	18	200	2515*	$x^5 - 10x^4 + 50x^2 - 25x$
18	3/4	1	8	80	894*	$x^5 - 2x^3 - x$
19	3/4	1	9	100	1222*	$x^5 - 1$
20	3/4	1	9	110	1501*	$11x^6 + 11x^3 - 4$
21	3/4	2	24	320	4474*	$x^5 + x$
22	13/16	1	9	100	1254*	$x^5 + 3x$
23	7/8	1	12	160	2237*	$x^5 + 2x$

TABLE 11. Moments of a_1 for genus 2 curves $y^2 = f(x)$ with $N = 2^{26}$.

Column $z(C)$ is the density of zero traces (a_p values). The starred values indicate uncertainty in the eighth moment statistic. In each case, if $T_8 = \lim_{N\to\infty} M_8$, we estimate that $-0.005 \le 1 - M_8/T_8 \le 0.01$ with very high probability (the larger uncertainty on the positive side is primarily due to an observed excess of zero traces for small values of N, we expect $M_8 \le T_8$ in most cases). See Table 13 for predicted values of T_8 and T_{10}.

statistics of a_1 for C to converge to

(A138552*)   1, 0, 2, 0, 11, 0, 90, 0, 889, 0, 9723, ...,

which is the binomial convolution of the sequences $(1, 0, 1, 0, 3, 0, 10, 0, 35, ...)$ and $(1, 0, 1, 0, 2, 0, 5, 0, 14, ...)$ mentioned in the introduction. These are the a_1 moment sequences of elliptic curves with and without complex multiplication. In terms of moment generating functions, we simply have

$$\mathcal{M}_C[a_1] = \mathcal{M}_{E_1}[a_1]\mathcal{M}_{E_2}[a_1].$$

Provided the covariance between the a_1 distributions of E_1 and E_2 is zero, one can prove the a_1 moment statistics of C converge to the sequence above using known results for genus 1 curves (note that E_1 has complex multiplication and E_2 has multiplicative reduction at $p = 107$).

Many of the distributions in Table 11 can be obtained from genus 1 moment sequences by constructing an appropriate genus 2 curve with split Jacobian, as shown in the next section. It is important to note that a curve whose Jacobian is simple (not split over $\overline{\mathbb{Q}}$) may still have an L-polynomial distribution matching that of a split Jacobian. Distribution #2, for example, corresponds to a Jacobian which splits as the product of two nonisogenous elliptic curves without complex multiplication. This distribution also arises for some genus 2 curves with simple Jacobians, including

$$y^2 = x^5 - x^4 + x^3 + x^2 - 2x + 1.$$

This is a modular genus 2 curve which appears as $C_{188,A}$ in [**17**], along with many similar examples.

We speculate that simple Jacobians with Distribution #2 are all of type I(2) in the classification of Moonen and Zarhin [**39**], corresponding to Jacobians whose endomorphism ring is isomorphic to the ring of integers in a real quadratic extension of $\mathbb{Q}$. A similar phenomenon occurs with Distribution #11, which arises for split Jacobians that are isogenous to the product of an elliptic curve and its twist, but also for simple Jacobians of type II(1) in the Moonen-Zarhin classification (these are QM-curves, see [**22**] for examples). The remaining two types of simple genus 2 Jacobians in the Moonen-Zarhin classification are type I(1), which is the typical case (Jacobians with endomorphism ring $\mathbb{Z}$), and type IV(2,1), which occurs for curves with complex multiplication over a quartic CM field (with no imaginary quadratic subfield). These correspond to Distributions #1 and #19 respectively (examples of the latter can be found in [**54**]). The remaining distributions appear to arise only for curves with split Jacobians.

7. Representation of Genus 2 Distributions in $USp(4)$

Conjecture 2 implies that each distribution in Table 11 is represented by the distribution of characteristic polynomials in some infinite compact subgroup H of $USp(4)$. In this section we will exhibit such an H for each distribution. We do not claim that each H we give is the "correct" subgroup for every curve with the corresponding distribution, in the sense of corresponding to the Galois image G_ℓ, as discussed in Section 3.[12] Rather, for each H we show that its density of zero traces and moment sequence are compatible with the corresponding data in Table 11. In most cases we also have evidence that suggests H is the correct subgroup for the particular corresponding curve listed in Table 11 (see Section 7.1).

For all but two cases we will construct H using two subgroups of $USp(2)$ that represent distributions of genus 1 curves: $G_1 = USp(2)$ for an elliptic curve without complex multiplication, and $G_2 = N(SO(2))$, the normalizer of $SO(2)$ in $SU(2)$, for an elliptic curve with complex multiplication. We will construct each $H \subset USp(4)$ from G_1 and/or G_2 explicitly as a group of matrices, however the motivation behind these constructions are split Jacobians.

---

[12]Indeed, a single H for each distribution would not suffice. As noted above, two curves may share the same $\overline{L}_p(T)$ distribution for different reasons.

The most obvious cases correspond to the product of two nonisogenous elliptic curves, and we have the groups $G_1 \times G_1$, $G_1 \times G_2$, and $G_2 \times G_2$ as groups of block diagonal 2×2 matrices in $USp(4)$. These correspond to Distributions #2, #3, and #8 in Table 11, and their moment sequences are easily computed via binomial convolutions of the appropriate genus 1 moment sequences (the example of the previous section corresponds to $G_1 \times G_2$).

To obtain additional distributions, we also consider the product of two isogenous elliptic curves. We may, for example, pair an elliptic curve with an isomorphic copy of itself, or with one of its twists.[13] For two isomorphic curves, the corresponding subgroup H_i contains block diagonal matrices of the form

$$B = \begin{pmatrix} A & 0 \\ 0 & A \end{pmatrix}$$

where A is an element of G_i ($i = 1$ or 2). To pair a curve with its twist we also include block diagonal matrices with A and $-A$ on the diagonal to obtain the subgroup H_i^-.

We now generalize this idea. Let $G = G_1$ (resp. G_2) be a compact subgroup of $USp(2)$, and let G^* be the subgroup of $U(2)$ obtained by extending G by scalars and taking the subgroup of elements whose determinants are kth roots of unity (for some positive integer k). For $A \in G^*$, let $\overline{A}$ denote the complex conjugate of A and define the block diagonal matrix

$$B = \begin{pmatrix} A & 0 \\ 0 & \overline{A} \end{pmatrix}.$$

The matrix B is clearly unitary, and one easily verifies that it is also symplectic, hence $B \in USp(4)$. The set of all such B forms our subgroup H. As a topological group, H has k (resp. $2k$) connected components, each a closed set consisting of elements with A having a fixed determinant, thus H is compact. The identity component is isomorphic to $USp(2)$ (resp. $SO(2)$), embedded diagonally.

We may write $A \in G^*$ as $A = \omega^j A_0$ with $A_0 \in G \subseteq USp(2)$, ω a primitive $2k$th root unity, and $1 \le j \le k$. We then have

$$\mathrm{tr}(B) = \mathrm{tr}(A) + \mathrm{tr}(\overline{A}) = \omega^j \mathrm{tr}(A_0) + \omega^{-j} \mathrm{tr}(\overline{A}_0) = (\omega^j + \omega^{-j}) \mathrm{tr}(A_0),$$

since $\mathrm{tr}(\overline{A}_0) = \mathrm{tr}(A_0)$ for any $A_0 \in USp(2)$. It follows that

(25) $$\mathbf{E}_H[(\mathrm{tr}(B))^n] = \left(\frac{1}{k} \sum_{j=1}^{k} (\omega^j + \omega^{-j})^n \right) \mathbf{E}_G[(\mathrm{tr}(A_0))^n]$$

where $\mathbf{E}_G[X]$ denotes the expectation of a random variable X over the Haar measure on G.

The term $(\omega^j + \omega^{-j})^n \mathbf{E}_G[(\mathrm{tr}(A_0))^n]$ corresponds to the nth moment of the trace distribution on a component of H. These moments will all be integers precisely when $k \in \{1, 2, 3, 4, 6\}$ (they are zero for n odd). In fact, these are the only values of k for which H plausibly represents the distribution of $\overline{L}_p(T)$ for a genus 2 curve defined over $\mathbb{Q}$, as we now argue.

---

[13]See [**7**, Ch. 14] and [**33**] for explicit methods of constructing such curves.

H	$k=1$	$k=2$	$k=3$	$k=4$	$k=6$
H_1^k	5	11	4	7	6
$J(H_1^k)$	11	17	10	16	14
H_2^k	13	21	12	18	15
$J(H_2^k)$	21	23	20	22	*

TABLE 12. Distributions matching H_i^k or $J(H_i^k)$.

Consider an L-polynomial $L_p(T)$ for which $\bar{L}_p(T) = L_p(p^{-1/2}T)$ is the characteristic polynomial of some $B \in H$. We may factor $L_p(T)$ as

$$(26) \qquad L_p(T) = (\chi(p)pT^2 - \alpha T + 1)(\overline{\chi}(p)pT^2 - \overline{\alpha}T + 1),$$

where $\alpha \in \mathbb{Z}[\zeta]$, with ζ a primitive kth root of unity, and $\chi(p)$ is a kth root of unity equal to the determinant of A in the component of H containing B. For the elliptic curve we have in mind as a factor of the Jacobian, $\chi(p)p$ is the determinant of its Frobenius element and α is the trace. Since $L_p(T)$ has integer coefficients, α must lie in a quadratic extension of $\mathbb{Q}$, giving $k \in \{1,2,3,4,6\}$.

For each of these k we can easily compute closed forms for the parenthesized expression in (25). Assuming $n > 0$ is even, we obtain

$$2^n, \quad 2^n/2, \quad (2^n+2)/3, \quad (2^n+2^{n/2+1})/4, \quad (2^n+2\cdot 3^{n/2}+2)/6,$$

for $k = 1,2,3,4,6$, respectively. Since $\mathbf{E}_G[(\mathrm{tr}(A_0))^n]$ is zero for odd values of n, these expressions may be used in (25) for all positive n. For even values of k, the density $z(H)$ of zero traces in H is $1/k$ (resp. $1/2+1/k$), and $z(H)$ is zero (resp. $1/2$) for k odd.

For any of the subgroups H constructed as above, we may also consider the group $J(H)$ generated by H and the block diagonal matrix

$$(27) \qquad J = \begin{pmatrix} 0 & I \\ -I & 0 \end{pmatrix}.$$

The group $J(H)$ contains H as a subgroup of index 2, with the nonidentity coset having all zero traces. For $n > 0$, the nth moment of the trace distribution in $J(H)$ is simply half that of H, and $z(J(H)) = (z(H)+1)/2$.

For $i = 1$ or 2 and $k \in \{1,2,3,4,6\}$, let H_i^k denote the group H constructed using G_i and k. By also considering $J(H_i^k)$, we can construct a total of 20 groups, 18 of which have distinct eigenvalue distributions. With the sole exception of $J(H_2^6)$, each of these matches one of the distributions in Table 11 to a proximity well within the accuracy of our computational methods. Note that $H_i^1 = H_i$, and the group H_i^2 has the same eigenvalue distribution as H_i^- (but is not conjugate).

We may also consider $J(G_1 \times G_1)$, which corresponds to Distribution #9. This construction does not readily apply to $G_1 \times G_2$ (in fact $J(G_1 \times G_2) = J(G_1 \times G_1)$). The group $J(G_2 \times G_2)$ does give a new distribution (it is the normalizer of $SO(2) \times SO(2)$ in $USp(4)$), but it does not correspond to any we have found. However, the group $J(G_2 \times G_2)$ contains a subgroup K not equal to $G_2 \times G_2$ which matches Distribution #19. The group K has identity component $SO(2) \times SO(2)$ and a cyclic component group, of order 4 (this determines K).

Of all the groups we have constructed, only K does not correspond, in some fashion, to a split Jacobian. As noted above, Distribution #19 arises for curves with simple Jacobians and complex multiplication over a quartic CM field, and in fact all nineteen such curves documented by van Wamelen have this distribution [54, 55]. The only remaining group to consider is $USp(4)$ itself, which of course gives Distribution #1.

Table 13 gives a complete list of the subgroups of $USp(4)$ we have identified, one for each distribution in Table 11, and also entries for $J(H_2^6)$ and $J(G_2 \times G_2)$. These last two distributions appear to be spurious; see § 7.2.

7.1. Supporting Evidence. Aside from closely matching the trace distributions we have found, there is additional data that supports our choice of the subgroups H appearing in Table 13. First, we find that these H not only match the distribution of the a_1 coefficient in $\bar{L}_p(T)$, they also appear to give the correct distribution of a_2. We should note that for the three curves in Table 11 where $f(x)$ has degree 6, the available methods for computing $L_p(T)$ are much less efficient, so we did not attempt this verification in these three cases.[14]

More significantly, the disconnected groups in Table 13 also appear to give the correct distribution of a_1 (and a_2 for $f(x)$ of degree 5) on each of their components. If we partition the components of H according to their distributions of characteristic polynomials, for a given curve we can typically find a partitioning of primes into sets of corresponding density with matching $\bar{L}_p(T)$ distributions.

Taking Distribution #10 as an example, we have $H = J(H_1^3)$ in Table 13 and the corresponding curve in Table 11 is given by

$$y^2 = f(x) = x^5 - 2x^4 - 3x^3 + 2x^2 + 8x = x(x-2)(x^3 - 3x - 4).$$

The cubic $g(x) = x^3 - 3x - 4$ has Galois group S_3. The set of primes P_3 for which $g(x)$ splits into three factors in $\mathbb{F}_p[x]$ has density $1/6$, and corresponds to the identity component of $J(H_1^3)$.[15] The set of primes P_2 where $g(x)$ splits into two factors has density $1/2$, and corresponds to the nonidentity coset of H_1^3 in $J(H_1^3)$, containing three components with identical eigenvalue distributions. The remaining set of primes P_1 for which $g(x)$ is irreducible has density $1/3$ and corresponds to the set of elements of H_1^3 for which the determinant of the block diagonal matrix A is not 1 (this includes two of the six components of $J(H_1^3)$). Table 14 lists the moment statistics for a_1 and a_2, restricted to the sets P_1, P_2, and P_3, with values for the corresponding subset of $J(H_1^3)$ beneath.

A similar analysis can be applied to the other disconnected groups in Table 13, however the definition of the sets P_i varies. For Distribution #11 the group H_1^- has two components, and for the curve $y^2 = f(x) = x^5 + x^3 = x$ the correct partitioning of primes simply depends on the value of p modulo 4, not on how $f(x)$ splits in $\mathbb{F}_p[x]$ (in fact, the set of primes where $f(x)$ splits intersects both partitions). In other cases both a modular constraint and a splitting condition may apply. In general there is some partitioning of primes which corresponds to a partitioning of the components of H, and the corresponding distributions appear to agree.

In addition to verifying the distribution of $\bar{L}_p(T)$ over sets of primes, for each group corresponding to a split Jacobian, one can also check whether $L_p(T)$ admits

---

[14]The a_1 coefficient can be computed reasonably efficiently in the degree 6 case by counting points on C in $\mathbb{F}_p$, but the group law on the Jacobian is much slower.

[15]We thank Dan Bump for suggesting this approach.

#	H	d	$c(H)$	$z(H)$	M_2	M_4	M_6	M_8	M_{10}
1	$USp(4)$	10	1	0	1	3	14	84	594
2	$G_1 \times G_1$	6	1	0	2	10	70	588	5544
3	$G_1 \times G_2$	4	2	0	2	11	90	889	9723
4	H_1^3	3	3	0	2	12	110	1204	14364
5	H_1	3	1	0	4	32	320	3584	43008
6	H_1^6	3	6	1/6	2	12	100	980	10584
7	H_1^4	3	4	1/4	2	12	100	1008	11424
8	$G_2 \times G_2$	2	4	1/4	2	12	110	1260	16002
9	$J(G_1 \times G_1)$	6	2	1/2	1	5	35	294	2772
10	$J(H_1^3)$	3	6	1/2	1	6	55	602	7182
11	H_1^-	3	2	1/2	2	16	160	1792	21504
12	H_2^3	1	6	1/2	2	18	220	3010	43092
13	H_2	1	2	1/2	4	48	640	8960	129024
14	$J(H_1^6)$	3	12	7/12	1	6	50	490	5292
15	H_2^6	1	12	7/12	2	18	200	2450	31752
16	$J(H_1^4)$	3	8	5/8	1	6	50	504	5712
17	H_2^4	1	8	5/8	2	18	200	2520	34272
18	$J(H_1^-)$	3	4	3/4	1	8	80	896	10752
19	K	2	4	3/4	1	9	100	1225	15876
20	$J(H_2^3)$	1	12	3/4	1	9	110	1505	21546
21	H_2^-	1	4	3/4	2	24	320	4480	64512
22	$J(H_2^4)$	1	16	13/16	1	9	100	1260	17136
23	$J(H_2^-)$	1	8	7/8	1	12	160	2240	32256
*	$J(G_2 \times G_2)$	2	8	5/8	1	6	55	630	8001
*	$J(H_2^6)$	1	24	19/24	1	9	100	1225	15876

TABLE 13. Candidate subgroups of $USp(4)$.

Row numbers correspond to the distributions in Table 11. Column d lists the real dimension of H, and $c(H)$ counts its components. The column $z(H)$ gives the density of zero traces, and $M_n = \mathbf{E}_H[(\mathrm{tr}(B))^n]$ for a random $B \in H$. The last two rows are not known to match the L-polynomial distribution of a genus 2 curve.

a factorization of the expected type. For the product groups we expect $L_p(T)$ to factor into two quadratics with coefficients in $\mathbb{Z}$. For the groups H_i^k we expect a factorization of the form given in (26), and a similar form applies to H_i^-.[16] For each curve in Table 11 with $f(x)$ of degree 5 corresponding to such a group, we

---

[16]As previously noted, there are curves with simple Jacobians matching distributions #2 and #11 for which this would not apply, but they don't appear in Table 11.

Set	X	M_1	M_2	M_3	M_4	M_5	M_6	M_7	M_8
P_1	a_1	0.000	1.001	-0.002	2.002	-0.007	5.005	-0.027	14.01
		0	1	0	2	0	5	0	14
	a_2	0.001	1.001	1.001	3.002	6.005	15.01	36.03	91.10
		0	1	1	3	6	15	36	91
P_2	a_1	0.000	0.000	0.000	0.000	0.000	0.000	0.000	0.000
		0	0	0	0	0	0	0	0
	a_2	1.000	2.010	3.001	6.003	10.00	20.01	35.02	70.05
		1	2	3	6	10	20	35	70
P_3	a_1	-0.002	3.999	-0.045	31.95	-0.590	319.3	-7.69	3574
		0	4	0	32	0	320	0	3584
	a_2	2.999	9.995	36.97	149.8	652.7	3005	14404	71160
		3	10	37	150	654	3012	14445	71398

TABLE 14. Component distributions of a_1 and a_2 for curve #10.

verified the existence of the expected factorization for primes $p \leq 10^6$. For groups with multiple components, we partitioned the primes appropriately (note that for groups of the form $J(H)$ we do not expect a factorization for primes corresponding to components with off-diagonal elements).

The final piece of evidence we present is much less precise, and of an entirely different nature. By computing the group structure of the Jacobian and determining the rank of its ℓ-Sylow subgroup for many p, one can estimate the size of the mod ℓ Galois image $G[\ell]$ in $GSp(2g, \mathbb{Z}/\ell\mathbb{Z})$. The primes for which the ℓ-Sylow subgroup has rank $2g$ correspond to the identity element in $G[\ell]$. The group $GSp(2g, \mathbb{Z}/\ell\mathbb{Z})$ has size $O(\ell^{11})$, and this corresponds to the real dimension of $USp(2g)$ which is 10 (the reduction in dimension arises from unitarization). For all but the first group in Table 13. the real dimension is at most 6. We should expect correspondingly small $G[\ell]$, at most $O(\ell^7)$ in size. By computing

$$(28) \qquad d = \frac{\log(\#G[\ell_1]) - \log(\#G[\ell_2])}{\log \ell_1 - \log \ell_2}$$

for various $\ell_1 \neq \ell_2$ we obtain a general estimate for $\#G[\ell] = O(\ell^d)$. The value $d - 1$ is then an estimate of the real dimension of H. This is necessarily a rather crude approximation, and one must take care to avoid exceptional values of ℓ. We performed this computation for each exceptional curve in Table 11 where $f(x)$ has degree 5 for $p \leq N = 2^{24}$ and ℓ ranging from 3 to 19. The results agreed with the corresponding dimensions in Table 11 to within ± 1.

7.2. Nonexistence Arguments.
We wish to thank Jean-Pierre Serre (private communication) for suggesting the following argument to rule out the case $H = J(G_2 \times G_2)$. Let E be the endomorphism algebra of the Jacobian of the curve over $\overline{\mathbb{Q}}$. Then $E \otimes \mathbb{Q}$ must be a commutative $\mathbb{Q}$-algebra of rank 4; more precisely, it must be either $K \otimes_{\mathbb{Q}} K'$ for some nonisomorphic imaginary quadratic fields K, K',

or a quartic CM-field. (We cannot have $K \cong K'$ or else H would be forced to have dimension 1 rather than 2.) In either case, $\text{Aut}(E \otimes \mathbb{Q})$ has at most 4 elements, so the elements of H are defined over a field of degree at most 4 over $\mathbb{Q}$. However, $J(G_2 \times G_2)$ has 8 connected components, so this is impossible.

We do not yet have an analogous argument to rule out $H = J(H_2^6)$. However, Serre suggests that it should be possible to give such an argument based on the following data: the algebra $E \otimes \mathbb{Q}$ must be a (possibly split) quaternion algebra over a quadratic field, and the image of Galois in $\text{Aut}(E \otimes \mathbb{Q})$ must have order 24.

8. Conclusion

Based on the results presented in Section 6, we now state a more explicit form of Conjecture 2 for genus 2 curves.

CONJECTURE 3. *Let C be a genus 2 curve. The distribution of $\bar{L}_p(T)$ over $p \leq N$ converges (as $N \to \infty$) to the distribution of $\chi(T)$ in one of the first 23 subgroups of $USp(4)$ listed in Table 13. For almost all curves, this group is $USp(4)$.*

It would be interesting to carry out a similar analysis in genus 3, but it is not immediately clear from the genus 2 results how many exceptional distributions one should expect. An exhaustive search of the type undertaken in genus 2 may not be computationally feasible in genus 3.

We end by once again thanking Nicholas Katz for his invaluable support throughout this project, and David Vogan for several helpful conversations. We also thank Zeev Rudnick for his feedback on an early draft of this paper, and Jean-Pierre Serre for his remarks in § 7.2.

9. Appendix I - Distributions of s_k

This appendix is a gallery of distributions of s_k for hyperelliptic curves of genus 1, 2, and 3 with large Galois image. The s_k are the kth power sums (Newton symmetric function) of the roots of the unitarized L-polynomial $\bar{L}_p(T)$ defined in Section 2.

Each figure represents a histogram of approximately $\pi(N)$ values derived from $\bar{L}_p(T)$ for $p \leq N$ where the curve has good reduction. The horizontal axis ranges from $-\binom{2g}{k}$ to $\binom{2g}{k}$, divided into approximately $\sqrt{\pi(N)}$ buckets. The vertical axis is scaled to fit the data, with the height of the uniform distribution indicated by a dotted line.

9.1. Genus 1. $N = 2^{35}$; $y^2 = x^3 + 314159x + 271828$.

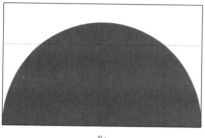

s_1

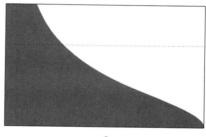

s_2

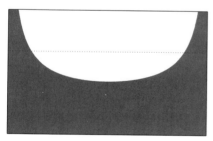

s_3

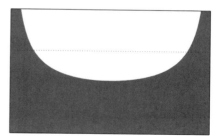

s_4

9.2. Genus 2. $N = 2^{26}$;
$$y^2 = x^5 + 314159x^3 + 271828x^2 + 1644934x + 57721566.$$

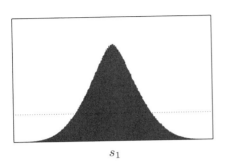

s_1

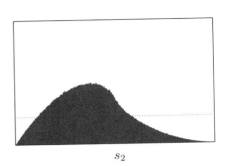

s_2

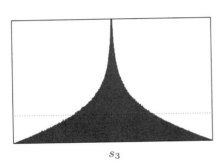

s_3

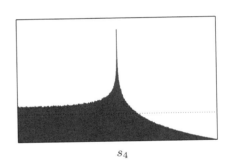

s_4

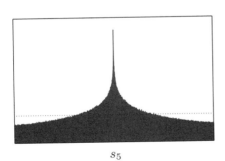

s_5

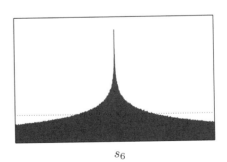

s_6

9.3. Genus 3. $N = 2^{25}$;
$$y^2 = x^7 + 314159x^5 + 271828x^4 + 1644934x^3 + 57721566x^2 + 1618034x + 141021.$$

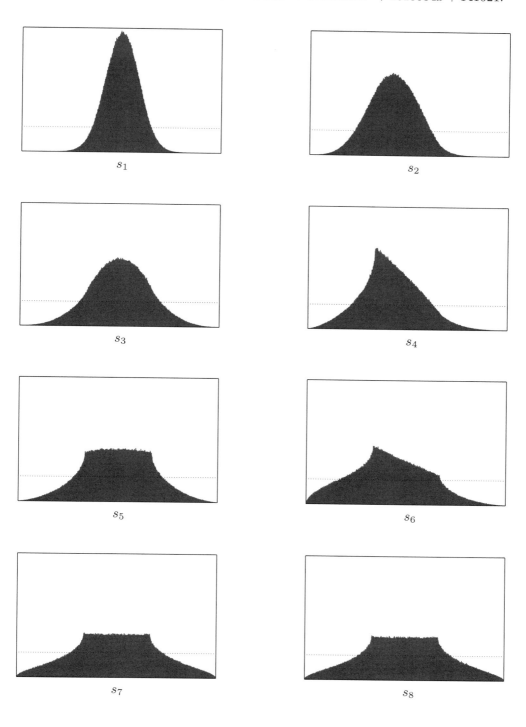

10. Appendix II - Distributions of a_1 in Genus 2

This appendix contains a_1 histograms for each of the first 12 curves listed in Table 11, computed with $N = 2^{26}$. The approximate area of the central spike is given by $z(C)$ in Table 11, corresponding to primes for which $a_p = 0$. The secondary spikes at $a_1 = \pm 2$ appearing in distributions #8 and #12 have approximately zero area.

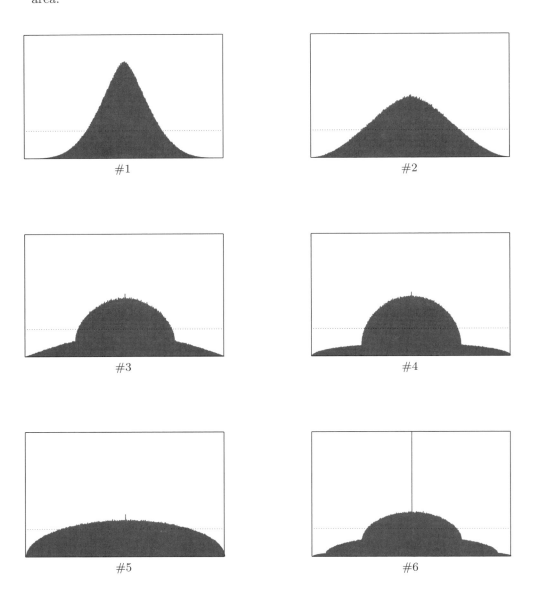

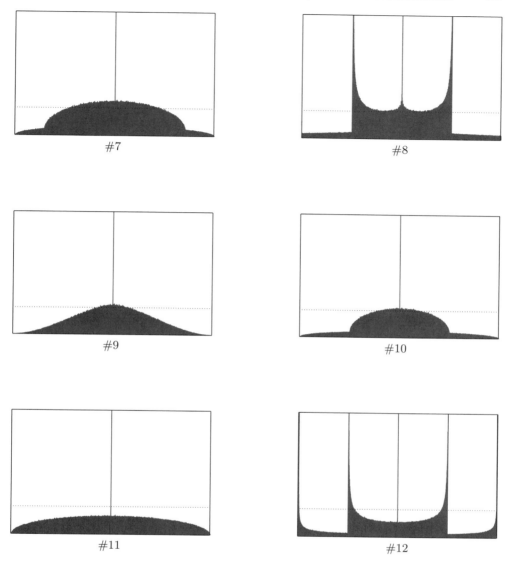

#7

#8

#9

#10

#11

#12

References

1. Martin Aigner, *Catalan and other numbers: a recurrent theme*, Algebraic combinatorics and computer science, a tribute to Gian-Carlo Rota, Springer, 2001, pp. 347–390.
2. Shigeki Akiyama and Yoshio Tanigawa, *Calculation of values of L-functions associated to elliptic curves*, Mathematics of Computation **68** (1999), no. 227, 1201–1231.
3. Emil Artin, *Quadratische Körper in gebiete der höheren Kongruenzen. I, II*, Mathematische Zeitschrift **19** (1924), 153–246.
4. Stephan Baier, *A remark on the conjectures of Lang-Trotter and Sato-Tate on average*, 2007, preprint, `http://arxiv.org/abs/0708.2535v3`.
5. Stephan Baier and Lianyi Zhao, *The Sato-Tate conjecture on average for small angles*, 2007, to appear, `http://arxiv.org/abs/math/0608318v4`.
6. William D. Banks and Igor E. Shparlinski, *Sato-Tate, cyclicity, and divisibility statistics on average for elliptic curves of small height*, 2007, to appear, `http://arxiv.org/abs/math/0609144`.
7. J. W. S. Cassels and E. V. Flynn, *Prolegomena to a middlebrow arithmetic of curves of genus 2*, London Mathematical Society Lecture Note Series, vol. 230, Cambridge University Press, 1996.
8. William Y.C. Chen, Eva Y.P. Deng, Rosena R.X. Du, Richard P. Stanley, and Catherine H. Yan, *Crossings and nestings of matchings and partitions*, Transactions of the American Mathematical Society **359** (2007), 1555–1575.
9. Laurent Clozel, Michael Harris, and Richard Taylor, *Automorphy for some l-adic lifts of automorphic mod ℓ Galois representations*, 2006, preprint, `http://www.math.harvard.edu/~rtaylor/twugnew.pdf`.
10. Alina Carmen Cojocaru, *On the surjectivity of the Galois representations associated to non-CM elliptic curves*, Canadian Mathematics Bulletin **48** (2005), no. 1, 16–31, With an appendix by Ernst Kani.
11. John Cremona, *The elliptic curve database for conductors to 130000*, Algorithmic Number Theory Symposium–ANTS VII, Lecture Notes in Computer Science, vol. 4076, 2006, pp. 11–29.
12. Max Deuring, *Die Klassenkörper der komplexen Multiplication*, Enzyklopädie der mathematischen Wissenschaften, vol. 12 (Book 10, Part II), B.G. Teubner Verlagsgesellschaft, Stuttgart, 1958.
13. Henri Cohen (Ed.) et al., *Handbook of elliptic and hyperelliptic curve cryptography*, Chapman and Hall, 2006.
14. Gerd Faltings, *Endlichkeitssätze für abelsche Varietäten über Zahlkörpern. [Finiteness theorems for abelian varieties over number fields]*, Inventiones Mathematicae **73** (1983), no. 3, 349–366. MR 718935 (85g:11026a) reviewed by James Milne.
15. Gerhard Frey and Michael Müller, *Arithmetic of modular curves and applications*, Algorithmic algebra and number theory (Matzat et al., ed.), Springer-Verlag, 1999, pp. 11–48.
16. Ira M. Gessel, *Symmetric functions and P-recursiveness*, Journal of Combinatorial Theory A **53** (1990), 257–285.
17. Enrique González-Jiménez and Josep González, *Modular curves of genus 2*, Mathematics of Computation **72** (2003), no. 241, 397–418.
18. David J. Grabiner and Peter Magyar, *Random walks in Weyl chambers and the decomposition of tensor powers*, Journal of Algebraic Combinatorics **2** (1993), no. 3, 239–260.
19. Richard K. Guy, Christian Krattenthaler, and Bruce Sagan, *Lattice paths, reflections and dimension-changing bijections*, Ars Combinatorica **34** (1992), 3–15.
20. Chris Hall, *An open image theorem for a general class of abelian varieties*, 2008, preprint, `http://arxiv.org/abs/0803.1682v1`.
21. Michael Harris, Nick Shepherd-Barron, and Richard Taylor, *A family of Calabi-Yau varieties and potential automorphy*, May 2006, preprint, `http://www.math.harvard.edu/~rtaylor/`.
22. Ki-Ichiro Hashimoto and Hiroshi Tsunogai, *On the Sato-Tate conjecture for QM-curves of genus two*, Mathematics of Computation **68** (1999), no. 228, 1649–1662.
23. Everett W. Howe, Franck Leprévost, and Bjorn Poonen, *Large torsion subgroups of split jacobians of curves of genus two or three*, Forum Mathematicum **12** (2000), no. 3, 315–364.
24. Jun-Ichi Igusa, *Arithmetic variety of moduli for genus two*, Annals of Mathematics **2** (1960), no. 72, 612–649.

25. Nicholas M. Katz, *Larsen's alternative, moments, and the monodromy of Lefschetz pencils*, Contributions to automorphic forms, geometry, and number theory, Johns Hopkins University Press, 2004, pp. 521–560.
26. _____, 2008, e-mail correspondence.
27. Nicholas M. Katz and Peter Sarnak, *Random matrices, Frobenius eigenvalues, and monodromy*, American Mathematical Society, 1999.
28. Kiran S. Kedlaya and Andrew V. Sutherland, *Computing L-series of hyperelliptic curves*, Algorithmic Number Theory Symposium–ANTS VIII, Lecture Notes in Computer Science, vol. 5011, Springer, 2008, pp. 312–326.
29. Anthony W. Knapp, *Basic algebra*, Birkhäuser, 2006.
30. Donald E. Knuth, *Two notes on notation*, American Mathematical Monthly **99** (1992), no. 5, 403–422.
31. Paul Koosis, *The logarithmic integral I*, Cambridge University Press, 1998.
32. Alain Kraus, *Une remarque sur les points de torsion des courbes elliptiques*, C. R. Acad. Sci. Paris Sér. I Math. **321** (1995), no. 9, 1143–1146.
33. Masato Kuwata, *Quadratic twists of an elliptic curve and maps from a hyperelliptic curve*, Mathematics Journal of Okayama University **47** (2005), 85–97.
34. Serge Lang, *Introduction to modular forms*, Springer, 1976.
35. _____, *Algebraic number theory*, second ed., Springer, 1994.
36. Serge Lang and Hale Trotter, *Frobenius distributions in GL₂ extensions*, Lecture Notes in Mathematics, vol. 504, Springer-Verlag, 1976.
37. Michael Larsen, *The normal distribution as a limit of generalized Sato-Tate measures*, early 1990s, preprint, http://mlarsen.math.indiana.edu/~larsen/papers/gauss.pdf.
38. Barry Mazur, *Finding meaning in error terms*, Bulletin of the American Mathematical Society **45** (2008), no. 2, 185–228.
39. B.J.J. Moonen and Yu. G. Zarhin, *Hodge classes on abelian varieties of low dimension*, Mathematische Annalen **315** (1999), 711–733.
40. David Mumford, *A note of Shimura's paper "Discontinuous groups and abelian varities"*, Mathematische Annalen **181** (1969), 345–351.
41. Eric M. Rains, *High powers of random elements of compact Lie groups*, Probability Theory and Related Fields **107** (1997), 219–241.
42. _____, *Images of eigenvalue distributions under power maps*, Probability Theory and Related Fields **125** (2003), no. 4, 522–538.
43. Fernando Rodriguez-Villegas, *Explicit models of genus 2 curves with split CM*, Algorithmic Number Theory Symposium–ANTS IV, Lecture Notes in Computer Science, vol. 1838, Springer-Verlag, 2000, pp. 505–514.
44. Jean-Pierre Serre, *Groupes linéaires modulo p et points d'ordre fini des variétés abéliennes*, 1986, Collège de France course notes by Eva Bayer-Fluckiger, http://alg-geo.epfl.ch/~bayer/files/Serre-cours.pdf.
45. _____, *Propriétés conjecturales des groupes de Galois motiviques et des représentations l-adiques*, Motives (Seattle, WA, 1991), Proceedings of Symposia in Pure Mathematics, vol. 55, American Mathematical Society, 1994, pp. 377–400. MR 1265537 (95m:11059)
46. _____, *Lettre a Marie-France Vigneras*, Oeuvres, Collected Papers, Springer, 2000, pp. 38–55.
47. N. J. A. Sloane, *The on-line encyclopedia of integer sequences*, 2007, www.research.att.com/~njas/sequences/.
48. N.P. Smart, *S-unit equations, binary forms and curves of genus 2*, Proceedings of the London Mathematical Society **75** (1997), no. 2, 271–307.
49. Jerome Spanier and Keith B. Oldham, *An atlas of functions*, Hemisphere, 1987.
50. William A. Stein and Mark Watkins, *A database of elliptic curves–first report*, Algorithmic Number Theory Symposium–ANTS V, Lecture Notes in Computer Science, vol. 2369, Springer, 2002, pp. 267–275.
51. Andrew V. Sutherland, *Notes on the Hasse-Witt matrix and the density of zero Frobenius traces*, 2008, in preparation.
52. J. T. Tate, *Algebraic cycles and poles of zeta functions*, Arithmetical Algebraic Geometry, Harper and Row, 1965.
53. Richard Taylor, *Automorphy for some ℓ-adic lifts of automorphic mod ℓ Galois representations II*, 2006, preprint, http://www.math.harvard.edu/~rtaylor/twugk6.pdf.

54. Paul van Wamelen, *Examples of genus two CM curves defined over the rationals*, Mathematics of Computation **68** (1999), no. 225, 307–320.

55. _____, *Proving that a genus 2 curve has complex multiplication*, Mathematics of Computation **68** (1999), no. 228, 1663–1677.

56. André Weil, *Numbers of solutions of equations in finite fields*, Bulletin of the American Mathematical Society **55** (1949), 497–508.

57. Hermann Weyl, *Classical groups*, second ed., Princeton University Press, 1946.

58. Yuri G. Zarhin, *Hyperelliptic jacobians with complex multiplication*, Mathematical Research Letters **7** (2000), no. 1, 123–132.

DEPARTMENT OF MATHEMATICS, MASSACHUSETTS INSTITUTE OF TECHNOLOGY, 77 MASSACHUSETTS AVENUE, CAMBRIDGE, MA 02139

E-mail address: (kedlaya|drew)@math.mit.edu

Contemporary Mathematics
Volume **487**, 2009

On Special Finite Fields

Florian Luca and Igor E. Shparlinski

ABSTRACT. It has been shown by J.-P. Serre that the largest possible number of $\mathbb{F}_q$-rational points on curves of small genus over the finite field $\mathbb{F}_q$ of q elements depends on the divisibility property $p \mid \lfloor 2q^{1/2} \rfloor$, where p is the characteristic of $\mathbb{F}_q$. In this paper, we obtain upper and lower bounds on the number of prime powers $q \leq Q$ which satisfy this condition and which are not perfect squares.

1. Introduction

In has been discovered by J.-P. Serre [**8, 9, 10**] that the largest possible number of $\mathbb{F}_q$-rational points on curves of small genus over a finite field $\mathbb{F}_q$ of q elements depends on whether the divisibility property

$$\tag{1} \left\lfloor 2q^{1/2} \right\rfloor \equiv 0 \pmod{p}$$

holds, where p the characteristic of $\mathbb{F}_q$. Certainly, if $q = p^{2k}$ is a perfect square, then relation (1) is trivial. However, the case of powers with odd exponents $q = p^{2k+1}$ is less understood.

Here, we obtain upper and lower bounds on the number $N(Q)$ of such prime powers $q = p^{2k+1} \leq Q$, where p is prime and $k \geq 1$ is an integer.

We recall that $A \ll B$ and $B \gg A$ are both equivalent to $A = O(B)$ (throughout the paper, the implied constants in symbols 'O' , '$\ll$' and '$\gg$' are absolute).

THEOREM 1. *For any sufficiently large Q, we have*

$$(\log Q)^{1/2} \ll N(Q) \ll Q^{17/140}.$$

Our upper bound is based on a combination of some rather deep tools, such as a bound on the number of integral points on *Mordell curves* due to H. A. Helfgott and A. Venkatesh [**6**], and an estimate on the number of integer points close to a curve which is due to M. Filaseta and O. Trifonov [**5**].

Our lower bound is based on some ideas of A. Dubickas and A. Novikas [**2**], combined with a result of D. H. Bailey, J. M. Borwein, R. E. Crandall and C. Pomerance [**1**].

1991 *Mathematics Subject Classification.* 11B50, 11G20, 11J25.

During the preparation of this paper, F. L. was supported in part by Grant SEP-CONACyT 46755, and I. S. by ARC Grant DP0556431.

2. Necessary Tools

2.1. Integral Points on Mordell Curves. Let us consider the family of Mordell elliptic curves

$$(2) \qquad Y^2 = X^3 + D,$$

where D is a nonzero integer.

Let $I(D)$ denote the number of integral points $(x, y) \in \mathbb{Z}^2$ on this curve. By the remark at the end of [6, Section 4], we have $I(D) = O\left(|D|^{0.22378}\right)$. However, using in the same argument the bound of J. S. Ellenberg and A. Venkatesh [4, Proposition 2] instead of [6, Theorem 4.2], one obtains the following stronger estimate.

LEMMA 1. *For every nonzero integer D, we have*

$$I(D) \ll |D|^{0.17}.$$

2.2. Integer Points Close to Curves. We make use of a special case of the following result of M. Filaseta and O. Trifonov [5, Theorem 6]. Let $\|\xi\|$ denote the distance between ξ and the closest integer. For a real function f and an integer $U > 1$, let $S_f(U, \delta)$ be the number of integers $u \in [U, 2U]$ with

$$\|f(u)\| \le \delta.$$

LEMMA 2. *Let $U > 1$ and $r \ge 3$ be integers and let $T > 0$ be real. Suppose that $f(x)$ is a function with at least r derivatives and such that*

$$TU^{-j} \ll f^{(j)}(x) \ll TU^{-j}$$

for all $x \in [U, 2U]$ and $j = r - 2, r - 1, r$. Then there is a positive constant κ depending only on r and the function $f(x)$, such that for any positive real number δ with

$$\delta < \kappa \min \left\{ TU^{-r+2}, T^{(r-4)/(r-2)}U^{-r+3} + TU^{-r+1} \right\},$$

we have

$$S_f(U, \delta) \ll T^{2/(r^2+r)}U^{(r-1)/(r+1)} + U\delta^{2/(r^2-3r+2)} + U\left(\delta TU^{1-r}\right)^{1/(r^2-3r+4)}.$$

3. Proof of the Upper Bound

3.1. Preparations. Let $q = p^{2k+1}$ satisfying the divisibility property (1). Then

$$2p^{(2k+1)/2} = mp + \alpha$$

holds with some integer m and a real number $\alpha \in [0, 1)$. Therefore,

$$(3) \qquad 2p^{(2k-1)/2} = m + \alpha/p.$$

Squaring the above relation we get

$$(4) \qquad 4p^{2k-1} = m^2 + 2m\alpha/p + (\alpha/p)^2.$$

Note also that

$$(5) \qquad 1 \le m \le 2p^{(2k-1)/2}.$$

We now denote by $N_{2k+1}(Q)$ the number of primes $p \le Q^{1/(2k+1)}$ satisfying the divisibility property (1). We shall treat separately the cases when $k = 1$, 2, 3 and $k \ge 4$.

3.2. Third Powers. If $k = 1$, we then derive from the inequality (5) that $m \leq 2p^{1/2}$. Thus,

$$2m\alpha/p + (\alpha/p)^2 < 2m/p + 1/p^2 \leq 4p^{-1/2} + 1/p^2 \leq 1,$$

where the last inequality above holds for all $p \geq 17$. The equation (4) now implies that $4p = m^2$, which is impossible. One also checks by hand that the divisibility relation (1) does not hold for any $p \leq 13$ with $k = 1$. Thus, there are no cubes of primes $q = p^3$ which satisfy relation (1), and we therefore obtained that

(6) $$N_3(Q) = 0.$$

3.3. Fifth Powers. If $k = 2$, then the equation (4) and the inequality (5) imply that

(7) $$4p^3 = m^2 + a$$

holds with some positive integer

$$a = O\left(m/p\right) = O\left(p^{1/2}\right) = O\left(Q^{1/10}\right).$$

Every integer solution (p, m) to the equation (7) leads to the integral point with coordinates $x = 4p$, $y = 4m$ on the elliptic curve

$$Y^2 = X^3 - 16a.$$

However, by Lemma 1, we know that the above curve has at most $O\left(a^{0.17}\right) = O\left(Q^{0.017}\right)$ integral points. Thus, there are at most

(8) $$N_5(Q) = O\left(Q^{1/10+0.017}\right) = O\left(Q^{0.117}\right)$$

fifth powers of primes $q = p^5$ which satisfy the divisibility relation (1).

3.4. Seventh Powers. For each of $\nu = 1, 2, \ldots$, we look at the primes p in the interval $[2^\nu, 2^{\nu+1}]$ which satisfy the divisibility relation (1) with $k = 3$. For such primes, we conclude from estimate (3) that the fractional parts $\{2p^{5/2}\}$ satisfy the bound

(9) $$\{2p^{5/2}\} \leq 1/p \leq 2^{-\nu}.$$

We now discard the condition that p is prime and look at all integers $u \in [2^\nu, 2^{\nu+1}]$ with $\{2u^{5/2}\} \leq 2^{-\nu}$. To count them, we apply Lemma 2 to the function $f(x) = 2x^{5/2}$ with the choices of parameters

$$r = 4, \qquad U = 2^\nu, \qquad T = U^{5/2}, \qquad \delta = U^{-1}.$$

We remark that with the above parameters we have

$$\min\left\{TU^{-r+2}, T^{(r-4)/(r-2)}U^{-r+3} + TU^{-r+1}\right\}$$
$$= \min\left\{U^{1/2}, U^{-1} + U^{-1/2}\right\} \gg U^{-1/2},$$

therefore the necessary condition of Lemma 2 is satisfied provided that ν is large enough. Thus, we get

$$S_f(2^\nu, 2^{-\nu}) \ll U^{17/20} + U^{2/3} + U^{13/16} \ll U^{17/20} = 2^{17\nu/20}.$$

In particular, defining the integer J by the conditions

$$2^{J-1} \leq Q^{1/7} \leq 2^J,$$

we derive that

(10) $$N_7(Q) \le \sum_{\nu=0}^{J} S_f(2^\nu, 2^{-\nu}) \ll 2^{17J/20} \ll Q^{17/140}.$$

3.5. Higher Powers. Trivially, we have $N_{2k+1}(Q) \le Q^{1/(2k+1)}$ and also $N_{2k+1}(Q) = 0$ for $k \ge (\log Q)/(2\log 2)$. Therefore,

$$\sum_{k\ge 4} N_{2k+1}(Q) = N_9(Q) + \sum_{k\ge 5} N_{2k+1}(Q)$$

(11)

$$= O\left(Q^{1/9} + Q^{1/11} \log Q\right) = O\left(Q^{1/9}\right).$$

3.6. Finishing the proof. Combining the bounds (6), (8), (10) and (11), and remarking that

$$\frac{17}{140} > 0.117 > \frac{1}{9},$$

we reach the upper bound of Theorem 1.

4. Proof of the Lower Bound

By a result of A. Dubickas and A. Novikas [**2**, Theorem 1], we see that $\lfloor 2^k \sqrt{2} \rfloor$ is even for infinitely many positive integers k. Thus $N(Q) \to \infty$ as $Q \to \infty$. In fact the argument of [**2**], combined with a result of D. H. Bailey, J. M. Borwein, R. E. Crandall and C Pomerance [**1**] leads to a more explicit lower bound on $N(Q)$. Indeed, the proof from [**2**] of the fact that $\lfloor 2^k \sqrt{2} \rfloor$ is even infinitely often is as follows. Since $\sqrt{2}$ is irrational, infinitely many of its binary digits are zero. Hence, the statement. To make it explicit, we note that in [**1**] it is shown that if α is a quadratic irrationality, then there is a constant $c > 0$ such that for a large integer K, among the first K binary digits of α there are at least $cK^{1/2}$ of them which are zero. Taking $K = \lfloor (\log Q/\log 2) - 1/2 \rfloor$, we get the lower bound of Theorem 1.

5. Final Remarks

We remark that the bound (8) can be improved as

$$N_5(Q) = O\left(Q^{1/10} \log Q\right),$$

if one uses instead the estimate $O\left(L^{1/2} \log L\right)$ on the number of solutions to the inequality $\{(rn^3)^{1/2}\} \le cL^{-1/2}$ in positive integers $n \le L$. Here, r is a positive rational number and c is an arbitrary positive real number. The proof of this estimate has been sketched by N. D. Elkies [**3**] (in fact, the result there is more general). Unfortunately, this does not improve the final result, but obtaining its generalization for $\{(rn^5)^{1/2}\}$ may be a way of improving upon estimate (10), and thus upon the exponent on Q in Theorem 1.

Certainly, the result of M. Filaseta and O. Trifonov [**5**, Theorem 6] can be used to give nontrivial estimates for $N_{2k+1}(Q)$ for $k \ge 4$ as well. Thus, if necessary, the bound (11) can be easily improved. One can appeal to a variety of other results of this type (see, for example, [**7**] and references therein).

The same argument as used in Section 4 shows that it is natural to expect that

$$N(Q) \sim A \log Q,$$

where

$$A = \sum_p \frac{1}{p \log p}.$$

Acknowledgements

The authors are grateful to Akshay Venkatesh for very useful comments and, in particular, for pointing out that the combination of the results from [4] and [6] yields Lemma 1.

References

[1] D. H. Bailey, J. M. Borwein, R. E. Crandall and C. Pomerance, 'On the binary expansions of algebraic numbers', *J. Théorie des Nombres Bordeaux* **16** (2004), 487-518.

[2] A. Dubickas and A. Novikas, 'Integer parts of powers of rational numbers', *Math. Zeitschr.* **251** (2005), 635–648.

[3] N. D. Elkies, 'Rational points near curves and small nonzero $|x^3 - y^2|$ via lattice reduction', *Proc. ANTS-4*, Lecture Notes in Comp. Sci., vol. 1838, Springer-Verlag, Berlin, 2000, 33-63.

[4] J. S. Ellenberg and A. Venkatesh, 'Reflection principles and bounds for class group torsion', *Int. Math. Res. Notices* **2007** (2007), Article ID rnm002, 1–18.

[5] M. Filaseta and O. Trifonov, 'The distribution of fractional parts with applications to gap results in number theory', *Proc. London Math. Soc.* **73** (1996), 241–278.

[6] H. A. Helfgott and A. Venkatesh, 'Integral points on elliptic curves and 3-torsion in class groups', *J. Amer. Math. Soc.* **19** (2006), 527–550.

[7] M. N. Huxley and P. Sargos, 'Points entiers au voisinage d'une courbe plane de classe C^n, II', *Funct. Approx. Comment. Math.* **35** (2006), 91–115.

[8] J.-P. Serre, 'Sur le nombre des points rationnels d'une courbe algébrique sur un corps fini', *C.R. Acad. Sci. Paris, Sér. I Math.* **296** (1983), 397–402.

[9] J.-P. Serre, 'Nombres de points des courbes algébriques sur $\mathbb{F}_q$', *Sém. Théorie Nombres de Bordeaux* **1982/83**, exp. no. 22, 1–8.

[10] J.-P. Serre, 'Appendix to: K. Lauter, The maximum or minimum number of rational points on genus three curves over finite fields', *Compositio Math.* **134** (2002), 87–111.

Instituto de Matemáticas, Universidad Nacional Autonoma de México, C.P. 58089, Morelia, Michoacán, México

E-mail address: `fluca@matmor.unam.mx`

Department of Computing, Macquarie University, North Ryde, NSW 2109, Australia

E-mail address: `igor@ics.mq.edu.au`

Contemporary Mathematics
Volume **487**, 2009

Borne sur le degré des polynômes
presque parfaitement non-linéaires

François Rodier

ABSTRACT. The vectorial Boolean functions are employed in cryptography to build block coding algorithms. An important criterion on these functions is their resistance to the differential cryptanalysis. Nyberg defined the notion of almost perfect non-linearity (APN) to study resistance to the differential attacks. Up to now, the study of APN functions was especially devoted to the power functions. Recently some people showed that certain quadratic polynomials were APN.

Here, we will give a criterion so that a function is not almost perfectly non-linear.

Janwa showed, by using Weil's bound, that certain cyclic codes could not correct two errors. Canteaut showed by using the same method that the power functions were not APN for a too large value of the exponent. We use Lang and Weil's bound and a result of Deligne (or more exactly an improvement given by Ghorpade and Lachaud) on the Weil's conjectures about surfaces on finite fields to generalize this result to many polynomials. We show that these polynomials cannot be APN if their degrees are too large. We study many examples, and make some computation, showing that one does not get any new APN function for a polynomial of degree not greater than 9.

1. Introduction

Les fonctions booléennes vectorielles sont employées en cryptographie pour construire des algorithmes de chiffrement par bloc. Un critère important sur ces fonctions est une résistance élevée à la cryptanalyse différentielle. Nyberg [**20**] a défini la notion de la non-linéarité presque parfaite (APN) qui caractérise les fonctions qui ont la meilleure résistance aux attaques différentielles.

Une fonction booléenne vectorielle f à m variables est d'autant plus résistante aux attaques différentielles que la valeur de δ est plus petite où

$$\delta = \sup_{\alpha \neq 0, \beta} \#\{x \in \mathbb{F}_2^m \mid f(x + \alpha) + f(x) = \beta\}.$$

Les fonctions APN sont celles qui atteignent la plus petite valeur de δ, c'est-à-dire 2.

2000 *Mathematics Subject Classification*. Primary 94A60, Secondary 11T71, 14G50, 14Q10.

Jusqu'ici, l'étude des fonctions APN a surtout porté sur les fonctions puissances (voir par exemple [**6**, **11**, **12**, **13**]). Récemment elle s'est étendue à d'autres fonctions, en particulier aux polynômes quadratiques (Pott, Carlet et al.[**14**, **2**]) ou à des polynômes sur des petits corps (Dillon [**10**]).

Divers auteurs (Berger, Canteaut, Charpin, Laigle-Chapuy [**1**], Byrne et McGuire [**3**], Jedlicka [**16**] ou Voloch [**22**]) ont démontré l'impossibilité pour une fonction d'être APN dans certains cas.

Nous montrons ici que pour beaucoup de fonctions polynomiales APN f sur $\mathbb{F}_{2^m}$, le nombre m est borné par une expression dépendant du degré de f.

Nous avons recours pour cela à une méthode déjà utilisée par Janwa qui a montré, en employant la borne de Weil, que certains codes cycliques ne pouvaient pas corriger deux erreurs [**18**]. Canteaut a montré par la même méthode que certaines fonctions puissances n'étaient pas APN pour une valeur trop grande de l'exposant [**5**]. Nous avons pu généraliser ce résultat à tous les polynômes en appliquant des résultats de Lang-Weil et de Deligne (ou plus exactement des améliorations, due à Ghorpade et Lachaud) sur les conjectures de Weil. Nous terminons cet article par l'étude des polynômes de petit degré.

Je remercie Felipe Voloch pour ses commentaires sur une ancienne version de cet article et Gregor Leander qui m'a aidé à faire des recherches exhaustives.

2. Préliminaires

Pour $a \in \mathbb{F}_2^{m*}$, l'équation $f(x+a) + f(x) = b$ a évidemment un nombre pair de solutions, et il existe $a \in \mathbb{F}_2^{m*}$ et b dans $\mathbb{F}_2^m$ tel que cette équation ait au moins une solution. D'où la définition suivante.

DÉFINITION 2.1. Une fonction $f : \mathbb{F}_2^m \longrightarrow \mathbb{F}_2^m$ est APN si et seulement si pour tout $a \in \mathbb{F}_2^{m*}$ et tout $b \in \mathbb{F}_2^m$, l'équation

$$f(x+a) + f(x) = b$$

n'a au plus que 2 solutions.

2.1. Polynômes équivalents. Posons $q = 2^m$. Au lieu de fonctions dans $\mathbb{F}_2^m$, on parlera de polynômes dans $\mathbb{F}_q$, ce qui est équivalent pour les polynômes de degré au plus $2^m - 1$.

Un polynôme q-affine est un polynôme dont les monômes sont de degré 0 ou une puissance de 2. Les propositions suivantes sont claires.

PROPOSITION 2.1. *La classe des fonctions APN est invariante par addition d'un polynôme q-affine.*

Nous prendrons désormais pour f une application polynomiale de $\mathbb{F}_{2^m}$ dans lui-même qui n'ait pas de termes de degré une puissance de 2 ni de terme constant.

PROPOSITION 2.2. *Pour tout a, b et c dans $\mathbb{F}_q$ le polynôme $cf(ax+b)$ est APN si et seulement si le polynôme f l'est.*

2.2. L'équivalence au sens de Carlet-Charpin-Zinoviev. Carlet, Charpin et Zinoviev ont défini une relation d'équivalence entre fonctions booléennes [**8**]. Pour une fonction f de $\mathbb{F}_2^m$ dans lui-même on note G_f le graphe de la fonction f :

$$G_f = \{(x, f(x)) \mid x \in \mathbb{F}_2^m\}.$$

DÉFINITION 2.2. On dit que les fonctions $f, f' : \mathbb{F}_2^m \longrightarrow \mathbb{F}_2^m$ sont équivalentes au sens de Carlet-Charpin-Zinoviev (on parle aussi de CCZ équivalence) s'il existe une permutation linéaire $L : \mathbb{F}_2^{2m} \longrightarrow \mathbb{F}_2^{2m}$ telle que $L(G_f) = G_{f'}$.

D'après [**8**] si f et f' sont CCZ équivalentes, alors f est APN si et seulement si f' l'est.

2.3. Les monômes APN connus. Ils sont décrits dans la monographie de Carlet [**7**] où on trouvera les références des démonstrations.

Les fonctions $f(x) = x^d$ suivantes sont APN sur $\mathbb{F}_{2^m}$, où d est donné par :
- $d = 2^h + 1$ où $pgcd(h, m) = 1$ (fonctions de Gold) ;
- $d = 2^{2h} - 2^h + 1$ où $pgcd(h, m) = 1$ (fonctions de Kasami) ;
- $d = 2^{(m-1)/2} + 3$ avec m impair (fonctions de Welch) ;
- $d = 2^{(m-1)/2} + 2^{(m-1)/4} - 1$, où $m \equiv 1 \mod 4$,
  $d = 2^{(m-1)/2} + 2^{(3m-1)/4} - 1$, où $m \equiv 3 \mod 4$ (fonctions de Niho) ;
- $d = 2^m - 2$, pour m impair ;
- $d = 2^{4m/5} + 2^{3m/5} + 2^{2m/5} + 2^{m/5} - 1$, où m est divisible par 5 (fonctions de Dobbertin).

Les seules fonctions connues qui sont APN pour une infinité de valeurs de m et où le degré est indépendant de m sont les fonctions de Gold et de Kasami.

3. Caractérisation des fonctions APN

PROPOSITION 3.1. *La fonction* $f : \mathbb{F}_{2^m} \longrightarrow \mathbb{F}_{2^m}$ *est APN si et seulement si la surface affine*

$$f(x_0) + f(x_1) + f(x_2) + f(x_0 + x_1 + x_2) = 0$$

a tous ses points rationnels contenus dans la surface $(x_0 + x_1)(x_2 + x_1)(x_0 + x_2) = 0$.

DÉMONSTRATION. Pour que la fonction $f : \mathbb{F}_{2^m} \longrightarrow \mathbb{F}_{2^m}$ soit APN il faut et il suffit que pour tout $\alpha \in \mathbb{F}_2^m$, $\alpha \neq 0$ et pour tout $\beta \in \mathbb{F}_2^m$,

$$\#\{x_0 \in \mathbb{F}_2^m : f(x_0) + f(x_1) = \beta \ , \ x_0 + x_1 = \alpha\} \leq 2$$

c'est-à-dire que pour tout $\alpha \in \mathbb{F}_2^m$, $\alpha \neq 0$ et pour tout $\beta \in \mathbb{F}_2^m$, il n'y ait pas 4 éléments distincts x_0, x_1, x_2, x_3 de $\mathbb{F}_2^m$ qui vérifient

$$\begin{cases} x_0 + x_1 = \alpha, f(x_0) + f(x_1) &=& \beta \\ x_2 + x_3 = \alpha, f(x_2) + f(x_3) &=& \beta \end{cases}$$

Cela revient à dire qu'il n'y a pas 3 éléments distincts x_0, x_1, x_2 de $\mathbb{F}_2^m$ qui vérifient

$$f(x_0) + f(x_1) + f(x_2) + f(x_0 + x_1 + x_2) = 0$$

autrement dit que la surface affine

$$f(x_0) + f(x_1) + f(x_2) + f(x_0 + x_1 + x_2) = 0$$

a tous ses points rationnels contenus dans la surface $(x_0 + x_1)(x_2 + x_1)(x_0 + x_2) = 0$. $\square$

Avant d'énoncer un corollaire, remarquons que le polynôme $f(x_0) + f(x_1) + f(x_2) + f(x_0 + x_1 + x_2)$ est divisible par $(x_0 + x_1)(x_2 + x_1)(x_0 + x_2)$, par conséquent le quotient

$$\frac{f(x_0) + f(x_1) + f(x_2) + f(x_0 + x_1 + x_2)}{(x_0 + x_1)(x_2 + x_1)(x_0 + x_2)}$$

définit bien un polynôme de degré $d-3$ où d est le degré de f. On vérifie facilement que ce polynôme est nul si et seulement si f est un polynôme q-affine.

COROLLAIRE 3.1. *Si l'application polynomiale f (de degré $d \geq 5$) est APN et si la surface affine X*

$$\frac{f(x_0) + f(x_1) + f(x_2) + f(x_0 + x_1 + x_2)}{(x_0 + x_1)(x_2 + x_1)(x_0 + x_2)} = 0$$

est absolument irréductible, alors la surface projective correspondante $\overline{X}$ admet au plus $4((d-3)q+1)$ points rationnels, où d est le degré de f et $q = 2^m$.

DÉMONSTRATION. Si f est APN, alors f n'est pas un polynôme q-affine et l'équation ci-dessus définit bien une surface de degré plus grand que 2. Si la surface $\overline{X}$ contenait le plan $x_0 + x_1 = 0$, elle serait égale à ce plan puisque la surface $\overline{X}$ est irréductible et elle serait de degré 1, ce qui est contraire à l'hypothèse. Par conséquent, elle coupe le plan $x_0 + x_1 = 0$ suivant une courbe de degré $d-3$. Cette courbe admet au plus $(d-3)q+1$ points rationnels d'après Serre [21]. De même pour le plan à l'infini.

Si f est APN, la surface $\overline{X}$ n'a pas d'autres points rationnels que ceux de la surface $(x_0 + x_1)(x_2 + x_1)(x_0 + x_2) = 0$, qui est réunion du plan $x_0 + x_1 = 0$ et de ses symétriques, ou du plan à l'infini. Donc elle admet au plus $4((d-3)q+1)$ points rationnels.   □

4. Borne inférieure pour le degré d'un polynôme APN

4.1. Une première borne.

THEOREM 4.1. *Soit f une application polynomiale de $\mathbb{F}_{2^m}$ dans lui-même, d son degré. Supposons que la surface X d'équation affine*

(4.1) $$\frac{f(x_0) + f(x_1) + f(x_2) + f(x_0 + x_1 + x_2)}{(x_0 + x_1)(x_2 + x_1)(x_0 + x_2)} = 0$$

soit absolument irréductible. Alors, si $d < 0,45 q^{1/4} + 0,5$ et $d \geq 9$, f n'est pas APN.

DÉMONSTRATION. D'une amélioration d'un résultat de Lang et Weil [19] par Ghorpade et Lachaud [15, section 11], on déduit

$$|\overline{X}(\mathbb{F}_{2^m}) - q^2 - q - 1| \leq (d-4)(d-5)q^{3/2} + 18d^4 q.$$

D'où

$$\overline{X}(\mathbb{F}_{2^m}) \geq q^2 + q + 1 - (d-4)(d-5)q^{3/2} - 18d^4 q.$$

Par conséquent, si $q^2 + q + 1 - (d-4)(d-5)q^{3/2} - 18d^4 q > 4((d-3)q+1)$, alors $\overline{X}(\mathbb{F}_{2^m}) > 4((d-3)q+1)$, et donc f n'est pas APN.

Cette condition s'écrit

$$q - (d-4)(d-5)q^{1/2} + (-18d^4 - 4d + 13) - 3/q > 0$$

La condition est vérifiée pour $q^{1/2} > 13,51 - 5d + 4,773d^2$ si $d \geq 2$. Ou encore pour $d < 0,45 q^{1/4} + 0,5$ et $d \geq 9$.   □

4.1.1. *Irréductibilité de X.* La proposition suivante donne un critère pour que la surface X soit irréductible.

PROPOSITION 4.1. *Soit f une application polynomiale de $\mathbb{F}_{2^m}$ dans lui-même, d son degré. Supposons que la courbe X_∞ d'équation*

$$\frac{x_0^d + x_1^d + x_2^d + (x_0 + x_1 + x_2)^d}{(x_0 + x_1)(x_2 + x_1)(x_0 + x_2)} = 0$$

soit absolument irréductible. Alors la surface X d'équation (4.1) est absolument irréductible.

DÉMONSTRATION. L'intersection X_∞ de la surface $\overline{X}$ avec le plan à l'infini a comme équation

$$\frac{x_0^d + x_1^d + x_2^d + (x_0 + x_1 + x_2)^d}{(x_0 + x_1)(x_2 + x_1)(x_0 + x_2)} = 0$$

puisque x^d est la composante de degré d du polynôme $f(x)$. Puisque la courbe X_∞ est absolument irréductible il en va de même de la surface $\overline{X}$ donc de X. $\qquad\square$

Janwa, McGuire et Wilson [17] ont étudié la courbe X_∞ et ont déduit un certain nombre de cas où elle est absolument irréductible.

PROPOSITION 4.2. *La courbe X_∞ est absolument irréductible pour les valeurs de $d \equiv 3 \mod 4$ et pour les valeurs de $d \equiv 5 \mod 8$ et $d > 13$.*

REMARQUE 4.2. La proposition 4.1 donne une condition suffisante mais pas nécessaire. Voir les exemples dans la section 5.

On peut aussi regarder l'intersection de X avec le plan $x_2 = a$ ou avec le plan $x_1 + x_2 = a$.

PROPOSITION 4.3. *Soit f une application polynomiale de $\mathbb{F}_{2^m}$ dans lui-même, d son degré. Supposons que la courbe X_a d'équation*

$$\frac{f(x_0) + f(x_1) + f(a) + f(x_0 + x_1 + a)}{(x_0 + x_1)(a + x_1)(x_0 + a)} = 0$$

soit absolument irréductible. Alors la surface X d'équation (4.1) est absolument irréductible.

PROPOSITION 4.4. *Soit f une application polynomiale de $\mathbb{F}_{2^m}$ dans lui-même, d son degré. Supposons que la courbe Y_a d'équation*

$$F_a(x, y) = \frac{f(x_0) + f(x_1) + f(x_1 + a) + f(x_0 + a)}{(x_0 + x_1)(x_0 + x_1 + a)} = 0$$

soit absolument irréductible. Alors la surface X d'équation (4.1) est absolument irréductible.

C'est la même courbe que Voloch considère dans [22].

4.2. Une deuxième borne. Sous certaines conditions, on peut obtenir une borne meilleure pour la dimension.

THEOREM 4.3. *Soit f une application polynomiale de $\mathbb{F}_{2^m}$ dans lui-même, d son degré. Supposons que la surface projective $\overline{X}$ associée à la surface X d'équation affine*

$$\frac{f(x_0) + f(x_1) + f(x_2) + f(x_0 + x_1 + x_2)}{(x_0 + x_1)(x_2 + x_1)(x_0 + x_2)} = 0$$

n'ait que des points singuliers isolés. Alors, si $d \geq 10$ et $d < q^{1/4} + 4$, f n'est pas APN.

DÉMONSTRATION. of dimension n and degree d

Notons $b_i(N, \delta)$ le $i^{\text{ème}}$ nombre de Betti d'une hypersurface non-singulière de degré δ dans un expace projectif de dimension N, c'est-à-dire la dimension de l'espace de cohomologie ℓ-adique de cette hypersurface à support compact, et $b_i'(N, \delta)$ le $i^{\text{ème}}$ nombre de Betti primitif, égal à $b_i(N, \delta)$ pour i impair et à $b_i(N, \delta) - 1$ pour i pair.

D'après une amélioration d'un résultat de Deligne [**9**] par Ghorpade-Lachaud [**15**, corollaire 7.2], on en déduit

$$\begin{aligned} |X(\mathbb{F}_{2^m}) - q^2 - q - 1| &\leq b_1'(2, d-3)q^{3/2} + (b_2(3, d-3) + 1)q \\ &\leq (d-4)(d-5)q^{3/2} + (d^3 - 13d^2 + 57d - 82)q \end{aligned}$$

D'où

$$X(\mathbb{F}_{2^m}) \geq q^2 + q + 1 - (d-4)(d-5)q^{3/2} - (d^3 - 13d^2 + 57d - 82)q.$$

Par conséquent, si

$$q^2 + q + 1 - (d-4)(d-5)q^{3/2} - (d^3 - 13d^2 + 57d - 82)q > 4((d-3)q + 1),$$

alors $X(\mathbb{F}_{2^m}) > 4((d-3)q + 1)$, et donc f n'est pas APN. Cette condition s'écrit

$$q + (-d^2 + 9d - 20)q^{1/2} + (-d^3 + 13d^2 - 61d + 95) - \frac{2}{q} > 0.$$

La condition est vérifiée pour $q > d^4 - 16d^3 + 94d^2 - 228d + 173$ dès que $d \geq 6$. Ou encore pour $d < q^{1/4} + 4$ dès que $d \geq 10$.   □

4.2.1. *Non-singularité de X_∞.*

THEOREM 4.4. *Soit f une application polynomiale de $\mathbb{F}_{2^m}$ dans lui-même, d son degré. Supposons que la courbe X_∞ d'équation*

$$\frac{x_0^d + x_1^d + x_2^d + (x_0 + x_1 + x_2)^d}{(x_0 + x_1)(x_2 + x_1)(x_0 + x_2)} = 0$$

soit lisse. Alors la surface $\overline{X}$ n'a que des points singuliers isolés.

DÉMONSTRATION. La démonstration du théorème 4.1 montre que X_∞ est l'intersection de X avec l'hyperplan à l'infini. Puisque la courbe X_∞ est absolument irréductible il en va de même de la surface $\overline{X}$.

On peut déduire que la surface $\overline{X}$ est régulière en codimension 1 (c'est-à-dire n'a que des points singuliers isolés) si la courbe X_∞ est non singulière (cf. Ghorpade-Lachaud [**15**], corollaire 1.4).   □

Janwa, et Wilson [**18**] ont étudié la courbe X_∞ et ont déduit un certain nombre de cas où elle est non singulière.

PROPOSITION 4.5. *La courbe X_∞ est non singulière pour les valeurs de $d = 2l + 1$ où*

- *l est un entier impair tel qu'il existe un entier r avec $2^r \equiv -1 \mod l$.*
- *l est un nombre premier plus grand que 17 tel que l'ordre de 2 modulo l soit $(l-1)/2$.*

En particulier la première condition est satisfaite si l est un nombre premier congru à ± 3 modulo 8. Parmi les degrés inférieurs à 100, on trouve 7, 11, 19, 23, 27, 35, 39, 47, 51, 55, 59, 67, 75, 83, 95.

4.3. La surface X ne peut pas être lisse. On pourrait se poser la question de savoir si la surface $\overline{X}$ peut être lisse, ce qui améliorerait encore les bornes sur le degré des fonctions APN. Ce ne peut être le cas. Soit en effet

$$\phi(x_0, x_1, x_2) = \frac{f(x_0) + f(x_1) + f(x_2) + f(x_0 + x_1 + x_2)}{(x_0 + x_1)(x_2 + x_1)(x_0 + x_2)}$$

l'équation affine de la surface X. Les points singuliers de cette surface X sont sur les surfaces d'équation $\phi'_{x_i}(x_0, x_1, x_2) = 0$ où on note ϕ'_{x_i} la dérivée de ϕ par rapport à x_i.

LEMME 4.5. *Le polynôme $x_1 + x_2$ divise $\phi'_{x_0}(x_0, x_1, x_2) = 0$.*

DÉMONSTRATION. Il suffit de démontrer le lemme pour chaque monôme de f. Si on fait le changement de variable $x_1 + x_2 = t$, le polynôme ϕ devient, pour un monôme de degré r

$$\begin{aligned}
\psi(x_0, x_1, t) &= \frac{x_0^r + x_1^r + (x_1 + t)^r + (x_0 + t)^r}{(x_0 + x_1)(x_0 + x_1 + t)t} \\
&= \frac{r(x_1^{r-1} + x_0^{r-1}) + tg(x_0, x_1, t)}{(x_0 + x_1)(x_0 + x_1 + t)}
\end{aligned}$$

où $g(x_0, x_1, t)$ est un polynôme. En dérivant par rapport à x_0, cela donne

$$\psi'_{x_0}(x_0, x_1, t) = \frac{r(r-1)(x_0^r + x_0^{r-2}x_1^2) + tg_1(x_0, x_1, t)}{(x_0 + x_1)^2(t + x_0 + x_1)^2}$$

où $g_1(x_0, x_1, t)$ est un polynôme. Cela montre le lemme car $r(r-1) \equiv 0 \mod 2$. $\square$

Par conséquent l'intersection de la droite $x_0 = x_1 = x_2$ avec la surface affine X est formée de points singuliers de X.

Si la surface projective $\overline{X}$ ne rencontre pas la droite en des points à distance finie, elles ont un point commun d'ordre $d - 3$ à l'infini. L'équation $\phi(u, u, u) = 0$ n'a pas de solution, donc $\phi(u, u, u)$ est une constante non-nulle a_0. Soit

$$\Phi(x_0 : x_1 : x_2 : z)$$

le polynôme homogène annulateur de la surface projective $\overline{X}$. Il est égal à

$$z^{d-3}\phi(\frac{x_0}{z}, \frac{x_1}{z}, \frac{x_2}{z}),$$

donc on a $\Phi(1 : 1 : 1 : z) = z^{d-3}\phi(\frac{1}{z}, \frac{1}{z}, \frac{1}{z}) = z^{d-3}a_0$. La formule

$$x_0\Phi'_{x_0} + x_1\Phi'_{x_1} + x_2\Phi'_{x_2} + z\Phi'_z = (d-3)\Phi(x_0, x_1, x_2, z)$$

restreinte à la droite $x_0 = x_1 = x_2$ devient donc $z\Phi'_z = (d-3)z^{d-3}a_0$, d'où $\Phi'_z = (d-3)z^{d-4}a_0$ sur cette droite. Par conséquent $\Phi'_z(1,1,1,0) = 0$ et le point $(1,1,1,0)$ est bien singulier si $d \geq 5$.

5. Autres exemples

5.1. Binômes.

PROPOSITION 5.1. *Soient d et r deux entiers tels que $d > r \geq 3$, et soit la fonction $f(x) = x^d + ax^r$, avec $a \in \mathbb{F}_q$. Soit ϕ_s le polynôme*

$$\frac{x_0^s + x_1^s + x_2^s + (x_0 + x_1 + x_2)^s}{(x_0 + x_1)(x_2 + x_1)(x_0 + x_2)}$$

pour $s = d$ ou r. Supposons que $(\phi_d, \phi_r) = 1$ et que
– ou bien ϕ_d se décompose en facteurs distincts sur $\overline{\mathbb{F}}_{2^m}$ et $r \geq 5$;
– ou bien ϕ_r se décompose en facteurs distincts sur $\overline{\mathbb{F}}_{2^m}$.
Alors, si $d < 0,45q^{1/4} + 0,5$ et $d \geq 9$, f n'est pas APN.

DÉMONSTRATION. L'équation de la surface X associée à f est $\phi_d + a\phi_r = 0$. Le lemme suivant montre que cette surface est irréductible sous les hypothèses faites.   □

LEMME 5.1. *Soit $\Phi(x,y,z) \in \mathbb{F}_{2^m}[x,y,z]$ la somme de deux polynômes homogènes, c'est-à-dire $\Phi = \Phi_r + \Phi_d$ où Φ_i est homogène de degré i, et $r < d$. Supposons que $(\Phi_r, \Phi_d) = 1$ et que*
– ou bien Φ_r se décompose en facteurs distincts sur $\overline{\mathbb{F}}_{2^m}$ et $r \geq 1$;
– ou bien Φ_d se décompose en facteurs distincts sur $\overline{\mathbb{F}}_{2^m}$ et $r \geq 0$.
Alors Φ est absolument irréductible sur $\mathbb{F}_{2^m}$.

DÉMONSTRATION. On fait le même raisonnement que Byrne et McGuire dans l'article [3, lemme 2]. La preuve fonctionne aussi dans le cas des polynômes à trois variables. Elle fonctionne aussi dans le deuxième cas du lemme (Φ_d se décompose en facteurs distincts sur $\overline{\mathbb{F}}_{2^m}$ et $r \geq 0$).   □

Par exemple, la proposition 5.1 montre que le polynôme $x^{13} + ax^7$ avec $a \neq 0$ ne peut être APN que si $m \leq 19$, car le polynôme ϕ_7 est irréductible et ne divise pas ϕ_{13} d'après [18].

On a aussi la proposition suivante due pour l'essentiel à Voloch.

PROPOSITION 5.2. *Soit $f(x) = x^d + cx^r$, où $c \in \overline{\mathbb{F}}_2^*$, $r < d$ sont des entiers, non tous les deux pairs, et non plus une puissance de 2 et tels que $(d-1, r-1)$ soit une puissance de 2. Alors, si $d < 0,45q^{1/4} + 0,5$ et $d \geq 9$, f n'est pas APN.*

DÉMONSTRATION. Voloch a montré dans [22] que ces hypothèses impliquaient que la courbe

$$F_a(x,y) = \frac{f(x_0) + f(x_1) + f(x_1 + a) + f(x_0 + a)}{(x_0 + x_1)(x_0 + x_1 + a)}$$

est irréductible dans $\mathbb{F}_2[x, y, a]$.   □

REMARQUE 5.2. L'énoncé de cette proposition [22, Theorem 3] supposait que r et d étaient premiers entre eux. Voloch m'a communiqué que ce n'était pas nécessaire.

5.2. Polynômes de degré 3 ou 5. D'après la proposition 2.1, il suffit de regarder les polynômes de la forme $a_5 x^5 + a_3 x^3$. Ces polynômes sont combinaisons linéaires de plusieurs monômes de la forme x^{2^i+1} et ils ne peuvent pas être APN d'après [1] sauf si a_3 ou a_5 est nul, auquel cas ce sont des fonctions de Gold.

5.3. Polynômes de degré 6.

PROPOSITION 5.3. *Soit $f(x) = x^6 + a_5 x^5 + a_3 x^3$ un polynôme de degré 6. Le polynôme f est APN si et seulement si $a_3 = a_5 = 0$. Il est alors équivalent à une fonction de Gold.*

DÉMONSTRATION. Il est facile de vérifier que la surface X associée ne contient un hyperplan que si $a_3 = a_5^3$. Donc elle est absolument irréductible, sauf si $a_3 = a_5^3$. Dans ce cas-là, on vérifie que la surface X se décompose en 3 hyperplans d'équation $x_0 + x_2 + a_5 = 0$, $x_0 + x_1 + a_5 = 0$, $a_5 + x_1 + x_2 = 0$ et la fonction f n'est pas APN si $a_5 \neq 0$, d'après la proposition 3.1.

Si $a_3 \neq a_5^3$ et $a_3 = 0$, on a donc $f(x) = x^6 + a_5 x^5$. Comme $a_5 \neq 0$, on se ramène par une transformation affine (cf. proposition 2.2) à $a_5^{-6} f(a_5 x) = x^6 + x^5$. Des points de la surface affine X sont alors

$$\left(\frac{1}{\lambda(1+\lambda)}, \frac{\lambda^3}{\lambda(1+\lambda)}, 1 \right)$$

avec $\lambda \in \mathbb{F}_q - \mathbb{F}_2$. Si $m \geq 3$, ils ne sont pas sur la surface d'équation

$$(x_0 + x_1)(x_2 + x_1)(x_0 + x_2) = 0,$$

donc la fonction $f(x) = x^6 + a_5 x^5$ avec $a_5 \neq 0$ n'est jamais APN, d'après la proposition 3.1.

Si $a_3 \neq a_5^3$ et si, de plus, $a_3 \neq 0$, alors la surface $\overline{X}$ n'a que des singularités isolées et la fonction f ne peut être APN que si $m \leq 4$ d'après la démonstration du théorème 4.3. D'après [4], ces fonctions ne peuvent pas être APN. $\square$

5.4. Polynômes de degré 7.

PROPOSITION 5.4. *Soit f un polynôme de degré 7. Pour $m \geq 3$, le polynôme f ne peut être APN que s'il est CCZ-équivalent au polynôme x^7 sur $\mathbb{F}_{32}$. Il est alors équivalent à une fonction de Welsh.*

DÉMONSTRATION. Le théorème 4.4 et la proposition 4.5 montrent que la surface $\overline{X}$ n'a que des points singuliers isolés. D'après la démonstration du théorème 4.3 on en déduit que f ne peut être APN que si $m \leq 6$. La proposition s'ensuit, d'après [4] et avec une recherche exhaustive pour $m = 6$. $\square$

5.5. Polynômes de degré 9.

PROPOSITION 5.5. *Soit f un polynôme de degré 9. Le polynôme f ne peut être APN pour une infinité de m que s'il est équivalent au polynôme x^9. Il est alors CCZ-équivalent à une fonction de Gold. Autrement le polynôme f ne peut être APN que pour $m = 6$ et il est égal à une fonction $f = x^9 + a_6 x^6 + a_3 x^3$ ou à une fonction CCZ-équivalente.*

DÉMONSTRATION. On peut se limiter grâce aux propositions 2.1 et 2.2 aux polynômes

$$f(x) = x^9 + a_7 x^7 + a_6 x^6 + a_5 x^5 + a_3 x^3.$$

Si $a_7 \neq 0$ on se réduit, d'après les mêmes propositions au polynôme

$$f(x) = x^9 + x^7 + a_5 x^5 + a_3 x^3.$$

On vérifie alors aisément que la surface $\overline{X}$ n'a que des singularités isolées (voir ci-dessous 5.5.1). La fonction f ne peut être APN que si $m \leq 13$, d'après la démonstration du théorème 4.3.

Si $a_7 = 0$, on obtient

$$f(x) = x^9 + a_6 x^6 + a_5 x^5 + a_3 x^3.$$

Si, de plus, $a_6 = 0$, on obtient un polynôme $f(x) = x^9 + a_5 x^5 + a_3 x^3$ dont tous monômes ont des degrés de la forme $2^r + 1$. D'après [1] ce polynôme ne peut pas être APN, sauf si c'est un monôme. Si $a_6 \neq 0$, on peut réduire l'étude, grâce à la proposition 2.2, aux polynômes $f(x) = x^9 + a_6 x^6 + x^5 + a_3 x^3$ et $f = x^9 + a_6 x^6 + a_3 x^3$.

La fonction x^9 est une fonction de Gold. On montre ci-dessous (voir 5.5.2) que pour $f(x) = x^9 + a_6 x^6 + a_3 x^3$ avec $a_3 = a_6^2 \neq 0$ la fonction f ne peut être APN ; pour $f(x) = x^9 + a_6 x^6 + a_5 x^5 + a_3 x^3$ avec a_3, a_5 ou a_6 non nuls et $a_3 \neq a_6^2$, la fonction f ne peut être APN que si $m \leq 8$. Un examen exhaustif des cas restants démontre la proposition. La seule fonction APN supplémentaire que l'on trouve est une fonction $f = x^9 + a_6 x^6 + a_3 x^3$ pour $m = 6$, déjà obtenue par Dillon [10].   □

5.5.1. *Points singuliers de X associé à $x^9 + x^7 + a_5 x^5 + a_3 x^3$.* La surface $\overline{X}$ a pour équation projective $B_6 + B_4 z^2 + a_5 B_2 z^4 + a_3 z^6$ où les B_i sont des polynômes homogènes de degré i en x_0, x_1, x_2 et ce sont les équations des courbes X_∞ correspondantes aux degrés i. Suivant [18], les polynômes B_6 et B_2 sont des équations de réunions de droites passant par le point $(1 : 1 : 1)$, définies sur $\mathbb{F}_8$ et $\mathbb{F}_4$ et le polynôme B_4 est l'équation d'une courbe irréductible. D'après le calcul des dérivées de X, on étudie plusieurs cas.

Cas n° 1 : $x_0 = x_1 = x_2$. Les points de la surface X qui sont sur cette droite sont ceux qui vérifient $x_0^4 + a_3 z^4 = 0$ ou $z = 0$.

Cas n° 2 : $x_1 = x_2$. Les points singuliers de la surface X qui sont sur ce plan sont ceux qui vérifient

$$\begin{cases} x_0^4 + x_2^4 + x_0 x_2 z^2 + a_5 z^4 = 0, \\ x_0^6 + x_0^4 x_2^2 + x_0^2 x_2^4 + x_2^6 + x_0^2 z^2 + x_0^2 x_2^2 z^2 + x_2^2 z^2 + a_5 x_0^2 z^4 + a_5 x_2^2 z^4 + a_3 z^6 = 0, \end{cases}$$

où la deuxième équation est 2 fois la courbe $(x_0 + x_2)^3 + (x_0^2 + x_0 x_2 + x_2^2) z + a_5 (x_0 + x_2) z^2 + a_3 z^3 = 0$. Un changement de variable $s = x_0 + x_2$ donne, pour $z = 1$: $s^4 + x_0(x_0 + s) + a_5 = 0$ et $s^3 + s^2 + x_0(s + x_0) + a_5 s + a_3 = 0$ qui sont deux courbes elliptiques distinctes. Les deux courbes ne se coupent qu'en un nombre fini de points.

Cas n° 3 : $x_0 \neq x_1 \neq x_2 \neq x_0$.
Les points singuliers de la surface X sont ceux qui vérifient

$$\begin{cases} a_5 + x_1 x_2 + B_2^2(x_0, x_1, x_2) & = & 0, \\ a_5 + x_0 x_2 + B_2^2(x_0, x_1, x_2) & = & 0, \\ a_5 + x_1 x_0 + B_2^2(x_0, x_1, x_2) & = & 0, \end{cases}$$

donc, par soustraction, qui vérifient

$$\begin{cases} (x_1 + x_0) x_2 & = & 0, \\ (x_2 + x_1) x_0 & = & 0, \\ (x_0 + x_2) x_1 & = & 0. \end{cases}$$

D'après l'hypothèse, on a donc $x_0 = x_1 = x_2$, donc on n'a rien de mieux.

5.5.2. *Surface associée à* $x^9 + a_6 x^6 + a_5 x_5 + a_3 x^3$. Si $a_3 = a_5 = a_6 = 0$, la fonction x^9 est une fonction de Gold. Supposons que l'un de ces coefficients soit non nul.

La surface X a pour équation projective $B_6 + a_6 B_3 z^3 + a_5 B_2 z^4 + a_3 z^6$ où les B_i sont comme plus haut. L'intersection de la surface X avec le plan $x_2 = 0$ est la courbe Y d'équation affine

$$\phi(x_0, x_1, 0) = B_6(x_0, x_1, 0) + a_6 B_3(x_0, x_1, 0) + a_5 B_2(x_0, x_1, 0) + a_3$$

avec

$$
\begin{aligned}
B_6(x_0, x_1, 0) &= \prod_{\beta \in \mathbb{F}_8 - \mathbb{F}_4} (x_0 + \beta x_1), \\
B_3(x_0, x_1, 0) &= x_0 x_1 (x_0 + x_1), \\
B_2(x_0, x_1, 0) &= (x_0 + \alpha x_1)(x_0 + \alpha^2 x_1),
\end{aligned}
$$

où $\alpha \in \mathbb{F}_4 - \mathbb{F}_2$.

Les raisonnements de Byrne et McGuire pour prouver le lemme 2 dans [3] peuvent s'appliquer. On trouve ainsi que le seul cas où $\phi(x_0, x_1, 0)$ pourrait être réductible est :

$$\phi(x_0, x_1, 0) = (P_3 + P_0)(Q_3 + Q_0) = P_3 Q_3 + P_3 Q_0 + P_0 Q_3 + P_0 Q_0$$

où les P_i et les Q_i sont des polynômes homogènes de degré i. De plus le fait que $P_3 Q_0 + P_0 Q_3 = a_6 B_3(x_0, x_1, 0)$ implique que $P_3 = (x_0 + \beta x_1)(x_0 + \beta^2 x_1)(x_0 + \beta^4 x_1)$, $Q_3 = (x_0 + \beta^3 x_1)(x_0 + \beta^5 x_1)(x_0 + \beta^6 x_1)$, et que $P_0 = Q_0$. Cela implique que le seul cas où la courbe Y est réductible est le cas où $a_3 = a_6^2$ et $a_5 = 0$.

Dans ce cas-là, la surface X se décompose en $X = X_1 \cup X_2$, où X_1 a pour équation $x_0^3 + x_0^2 x_1 + x_1^3 + x_1^2 x_2 + x_0 x_2^2 + x_2^3 + a_6 = 0$ et X_2 a pour équation $x_0^3 + x_0 x_1^2 + x_1^3 + x_0^2 x_2 + x_1 x_2^2 + x_2^3 + a_6 = 0$. L'intersection de la surface X_1 avec le plan $x_2 = 0$ est une courbe elliptique d'équation affine $1 + x_0^3 + x_0 x_1^2 + x_1^3 = 0$. Elle ne peut rencontrer chacune des droites $x_1 = 0$, $x_2 = 0$, $x_1 + x_3 = 0$ ou la droite à l'infini qu'en 3 points. Or elle a au moins $1 + q - 2\sqrt{q}$ points rationnels d'après la borne de Hasse-Weil, ce qui prouve que la surface affine X_1 a des points rationnels en dehors de la surface $(x_0 + x_1)(x_2 + x_1)(x_0 + x_2) = 0$ pour $1 + q - 2\sqrt{q} > 12$. Par conséquent, $f(x) = x^9 + a_6 x^6 + a_6^2 x^3$ n'est pas APN pour $m \geq 5$. Pour $m \leq 4$, le calcul exhaustif montre que l'on n'obtient pas de nouvelles fonctions APN (cf. [4]).

Dans le cas où la courbe Y est irréductible, elle a au plus $1 + q - 20\sqrt{q}$ points rationnels d'après le corollaire 7.4 de [15]. Si la fonction f est APN, la courbe projective $\overline{Y}$ doit avoir tous ses points rationnels contenus dans la courbe réunion de 4 droites $x_0 x_1 (x_0 + x_1) z = 0$. Elle recoupe chaque droite en 6 points au plus, donc elle doit avoir au plus 24 points. Cela n'est possible que si $m \leq 8$.

6. Applications numériques

Quand la surface X est irréductible, on obtient que la fonction f ne peut être APN que si $m \leq m_{max}$ où m_{max} est donné par le tableau suivant.

$d \leq$	7	9	10	12	15	17	21	23	29	36	41	49	50	70	83
m_{max}	15	16	17	18	19	20	21	22	23	24	25	26	27	28	29

Quand la surface $\overline{X}$ est de plus à singularités isolées, on obtient que la fonction f ne peut être APN que si $m \leq m_{max}$ où m_{max} est donné par le tableau suivant.

$d \leq$	7	9	10	12	13	15	17	20	23	26	30	36	42	49	57
m_{max}	6	9	10	11	12	13	14	15	16	17	18	19	20	21	22

Références

1. T. Berger, A. Canteaut, P. Charpin, Y. Laigle-Chapuy, *On almost perfect nonlinear functions over F_2^n*, IEEE Trans. Inform. Theory 52, n° 9 (2006), 4160–4170.

2. L. Budaghyan and C. Carlet and P. Felke and G. Leander, *An infinite class of quadratic APN functions which are not equivalent to power mappings*, Cryptology ePrint Archive, n° 2005/359

3. E. Byrne and G. McGuire, *On the Non-Existence of Quadratic APN and Crooked Functions on Finite Fields*, preprint.
http ://www.maths.may.ie/sta/gmg/APNniceWeilEBGMG.pdf.

4. M. Brinkman, G. Leander, *On the classification of APN functions up to dimension five*, International Workshop on Coding and Cryptography (WCC), Versailles, France, 2007.

5. A. Canteaut, *Differential cryptanalysis of Feistel ciphers and differentially δ-uniform mappings*, In Selected Areas on Cryptography, SAC'97, pp. 172-184, Ottawa, Canada, 1997.

6. A. Canteaut, P. Charpin, and H. Dobbertin, *Weight divisibility of cyclic codes, highly nonlinear functions on $GF(2^m)$ and crosscorrelation of maximum-length sequences*, SIAM Journal on Discrete Mathematics, 13(1), 2000.

7. C. Carlet, *Vectorial Boolean Functions for Cryptography*, Chapter of the monography Boolean Methods and Models, Y. Crama and P. Hammer eds, Cambridge University Press, to appear.

8. C. Carlet, P. Charpin and V. Zinoviev, *Codes, bent functions and permutations suitable for DES-like cryptosystems*, Designs, Codes and Cryptography, 15(2), pp. 125-156, 1998.

9. P. Deligne, *La conjecture de Weil : I*, Publications Mathématiques de l'IHES, 43 (1974), pp. 273-307

10. J. F. Dillon, *APN Polynomials and Related Codes*, Conference on Polynomials over Finite Fields and Applications, Banff International Research Station November 2006.

11. H. Dobbertin, *Almost perfect nonlinear power functions over $GF(2^n)$: the Niho case*, Inform. and Comput., 151, pp. 57-72, 1999.

12. H. Dobbertin, *Almost perfect nonlinear power functions over $GF(2^n)$: the Welch case*, IEEE Trans. Inform. Theory, 45, pp. 1271-1275, 1999.

13. H. Dobbertin, *Almost perfect nonlinear power functions over $GF(2^n)$: a new case for n divisible by 5*, Proceedings of Finite Fields and Applications FQ5, D. Jungnickel and H. Niederreiter eds., Augsburg, Germany, Springer, pp. 113-121, 2000.

14. Y. Edel, G. Kyureghyan and A. Pott, *A new APN function which is not equivalent to a power mapping*, Preprint, 2005, http ://arxiv.org/abs/math.CO/0506420

15. S. R. Ghorpade, G. Lachaud, *Étale cohomology, Lefschetz theorems and number of points of singular varieties over finite fields*, Mosc. Math. J. 2 n° 3, (2002), pp. 589–631.

16. D. Jedlicka, *APN monomials over GF(2^n) for infinitely many n*, Finite Fields Appl. 13 no. 4, (2007), pp. 1006–1028.

17. H. Janwa, G. McGuire, R. Wilson, *Double-error-correcting cyclic codes and absolutely irreducible polynomials over GF(2)*, J. Algebra 178 n° 2, (1995), pp. 665–676.

18. H. Janwa, R. Wilson, *Hyperplane sections of Fermat varieties in P^3 in char. 2 and some applications to cyclic codes*, Applied algebra, algebraic algorithms and error-correcting codes (San Juan, PR, 1993), pp. 180–194, Lecture Notes in Comput. Sci., n° 673, Springer, Berlin, 1993.

19. S. Lang, A. Weil, *Number of points of varieties in finite fields*, Amer. J. Math. n° 76, (1954), pp. 819–827.

20. K. Nyberg, *Differentially uniform mappings for cryptography*, Advances in cryptology— Eurocrypt '93 (Lofthus, 1993), 55–64, Lecture Notes in Comput. Sci., n° 765, Springer, Berlin, 1994.

21. J. -P. Serre, *Lettre à M. Tsfasman,* Astérisque n° 198-199-200 (1991), pp. 351-353.

22. F. Voloch, *Symmetric Cryptography and Algebraic Curves,* preprint,
    http ://www.ma.utexas.edu/users/voloch/preprint.html

INSTITUT DE MATHÉMATIQUES DE LUMINY – C.N.R.S. – MARSEILLE – FRANCE
E-mail address: `rodier@iml.univ-mrs.fr`

Contemporary Mathematics
Volume **487**, 2009

How to use finite fields for problems concerning infinite fields

Jean-Pierre Serre

As the title indicates, the purpose of the present lecture is to show how to use finite fields for solving problems on infinite fields. This can be done on two different levels: the elementary one uses only the fact that most algebraic geometry statements involve only finitely many data, hence come from geometry over a finitely generated ring, and the residue fields of such a ring are finite; the examples we give in §§1-4 are of that type. A different level consists in using Chebotarev's density theorem and its variants, in order to obtain results over non-algebraically closed fields; we give such examples in §§5-6. The last two sections were only briefly mentioned in the actual lecture; they explain how cohomology (especially the étale one) can be used instead of finite fields; the proofs are more sophisticated[1], but the results have a wider range.

1. Automorphisms of the affine n-space

Let us start with the following simple example:

THEOREM 1.1. *Let σ be an automorphism of the complex affine n-space $\mathbf{C}^n$, viewed as an algebraic variety. Assume that $\sigma^2 = 1$. Then σ has a fixed point.*

Surprisingly enough this theorem can be proved by "replacing $\mathbf{C}$ by a finite field".

More generally:

THEOREM 1.2. *Let G be a finite p-group acting algebraically on the affine space $\mathbf{A}^n$ over an algebraically closed field k with char $k \neq p$. Then the action of G has a fixed point.*

Proof of Theorem 1.2

a) The case $k = \overline{\mathbf{F}}_\ell$, where ℓ is a prime number $\neq p$

We may assume that the action of G is defined over some finite extension $\mathbf{F}_{\ell^m}$ of $\mathbf{F}_\ell$. Then the group G acts on the product $\mathbf{F}_{\ell^m} \times \cdots \times \mathbf{F}_{\ell^m}$. However, G is a

---

2000 *Mathematics Subject Classification.* Primary 20G30, 14L30, 14R20, 11E57; Secondary 14F20.

Key words and phrases. finite field, finite subgroup, algebraic group, automorphism, fixed point, Smith theory.

I want to thank A.Zykin who wrote a preliminary version of this lecture.

[1]Indeed, I would not have been able to give them without the help of Luc Illusie and of his two reports [**12**] and [**13**].

p-group and the number of elements of $\mathbf{F}_{\ell^m} \times \cdots \times \mathbf{F}_{\ell^m}$ is not divisible by p. Hence there is an orbit consisting of one element, i.e. there is a fixed point for the action of G.

b) Reduction to the case $k = \overline{\mathbf{F}}_\ell$

Since G is finite, we can find a ring $\Lambda \subset \mathbf{C}$ finitely generated over $\mathbf{Z}$, over which the action of G can be defined. This means that the action of G is given by

$$g(x_1, \ldots, x_n) = (P_{g,1}(x_1, \ldots, x_n), \ldots, P_{g,n}(x_1, \ldots, x_n)),$$

where the coefficients of the polynomials $P_{g,i}(x_1, \ldots, x_n)$ belong to Λ. Assume that there is no fixed point. The system of equations

$$x_i - P_{g,i}(x_1, \ldots, x_n) = 0$$

has no solution in $\mathbf{C}$. Thus, by Hilbert's Nullstellensatz, there exist polynomials $Q_{g,i}(x_1, \ldots, x_n)$ such that

(1) $$\sum_{g,i}(x_i - P_{g,i}(x_1, \ldots, x_n))Q_{g,i}(x_1, \ldots, x_n) = 1.$$

By enlarging Λ if necessary, we may assume that it contains $1/p$ and the coefficients of the $Q_{g,i}$'s. Let $\mathfrak{m}$ be a maximal ideal of Λ. Then the field $\Lambda/\mathfrak{m}$ is finite (see e.g. [**5**], p.68, cor.1), we have $\operatorname{char} \Lambda/\mathfrak{m} \neq p$ (since p is invertible in Λ) and by (1) the conditions of the theorem hold for the algebraic closure of $\Lambda/\mathfrak{m}$. So we can apply part a) of the proof to get a contradiction.

Remark. The technique of replacing a scheme X of finite type over k by a scheme over Λ is sometimes called "spreading out X"; its properties are described in [**10**], §10.4.11 and §17.9.7.

Question. Assume the hypotheses of Theorem 1.2. Let k_o be a subfield of k such that the action of G is defined over k_o. Does there exist a fixed point of G which is rational over k_o ? Even the case $k = \mathbf{C}$, $k_o = \mathbf{Q}$, $|G| = 2$, $n = 3$ does not seem to be known.

Exercises

1. Let L be an infinite set of prime numbers. For every $p \in L$, let $k(p)$ be a denumerable field of characteristic p. Let $A = \prod k(p)$ be the product of the $k(p)$'s. Show that there exists a quotient of A which is isomorphic to a subfield of $\mathbf{C}$.(*Hint.* Use an ultrafilter on L.)

2. Let $P_i(X_1, \ldots, X_n)$ be a family of polynomials with coefficients in $\mathbf{Z}$. Show that the following properties are equivalent:

a) The P_i's have a common zero in $\mathbf{C}$.

b) There exists an infinite set of primes p such that the P_i's have a common zero in $\mathbf{F}_p$.

c) For every prime p, except a finite number, there exists a field of characteristic p in which the P_i's have a common zero.

3. Assume the hypotheses of Theorem 1.2. Show that the number of fixed points of G is either infinite or $\equiv 1 \bmod p$. (*Hint.* Suppose the set S of fixed points is finite. Using the same argument as in the proof of Theorem 1.2, we may assume that the action of G is defined over a finite field k_1 with q elements, with $(q, p) = 1$, that the points of S are rational over k_1, and that k_1 contains the p-th roots of unity. We then get $|S| \equiv q^n \bmod p$, hence $|S| \equiv 1 \bmod p$ since $q \equiv 1 \bmod p$.)

[Smith's theory gives more: if S is finite, it has one element only, see §7.4.]

2. Fixed points for finite group actions

Consider a finite group G of order m acting on a k-variety X, where k is an algebraically closed field[2]. Let us assume that there are a finite number of fixed points in $X(k)$ and that G acts freely outside these points.

THEOREM 2.1. *Suppose we have two actions of G on X satisfying the above properties, the number of fixed points being a and a' respectively. Then $a \equiv a' \pmod{m}$.*

Sketch of proof

Assume first that X, the two actions of G, and the fixed points, are defined over a finite field $\mathbf{F}_q$. We then have

$$|X(\mathbf{F}_q)| \equiv a \pmod{m} \quad \text{and} \quad |X(\mathbf{F}_q)| \equiv a' \pmod{m},$$

hence $a \equiv a' \pmod{m}$.

The general case is reduced to this case by an argument similar to (but less obvious than) the one given in §1. [One replaces X by a separated scheme of finite type X_0 over a ring Λ which is finitely generated over $\mathbf{Z}$, in such a way that the the the two actions of G extend to X_0; one also needs that the corresponding fixed points Y and Y' are finite and étale over $\mathrm{Spec}(\Lambda)$, and that the two actions of G on $X_0 - Y$ and on $X_0 - Y'$ are free. That all these conditions can be met is a consequence of [10], *loc.cit.*] One can then reduce modulo a maximal ideal of Λ.

Remarks

1) Theorem 1.1 is a corollary of Theorem 2.1. Indeed the involution $z \mapsto -z$ of $\mathbf{C}^n$ has only one fixed point, hence the number of fixed points of any other involution is either odd or infinite; it cannot be 0.

2) The results of §§1, 2 can also be proved by topological arguments, see §7.4 below.

3. Injectivity and surjectivity of maps between algebraic varieties

The following theorem was proved independently by J.Ax and A.Grothendieck in the 60's (see [2], [8], p.184, and [10], §10.4.11), and has been rediscovered several times since.

THEOREM 3.1. *Let X be an algebraic variety over an algebraically closed field k. If a morphism $f : X \to X$ is injective then it is bijective.*

Sketch of proof

Assume first that k is an algebraic closure of a finite field k_1, and that f is defined over k_1. Then $X(k_1)$ is finite; since

$$f : X(k_1) \to X(k_1)$$

is injective, it is bijective. The same argument, applied to the finite extensions of k_1, shows that $f : X(k) \to X(k)$ is bijective. The case of an arbitrary algebraically closed field is reduced to the one above as in §§1-2, by choosing a ring Λ of finite type over $\mathbf{Z}$ over which X and f are defined, and reducing modulo a maximal ideal of Λ; for more details, see Grothendieck, [10], *loc.cit.*

---

[2]The reader may interpret the word "k-variety" in the sense of FAC, i.e. as meaning a separated and reduced scheme of finite type over $\mathrm{Spec}(k)$, cf. [10], §10.10. Since we are only interested in the k-points, the "reduced" assumption has no importance.

Remark. When $k = \mathbf{C}$ a topological proof of this theorem was given by Borel in 1969 (see [**4**]).

4. Nilpotent groups

In 1955 M.Lazard ([**17**]) proved the following theorem:

THEOREM 4.1. *Let G be an algebraic group over an algebraically closed field k. If the underlying variety of G is isomorphic to the affine space $\mathbf{A}^n$, for some $n \geqslant 0$, then G is nilpotent.*

Sketch of proof
    Lazard proves even more: he shows that G is *nilpotent of class $\leqslant n$*, i.e. that every iterated commutator of length $> n$ is equal to 1. As in §1, we may assume that k is an algebraic closure of a finite field k_1 and that G is defined over k_1. If $p = \operatorname{char} k$, the group $G(k_1)$ is a finite p-group, hence is nilpotent. By applying this to the finite extensions of k_1 one sees that $G(k)$ is a locally nilpotent group, i.e. is an increasing union of nilpotent groups. A further argument is needed to show that $G(k)$ is indeed nilpotent of class $\leqslant n$, see [**17**][3].

    Note that we may deduce Theorem 4.1 from Theorem 1.2 together with the following standard result:

THEOREM 4.2. *Let G be a connected linear algebraic group. Then either G is nilpotent or it contains a one dimensional torus $\mathbf{G}_m$ as a subgroup.*

Proof of Theorem 4.1 from Theorem 4.2 and Theorem 1.2
    Assume that the underlying variety of G is isomorphic to $\mathbf{A}^n$. In particular, G is an affine variety; this is known to imply that the group G can be embedded in some $\mathbf{GL}_N$, i.e. that G is a linear group. Assume further that G is not nilpotent; then $\mathbf{G}_m \subset G$ by Theorem 4.2. Choose an element σ of $\mathbf{G}_m$ of prime order $\ell \neq \operatorname{char} k$. The element σ acts on $G \simeq \mathbf{A}^n$ by left translation and thus by Theorem 1.2 has a fixed point: contradiction!

Exercise. Give a topological proof of Theorem 4.1 when $k = \mathbf{C}$ by using the fact that, for a non nilpotent connected group, either H^1 or H^3 is non zero. Extend this proof to arbitrary fields using ℓ-adic cohomology.

5. Finite subgroups of $\mathbf{GL}_n(\mathbf{Q})$

    The following theorem is a well-known result of Minkowski (see [**18**], [**20**], [**23**]); it gives a multiplicative upper bound for the order of a finite subgroup of $\mathbf{GL}_n(\mathbf{Q})$.

THEOREM 5.1. *Let ℓ be a prime number. If A is a subgroup of $\mathbf{GL}_n(\mathbf{Q})$ of order ℓ^a then*

$$(2) \qquad a \leqslant M(n,\ell) = \left[\frac{n}{\ell - 1}\right] + \left[\frac{n}{\ell(\ell - 1)}\right] + \left[\frac{n}{\ell^2(\ell - 1)}\right] + \cdots.$$

Proof for $\ell \neq 2$
    The idea is to reduce $\operatorname{mod} p$ for an appropriate choice of $p \neq \ell$. First, we have $A \subset \mathbf{GL}_n(\mathbf{Z}[1/N])$ for a suitable $N \geqslant 1$. If p is sufficiently large ($p > 2$ is enough) and does not divide N, then the reduction $\operatorname{mod} p$ gives a subgroup A'

---

[3]Note the following misprints in [**17**]: in the last part of the proof of Lemma 1, "$s \geqslant t$" should be "s divisible by t" and "$W = C_2(X, W)$" should be "$W = C_2(X, V)$".

of $\mathbf{GL}_n(\mathbf{F}_p)$ such that $|A| = |A'|$. Hence $|A|$ divides $|\mathbf{GL}_n(\mathbf{F}_p)|$, which is equal to $p^{n(n-1)/2} \prod_{i=1}^n (p^i - 1)$. Let us choose p such that its image in $(\mathbf{Z}/\ell^2\mathbf{Z})^*$ is a generator of this group. This is always possible by Dirichlet's theorem on primes in arithmetic progressions since $(\mathbf{Z}/\ell^2\mathbf{Z})^*$ is cyclic. Once p is chosen in this way, $p^i - 1$ is divisible by ℓ only if i is divisible by $\ell - 1$ and in this case the ℓ-adic valuation $v_\ell(p^i - 1)$ of $p^i - 1$ is equal to $1 + v_\ell(i)$. Hence $v_\ell(|A|) \leqslant \sum (1 + v_\ell(i))$, where the sum is over the integers i with $1 \leqslant i \leqslant n$ which are divisible by $\ell - 1$; a simple computation shows that this sum is equal to $M(n, \ell)$ if $\ell \neq 2$, cf. [23], §1.3.

REMARK 5.2. In the case $\ell = 2$ one has to replace the group $\mathbf{GL}_n$ by an orthogonal group $\mathbf{O}_n$ in order to get the desired bound.

REMARK 5.3. The result is optimal in the sense that, for every prime number ℓ, there exist subgroups of $\mathbf{GL}_n(\mathbf{Q})$ of order $\ell^{M(n,\ell)}$, see [18] and [23]. These subgroups have the following Sylow-type property:

THEOREM 5.4. Let A, P be two finite ℓ-subgroups of $\mathbf{GL}_n(\mathbf{Q})$. Suppose that $|P| = \ell^{M(n,\ell)}$. Then A is $\mathbf{GL}_n(\mathbf{Q})$-conjugate to a subgroup of P.

[In particular, if $|A| = |P|$, then A and P are conjugate.]

Sketch of proof

We only give the proof when $\ell \neq 2$ (otherwise we have to consider orthogonal groups). Let us reduce mod p for a prime p chosen as above. We get two ℓ-subgroups A and P of $\mathbf{GL}_n(\mathbf{F}_p)$; by construction, P is a Sylow subgroup of $\mathbf{GL}_n(\mathbf{F}_p)$. By Sylow's theorem, A is conjugate to a subgroup of P, i.e. there exists an embedding $i : A \to B$ which is the restriction of an inner automorphism of $\mathbf{GL}_n(\mathbf{F}_p)$. The linear representations $A \to \mathbf{GL}_n(\mathbf{Q})$ and $A \xrightarrow{i} P \to \mathbf{GL}_n(\mathbf{Q})$ become isomorphic after reduction mod p. Since $p \neq \ell$, a standard argument shows that they are isomorphic over $\mathbf{Q}_p$, hence also over $\mathbf{Q}$, and this completes the proof.

REMARK 5.5. In [20], Schur gave another proof of Minkowski's Theorem using an interesting lemma on characters of finite groups ([23], §2.1, prop.1). He also extended the theorem (by a different method) to arbitrary number fields (see [23], §2.2).

6. Generalizations to other algebraic groups and fields

A natural question is: what happens in Theorem 5.1 if $\mathbf{Q}$ is replaced by an arbitrary field k and $\mathbf{GL}_n$ by an arbitrary reductive group G ? This is answered in [23]: roughly speaking, one can give a sharp bound for the order of a finite ℓ-subgroup of $G(k)$ when one knows the root system of G and the Galois group of the ℓ-cyclotomic tower of k (one needs also to assume that G is "of inner type", but this is automatic for the most interesting examples, such as G_2, F_4, E_7 or E_8). The proof follows Minkowski's method; the main difference is that Dirichlet's theorem on arithmetic progressions is replaced by a variant of the Chebotarev's density theorem which applies to every normal domain which is finitely generated over $\mathbf{Z}$ (see [21], §2.7 and [23], §6.4). As a sample, here is the case where $k = \mathbf{Q}$ and G is of type E_8:

THEOREM 6.1. Let G be a group of type E_8 over $\mathbf{Q}$ and let A be a finite subgroup of $G(\mathbf{Q})$. Then $|A|$ divides the number

$$M(\mathbf{Q}, E_8) = 2^{30} \cdot 3^{13} \cdot 5^5 \cdot 7^4 \cdot 11^2 \cdot 13^2 \cdot 19 \cdot 31.$$

Sketch of proof

We may assume that G is defined over $\mathbf{Z}[1/N]$ for some $N \geqslant 1$ and that A is an ℓ-subgroup of $G(\mathbf{Z}[1/N])$ for some prime ℓ (the general bound is then obtained by multiplicativity). By reducing mod p for p large enough, we see that $|A|$ divides $|E_8(\mathbf{F}_p)|$. One knows (see e.g. [7], Theorem 9.4.10) that

$$|E_8(\mathbf{F}_p)| = p^{120}(p^2-1)(p^8-1)(p^{12}-1)(p^{14}-1)(p^{18}-1)(p^{20}-1)(p^{24}-1)(p^{30}-1).$$

By choosing p as in Minkowski's proof, one gets the desired bound. As an example, let us do the computation when ℓ is equal to 3:

Choose $p \equiv 2, 4, 5$ or $7 \pmod 9$. Then $p^2 - 1$ is divisible by 3 but not by 9. This implies that the 3-adic valuations of the eight factors $p^2 - 1$, $p^8 - 1$, ..., $p^{30} - 1$ are respectively: 1, 1, 2, 1, 3, 1, 2, 2. Their sum is 13, as claimed.

REMARK 6.2. The bound in Theorem 6.1 is optimal in the following sense: for every $\ell = 2, 3, ..., 31$, there exists a group G of type E_8 over $\mathbf{Q}$ such that $G(\mathbf{Q})$ contains a subgroup of order 2^{30}, 3^{13}, ..., 31, cf. [23], §13.5. [Recall that there are three different groups of type E_8 over $\mathbf{Q}$, up to isomorphism; they are characterized by the structure of their $\mathbf{R}$-points. For most values of ℓ, one can choose G such that $G(\mathbf{R})$ is compact.]

Exercise. Let K be a quadratic number field with discriminant d. Show that Theorem 6.1 is valid with $\mathbf{Q}$ replaced by K and $M(\mathbf{Q}, E_8)$ replaced by $c(d)M(\mathbf{Q}, E_8)$, where $c(d)$ is defined by:

$c(8) = 2^8$, $c(5) = 5^5$, $c(13) = 13^2$, $c(17) = 17^2$, $c(29) = 29$, $c(37) = 37$,

$c(41) = 41$, $c(61) = 61$ and $c(d) = 1$ for the other values of d (in particular those which are negative).

7. Proofs via Topology

As indicated at the end of §2, the results of §§1,2 can also be obtained - and sometimes improved - by using topological methods, based on cohomology (either standard, if the ground field is $\mathbf{C}$, or étale). There are several ways to do so; we shall summarize a few of them below.

The notation will be the following: X is an algebraic variety over an algebraicalled closed field k and G is a finite group which acts on X; the fixed point set of G is denoted by X^G. We assume that X is quasi-projective (in most applications, X is affine), so that the quotient variety X/G is well defined. The cohomology groups $H^i(X)$ will be understood as the étale ones, with arbitrary support; in case we need cohomology with proper support, we shall write $H^j_c(X)$, with a c in index position. The letter ℓ will denote a prime number distinct from char k.

7.1. *Using Cartan-Leray's spectral sequence*

Suppose that G acts freely on X. There is a Cartan-Leray spectral sequence (first defined in [6] in the context of standard sheaf cohomomology - for the case of étale cohomology, see [11], §8.5)

$$H^i(G, H^j(X, C)) \Longrightarrow H^{i+j}(X/G, C),$$

where C is any finite abelian group. If X is the affine n-space $\mathbf{A}^n$ and $|C|$ is prime to char k, then $H^j(X, C) = 0$ for $j > 0$ and $H^0(X, C) = C$. In that case the spectral sequence degenerates and gives $H^i(G, C) = H^i(X/G, C)$ for every i, i.e. X/G has the same cohomology as the classifying space of G. Take now $C = \mathbf{Z}/\ell\mathbf{Z}$

and suppose that ℓ divides $|G|$. It is well known that $H^j(G, C)$ is non zero for infinitely many j's, and that $H^j(X/G, C)$ is zero for $j > 2.\dim X$: contradiction!

Conclusion: *the only finite groups which can act freely on $\mathbf{A}^n$ are the p-groups,* with $p = \operatorname{char} k$. This gives another proof of Theorem 1.1.

Exercise. Let G be a finite p-group, with $p = \operatorname{char} k$. Show that there exists a free action of G on $\mathbf{A}^n$, provided that n is large enough. (*Hint*: embed G in a connected unipotent group.)

7.2. *Using Euler-Poincaré characteristics*

Let $\chi(X)$ be the Euler-Poincaré characteristic of X, relative to the ℓ-adic cohomology. It is known (see [13], [16]) that $\chi(X)$ does not depend on the choice of ℓ, and that it coincides with the Euler-Poincaré characteristic of X with proper support (Grothendieck-Laumon's theorem, cf. [16] - in case $k = \mathbf{C}$, see the Appendix of [15]). In other words, we have

$$\chi(X) = \sum (-1)^j \dim H^j(X, \mathbf{Q}_\ell) = \sum (-1)^j \dim H^j_c(X, \mathbf{Q}_\ell).$$

A useful property of χ is its *additivity* : if X is the disjoint union of locally closed subvarieties X_λ, then $\chi(X) = \sum \chi(X_\lambda)$; this follows from the definition of $\chi(X)$ via cohomology with proper support.

If $|G|$ is prime to $\operatorname{char} k$, and G acts freely on X, one has $\chi(X) = |G| . \chi(X/G)$, cf. [13]. (In particular $\chi(X) = 1$ implies $|G| = 1$; since $\chi(\mathbf{A}^n) = 1$, we recover the statement given at the end of §7.1.)

Assume now that G is an ℓ-group. We have:

$$\chi(X^G) \equiv \chi(X) \ (\operatorname{mod} \ell).$$

Indeed, using induction on $|G|$, one may assume that G is cyclic of order ℓ; in that case, G acts freely on $Y = X - X^G$ and we have

$$\chi(X) = \chi(X^G) + \chi(Y) = \chi(X^G) + \ell . \chi(Y/G) \equiv \chi(X^G) \ (\operatorname{mod} \ell).$$

In particular, *if $\chi(X)$ is not divisible by ℓ, then X^G is non empty.* This gives another proof of Theorem 1.2 (with the letter p replaced by ℓ). A similar argument applies to Theorem 2.1: with the notation of that theorem, one has $a \equiv \chi(X) \ (\operatorname{mod} m)$ and $a' \equiv \chi(X) \ (\operatorname{mod} m)$, hence $a \equiv a' \ (\operatorname{mod} m)$.

7.3. *Using Lefschetz numbers*

If s is an element of G, of order prime to $\operatorname{char} k$, let $t(s)$ be its Lefschetz number:

$$t(s) = \sum (-1)^j . \operatorname{Tr}(s, H^j(X, \mathbf{Q}_\ell)) = \sum (-1)^j . \operatorname{Tr}(s, H^j_c(X, \mathbf{Q}_\ell)), \quad \text{cf. [13]}.$$

It is known (*loc.cit.*) that $t(s) = 0$ if s has no fixed point. If $X = \mathbf{A}^n$, one has obviously $t(s) = 1$. Hence *every automorphism of $\mathbf{A}^n$, whose order is finite and prime to $\operatorname{char}(k)$, has a fixed point.* In the special case where $k = \mathbf{C}$, this was proved by D.Petrie and J.D.Randall [19] by a similar argument, based on standard cohomology.

Remark. For a general report on the possible actions of algebraic groups (not necessarily finite) on $\mathbf{A}^n$, see H.Kraft [14].

7.4. *Using Smith theory*

In the situation of Theorem 1.2, for $k = \mathbf{C}$, Smith theory (cf. P.A.Smith [24] and A.Borel [3]) gives more than the mere existence of a fixed point: it gives non trivial information on the cohomology of X^G. For instance, it shows that, if $\dim X^G = 0$, then X^G is reduced to one point. Similar results hold in any characteristic. More precisely, suppose that G is a finite p-group and let us write $H^j(X)$ instead of $H^j(X, \mathbf{Z}/p\mathbf{Z})$.

Theorem 7.5. *Let N be an integer such that $H^j(X) = 0$ for all $j \geqslant N$. Then:*

a) *$H^j(X^G) = 0$ for all $j \geqslant N$.*

b) *If $N = 1$ and $\dim H^0(X) = 1$, then $\dim H^0(X^G) = 1$.*

The proof when $|G| = p$ will be given in §8 below. The general case follows by induction on $|G|$: if $|G| > p$, one chooses a central subgroup H of G of order p and one applies the induction hypothesis to the action of G/H on X^H.

Corollary. *If X is p-acyclic, so is X^G.*

This is obvious, since "p-acyclic" means that $\dim H^j$ is equal to 0 for $j > 0$ and to 1 for $j = 0$. Note that this implies that, if X^G is finite, it has only one element.

Remark. The prime number p is allowed to be equal to char k. This case is of interest mainly when X is a projective variety. For instance, if X is a smooth projective surface in characteristic p and if X is rational, then X is p-acyclic [use the Artin-Schreier exact sequence] and the corollary above shows that the same is true for X^G; in particular, X^G is not empty.

8. Smith Theory: Proof of Theorem 7.5

Let G be a cyclic group of prime order p. The group algebra $\mathbf{F}_p[G]$ is isomorphic to the truncated polynomial ring $\mathbf{F}_p[t]/(t^p)$. This very simple fact is basic in Smith's proofs. It explains the rather artificial-looking definitions given below.

8.1. *R-abelian categories. Definitions*

Let F be a field and let n be an integer > 1. Let R be the F-algebra generated by an element t with the relation $t^n = 0$; it has for basis $1, ..., t^{n-1}$. Let C be an *R-abelian category*, i.e. an abelian category such that, for every pair of objects A, B of C, $\mathrm{Hom}_C(A, B)$ has an R-module structure, and the composition of maps is R-bilinear. In particular, t defines an endomorphism t_A of every object A of C, and we have $t_A^n = 0$. If i is a positive integer, the image of the endomorphism t_A^i of A will be denoted by $t^i A$, and its kernel will be written A_{t^i}.

We have $A \supset tA \supset t^2 A \supset ... \supset t^n A = 0$, and $A/A_{t^i} = t^i A$.

We shall say that A is *constant* if $tA = 0$, and that A is *free* if the morphism $A/tA \to t^{n-1} A$ given by t_A^{n-1} is an isomorphism, i.e. if $A_{t^{n-1}} = tA$; this implies that all the quotients $t^i A/t^{i+1} A$ are isomorphic to A/tA for $i = 1, ..., n - 1$.

Example. If C is the category Mod_R of all R-modules, the notion of freeness just defined coincides with the usual one. As for the "constant" R-modules, they are the F-vector spaces with zero t-action.)

8.2. *The (I,A,B) setting*

We now choose an exact sequence in C:

(8.2.1) $0 \to I \to A \to B \to 0$

such that I is free and B is constant.

Lemma 8.2.2. *For every i with $1 \leqslant i \leqslant n - 1$ we have:*

(8.2.3) *The natural map $A_{t^i}/t^{n-i}A \to B$ is an isomorphism.*

(8.2.4) *The natural map $A/t^i A \to B \oplus t^{n-i} A$ is an isomorphism.*

(In (8.2.4), the map $A/t^i A \to t^{n-i} A$ is induced by t_A^{n-i}.)

Proof. For every object Y of C, let us put $h_i(Y) = Y_{t^i}/t^{n-i} Y$. If

$$0 \to Y' \to Y \to Y'' \to 0$$

is an exact sequence in C, we have an hexagonal exact sequence

$$h_i(Y') \quad \rightarrow \quad h_i(Y)$$
$$\nearrow \qquad\qquad\qquad\qquad \searrow$$
$$h_{n-i}(Y'') \qquad\qquad\qquad\qquad\qquad\qquad h_i(Y'')$$
$$\searrow \qquad\qquad\qquad\qquad \nearrow$$
$$h_{n-i}(Y) \quad \leftarrow \quad h_{n-i}(Y')$$

Apply this to (8.2.1). Since B is constant, we have $h_i(B) = B$, and since I is free, we have $h_i(I) = 0$ for every i. We thus get the exact sequence

$$0 = h_i(I) \longrightarrow h_i(A) \longrightarrow h_i(B) = B \longrightarrow h_{n-i}(I) = 0,$$

which proves (8.2.3). As for (8.2.4), it follows from the fact that $A/t^i A \to t^{n-i} A$ is surjective and that its kernel, by (8.2.3), is $h_i(A) = B$.

Remark. When $C = Mod_R$, (8.2.1) implies that the R-module A is a direct sum[4] of indecomposable modules which are isomorphic to either F or R : in other words, the only "Jordan blocks" which can occur are of rank either 1 or n.

8.3. *The main statements*

We now consider another R-abelian category C' and a cohomological functor (H, δ) on C with values in C' (cf. [**9**], §2.1); we assume that this functor is compatible with the R-structures of C and C'. For every $j \in \mathbf{Z}$, and every object E of C, $H^j(E)$ is an object of C'; every exact sequence $0 \to E \to E' \to E'' \to 0$ in the category C gives rise to an infinite exact sequence in C':

$$\ldots \longrightarrow H^j(E) \longrightarrow H^j(E') \longrightarrow H^j(E'') \xrightarrow{\delta_j} H^{j+1}(E) \longrightarrow \ldots$$

We make the following assumptions:

(8.3.1) $H^j(E) = 0$ for every $j < 0$ and every $E \in Ob(C)$. This implies that the functor H^0 is left exact.

(8.3.2) For every $E \in Ob(C)$, one has $H^j(E) = 0$ for j large enough (which may depend on E).

We are now going to apply the cohomological functor (H, δ) to the C-objects I, A, B of the exact sequence (8.2.1). The first result is:

Proposition 8.3.3. *Let N be a positive integer. Assume that $H^j(A) = 0$ for all $j \geqslant N$. Then the same is true for B, i.e. $H^j(B) = 0$ for all $j \geqslant N$.*

In the special case $N = 1$ one can say more:

Proposition 8.3.4. *Assume that $H^j(A) = 0$ for all $j > 0$, and that $H^0(A)$ is constant, i.e. $t.H^0(A) = 0$. Then the same is true for B and the natural map $H^0(A) \to H^0(B)$ is an isomorphism.*

8.4. *Proof of Proposition 8.3.3*

We prove first:

Lemma 8.4.1. *Let m be a positive integer. Suppose that $H^j(A), H^j(B), H^j(t^i A)$ and $H^j(A_{t^i})$ are 0 for $j = m + 1$ and all $i = 1, ..., n - 1$. Suppose also that $H^m(A) = 0$. Then $H^m(B), H^m(t^i A)$ and $H^m(A_{t^i})$ are 0.*

Proof. Since $H^{m+1}(t^i A) = 0$ the map $H^m(A) \to H^m(A/t^i A)$ is surjective, and this shows that $H^m(A/t^i A) = 0$. By (8.2.4) this implies that $H^m(B)$ and $H^m(t^{n-i} A)$ are 0, and since $n - i$ takes all values between 1 and $n - 1$ we also have $H^m(t^i A) = 0$. By (8.2.3), A_{t^i} is an extension of B by $t^{n-i} A$; hence $H^m(A_{t^i})$ is also 0.

---

[4]possibly infinite : Smith theory does not require any finiteness assumption.

We can now prove Proposition 8.3.3. Indeed, by (8.3.2), we may choose an $m > N$ such that the hypotheses of Lemma 8.4.1 are satisfied. One then uses descending induction on m. By Lemma 8.4.1, this is possible until we reach $m = N$; the proposition follows.

Note that we obtain at the same time the vanishing of $H^j(t^i A)$ and $H^j(A_{t^i})$ for all $j \geqslant N$ and all $i = 1, ..., n-1$.

8.5. *Proof of Proposition 8.3.4*

We apply Proposition 8.3.3 with $N = 1$. As mentioned above, we obtain the vanishing, not only of $H^1(B)$, but also of $H^1(tA)$ and $H^1(A_t)$. Since H^0 is left exact, the sequence

$$0 \longrightarrow H^0(A_t) \longrightarrow H^0(A) \overset{t}{\longrightarrow} H^0(A)$$

is exact, and since $t.H^0(A) = 0$ we see that $H^0(A_t) \to H^0(A)$ is an isomorphism. Using the exact sequence

$$H^0(A_t) \longrightarrow H^0(A) \longrightarrow H^0(tA) \longrightarrow H^1(A_t),$$

we deduce that $H^0(tA) = 0$. By (8.2.4) this implies that the natural map $H^0(A/t) \to H^0(B)$ is an isomorphism. But the map $H^0(A) \to H^0(A/tA)$ is also an isomorphism, since $H^0(tA)$ and $H^1(tA)$ are both 0. Hence $H^0(A) \to H^0(B)$ is an isomorphism.

8.6. *Proof of Theorem 7.5*

We can now prove Theorem 7.5 by applying what we have done to the case where $R = \mathbf{F}_p[G] = \mathbf{F}_p[t]/(t^p)$ and C is the category of the $\mathbf{F}_p[G]$-étale sheaves over X/G, i.e. the sheaves which are killed by p, and are endowed with an action of G.

We take for C' the categorie Mod_R of all R-modules, and for cohomological functor the functor "étale cohomology over X/G"; hence, if Y is a sheaf belonging to C, $H^j(Y)$ is nothing else than $H^j(X/G, Y)$; condition (8.3.1) is obviously satisfied and condition (8.3.2) is a well known theorem of M.Artin, cf. [**1**], §5.

We take for sheaf A the direct image by the map $\pi : X \to X/G$ of the constant sheaf $\mathbf{Z}/p\mathbf{Z}$, i.e. $A = \pi_*(\mathbf{Z}/p\mathbf{Z})$; we have $H^j(A) = H^j(X, \mathbf{Z}/p\mathbf{Z}) = H^j(X)$, cf. [11, §5]. There is a natural action of G on A, coming from the action of G on the variety X.

We take for sheaf B the constant sheaf $\mathbf{Z}/p\mathbf{Z}$ on X^G "extended by zero on $X/G - X^G$" (direct image by the inclusion[5] $X^G \to X/G$); we have $H^j(B) = H^j(X^G)$. The sheaf B is constant (in the sense of 8.1) : the group G acts trivially on it.

There is a natural surjection $A \to B$. Its kernel I is free in the sense of §8.1. This is checked "fiber by fiber"; if x is a geometric point of X/G, and x belongs to X^G, the fiber of I at x is 0; if x belongs to $X - X^G$, the fiber of I at x is a free R-module of rank 1.

We can now apply Proposition 8.3.3 and Proposition 8.3.4. They give the two parts of Theorem 7.5.

Remark. The same proof applies in the usual context of sheaf theory, provided that the space X/G has finite cohomogical dimension (so that (8.3.2) holds). It also

---

[5] The natural map $X^G \to \pi(X^G)$ is a homeomorphism for the étale topology (it is even an isomorphism of schemes if p is distinct from char k); hence we may identify X^G with its image in X/G.

applies in a combinatorial context, with cohomology replaced by homology; this was already mentioned in [22], proof of Prop.2.10.

References

1. Artin, M. Dimension cohomologique: premiers résultats. In SGA 4, vol.3, Lect. Notes in Math. **305** (1973), pp. 43-63.
2. Ax, J. The elementary theory of finite fields. Ann. Math. **88** (1968), 239-271.
3. Borel, A. Nouvelle démonstration d'un théorème de P.A.Smith. Comment. Math. Helv. **29** (1955), 27-39 (= Oe., vol.I, n°34).
4. Borel, A. Injective endomorphisms of algebraic varieties. Arch. Math. **20** (1969), 531–537 (= Oe., vol.III, n°83).
5. Bourbaki, N. Algèbre Commutative. Chapitre V. Entiers, Hermann, Paris, 1964.
6. Cartan, H. and Leray, J. Relations entre anneaux de cohomologie et groupe de Poincaré. Colloque de Topologie Algébrique, Paris, CNRS (1947), pp. 83-85; see also H.Cartan's Oeuvres, vol.III, pp. 1226-1231.
7. Carter, R.W. Simple Groups of Lie Type. John Wiley & Sons, London-New York-Sydney, 1972, Pure and Applied Mathematics, vol. **28**.
8. Colmez, P. and Serre, J-P. Grothendieck-Serre Correspondence. A.M.S. - S.M.F., 2004.
9. Grothendieck, A. Sur quelques points d'algèbre homologique. Tôhoku Math.J. **9** (1957), 119-121.
10. Grothendieck, A. Eléments de Géométrie Algébrique (rédigés avec la collaboration de J.Dieudonné). IV. Etude Locale des Schémas et des Morphismes de Schémas (Troisième Partie), Publ. Math. I.H.E.S. **28** (1966), 5-255.
11. Grothendieck, A. Foncteurs fibres, supports, étude cohomologique des morphismes finis. In SGA 4, vol.2, Lect. in Math. **272** (1972), pp. 366-412.
12. Illusie, L. Miscellany on traces in ℓ-adic cohomology: a survey. Japanese J. Math. **1** (2006), 107-136.
13. Illusie, L. Odds and Ends on finite group actions and traces. Notes, Tokyo-Kyoto, 2008.
14. Kraft, H. Challenging problems on affine n-space. Sém. Bourbaki **802** (1994-1995), Astérisque **237** (1996), pp. 295-317.
15. Kraft, H. and Popov, V.L. Semisimple actions on the three-dimensional affine space are trivial. Comment. Math. Helv. **60** (1985), 466-479.
16. Laumon, G. Comparaison de caractéristiques d'Euler-Poincaré en cohomologie ℓ-adique, C. R. Acad. Sci. Paris **292** (1981), 209-212.
17. Lazard, M. Sur la nilpotence de certains groupes algébriques. C. R. Acad. Sci. Paris **241** (1955), 1687–1689.
18. Minkowski, H. Zur Theorie der positiven quadratischen Formen, J. Crelle **101** (1887), 196–202 (= Ges. Abh., Band I, n°VI).
19. Petrie, D. and Randall, J.D. Finite order automorphisms of affine varieties, Comment. Math. Helv. **61** (1986), 203-221.
20. Schur, I. Über eine Klasse von endlichen Gruppen linearer Substitutionen, Sitz. Preuss. Akad. Wiss. Berlin (1905), 77–91 (= Ges. Abh., Band I, n°6).
21. Serre, J-P. Zeta and L-functions, in "Arithmetical Algebraic Geometry" (Proc. Conf. Purdue Univ. 1963), pp. 82–92, Harper and Row, New York, 1965 (= Oe., vol.II, n°64).
22. Serre, J-P. Complète Réductibilité, Sém. Bourbaki **932** (2003-2004) (= Exposés de Séminaires 1950-1999, 2ème édition, S.M.F., Doc. Math. **1** , 2008, pp. 265-289).
23. Serre, J-P. Bounds for the orders of the finite subgroups of $G(k)$, in "Group Representation Theory" (edit. M.Geck, D.Testerman & J.Thévenaz), E.P.F.L. Press, 2007, pp. 405–450 [available on http://www.college-de-france.fr/default/EN/all/ins-pro/essai.htm].
24. Smith, P.A. A theorem on fixed points for periodic transformations, Ann. Math. **35** (1934), 572–578.

JEAN-PIERRE SERRE
COLLÈGE DE FRANCE
3, RUE D'ULM, F-75005 PARIS, FRANCE.
E-mail address: serre@noos.fr

Contemporary Mathematics
Volume **487**, 2009

On the generalizations of the Brauer–Siegel theorem

Alexey Zykin

ABSTRACT. The classical Brauer–Siegel theorem states that if k runs through the sequence of normal extensions of $\mathbb{Q}$ such that $n_k/\log|D_k| \to 0$, then $\log(h_k R_k)/\log\sqrt{|D_k|} \to 1$. In this paper we give a survey of various generalizations of this result including some recent developements in the study of the Brauer–Siegel ratio in the case of higher dimensional varieties over global fields. We also present a proof of a higher dimensional version of the Brauer–Siegel theorem dealing with the study of the asymptotic properties of the residue at $s = d$ of the zeta function in a family of varieties over finite fields.

1. Introduction

Let K be an algebraic number field of degree $n_K = [K : \mathbb{Q}]$ and discriminant D_K. We define the genus of K as $g_K = \log\sqrt{D_K}$. By h_K we denote the class-number of K, R_K denotes its regulator. We call a sequence $\{K_i\}$ of number fields a family if K_i is non-isomorphic to K_j for $i \neq j$. A family is called a tower if also $K_i \subset K_{i+1}$ for any i. For a family of number fields we consider the limit

$$\mathrm{BS}(\mathcal{K}) := \lim_{i\to\infty} \frac{\log(h_{K_i} R_{K_i})}{g_{K_i}}.$$

The classical Brauer–Siegel theorem, proved by Brauer (see [**3**]) can be stated as follows:

THEOREM 1.1 (Brauer–Siegel). *For a family $\mathcal{K} = \{K_i\}$ we have*

$$\mathrm{BS}(\mathcal{K}) := \lim_{i\to\infty} \frac{\log(h_{K_i} R_{K_i})}{g_{K_i}} = 1$$

if the family satisfies two conditions:

(i) $\displaystyle\lim_{i\to\infty} \frac{n_{K_i}}{g_{K_i}} = 0$;

(ii) *either the generalized Riemann hypothesis (GRH) holds, or all the fields K_i are normal over $\mathbb{Q}$.*

2000 *Mathematics Subject Classification.* Primary 11R42, 11G05, 11G25; Secondary 11M38, 14J27.

Key words and phrases. Brauer–Siegel theorem, infinite global field, number field, function field, varieties over finite fields.

This research was supported in part by the Russian-Israeli grant RFBR 06-01-72004-MSTIa, by the grants RFBR 06-01-72550-CNRSa, 07-01-00051a, and INTAS 05-96-4634.

The initial motivation for the Brauer–Siegel theorem can be traced back to a conjecture of Gauss:

CONJECTURE 1.2 (Gauss). *There are only 9 imaginary quadratic fields with class number equal to one, namely those having their discriminants equal to* -3, -4, -7, -8, -11, -19, -43, -67, -163.

The first result towards this conjecture was proven by Heilbronn in [11]. He proved that $h_K \to \infty$ as $D_K \to -\infty$. Moreover, together with Linfoot [12] he was able to verify that Gauss' list was complete with the exception of at most one discriminant. However, this "at most one" part was completely ineffective. The initial question of Gauss was settled independently by Heegner [10], Stark [28] and Baker [1] (initially the paper by Heegner was not acknowledged as giving the complete proof). We refer to [35] for a more thorough discussion of the history of the Gauss class number problem.

A natural question was to find out what happens with the class number in the case of arbitrary number fields. Here the situation is more complicated. In particular a new invariant comes into play: the regulator of number fields, which is very difficult to separate from the class number in asymtotic considerations (in particular, for this reason the other conjecture of Gauss on the infinitude of real quadratic fields having class number one is still unproven). A major step in this direction was made by Siegel [27] who was able to prove Theorem 1.1 in the case of quadratic fields. He was followed by Brauer [3] who actually proved what we call the classical Brauer–Siegel theorem.

Ever since a lot of different aspects of the problem have been studied. For example, the major difficulty in applying the Brauer–Siegel theorem to the class number problem is its ineffectiveness. Thus many attempts to obtain good explicit bounds on $h_K R_K$ were undertaken. In particular we should mention the important paper of Stark [29] giving an explicit version of the Brauer–Siegel theorem in the case when the field contains no quadratic subfields. See also some more recent papers by Louboutin [21], [22] where better explicit bounds are proven in certain cases. Even stronger effective results were needed to solve (at least in the normal case) the class-number-one problem for CM fields, see [15], [25], [2].

In another direction, assuming the generalized Riemann hypothesis (GRH) one can obtain more precise bounds on the class number then those given by the Brauer–Siegel theorem. For example in the case of quadratic fields we have $h_K << D_K^{1/2}(\log \log D_K / \log D_K)$. In particular they are known to be optimal in many cases (see [5], [6], [4]).

A full survey of the problems stemming from the study of the Brauer–Siegel type questions definitely lies beyond the scope of this article. Our goal is more modest. Here we survey the results that generalize the classical Brauer–Siegel theorem. In §2 the case of families of number fields violating one (or both) of the conditions (i) and (ii) of theorem 1.1 is discussed. In particular we introduce the notion of Tsfasman–Vlăduţ invariants of global fields that allow to express the Brauer–Siegel limit in general. In §3 we survey the known results and conjectures about the Brauer–Siegel type statements in the higher dimensional situation. Finally, in the last §4 we prove a Brauer–Siegel type result (theorem 3.2) for families of varieties over finite fields. This theorem expresses the asymptotic properties of the residue at $s = d$ of the zeta function of smooth projective varieties over finite fields via the asymptotics of the number of $\mathbb{F}_{q^m}$-points on them.

2. The case of global fields: Tsfasman–Vlăduţ approach

A natural question is whether one can weaken the conditions (i) and (ii) of theorem 1.1. The first condition seems to be the most restrictive one. Tsfasman and Vlăduţ were able to deal with it first in the function field case [31], [32] and then in the number field case [33] (which was as usual more difficult, especially from the analytical point of view). It turned out that one has to take in account non-archimedian place to be able to treat the general situation. Let us introduce the necessary notation in the number field case (for the function field case see §3).

For a prime power q we set

$$\Phi_q(K_i) := |\{v \in P(K_i) : \operatorname{Norm}(v) = q\}|,$$

where $P(K_i)$ is the set of non-archimedian places of K_i. Taking in account the archimedian places we also put $\Phi_{\mathbb{R}}(K_i) = r_1(K_i)$ and $\Phi_{\mathbb{C}}(K_i) = r_2(K_i)$, where r_1 and r_2 stand for the number of real and (pairs of) complex embeddings.

We consider the set $A = \{\mathbb{R}, \mathbb{C}; 2, 3, 4, 5, 7, 8, 9, \ldots\}$ of all prime powers plus two auxiliary symbols $\mathbb{R}$ and $\mathbb{C}$ as the set of indices.

DEFINITION 2.1. A family $\mathcal{K} = \{K_i\}$ is called asymptotically exact if and only if for any $\alpha \in A$ the following limit exists:

$$\phi_\alpha = \phi_\alpha(\mathcal{K}) := \lim_{i \to \infty} \frac{\Phi_\alpha(K_i)}{g_{K_i}}.$$

We call an asymptotically exact family $\mathcal{K}$ asymptotically good (respectively, bad) if there exists $\alpha \in A$ with $\phi_\alpha > 0$ (respectively, $\phi_\alpha = 0$ for any $\alpha \in A$). The ϕ_α are called the Tsfasman–Vlăduţ invariants of the family $\{K_i\}$.

One knows that any family of number fields contains an asymptotically exact subfamily so the condition on a family to be asymptotically exact is not very restrictive. On the other hand, the condition of asymptotical goodness is indeed quite restrictive. It is easy to see that a family is asymptotically bad if and only if it satisfies the condition (i) of the classical Brauer–Siegel theorem. In fact, before the work of Golod and Shafarevich [9] even the existence of asymptotically good families of number fields was unclear. Up to now the only method to construct asymptotically good families in the number field case is essentially based on the ideas of Golod and Shafarevich and consists of the usage of classfield towers (quite often in a rather elaborate way). This method has the disadvantage of beeing very inexplicit and the resulting families are hard to controll (ex. splitting of the ideals, ramification, etc.). In the function field case we dispose of a much wider range of constructions such as the towers coming from supersingular points on modular curves or Drinfeld modular curves ([16], [34]), the explicit iterated towers proposed by Garcia and Stichtenoth [7], [8] and of course the classfield towers as in the number field case (see [26] for the treatement of the function field case).

This partly explains why so little is known about the above set of invariants ϕ_α. Very few general results about the structure of the set of possible values of (ϕ_α) are available. For instance, we do not know whether the set $\{\alpha \mid \phi_\alpha \neq 0\}$ can be infinite for some family $\mathcal{K}$. We refer to [20] for an exposition of most of the known results on the invariants ϕ_α.

Before formulating the generalization of the Brauer–Siegel theorem proven by Tsfasman and Vlăduţ in [33] we have to give one more definition. We call a number field almost normal if there exists a finite tower of number fields $\mathbb{Q} = K_0 \subset K_1 \subset \cdots \subset K_m = K$ such that all the extensions K_i/K_{i-1} are normal.

THEOREM 2.2 (Tsfasman–Vlăduţ). *Assume that for an asymptotically good tower $\mathcal{K}$ any of the following conditions is satisfied:*

- *GRH holds*
- *All the fields K_i are almost normal over $\mathbb{Q}$.*

Then the limit $\mathrm{BS}(\mathcal{K}) = \lim\limits_{i \to \infty} \frac{\log(h_{K_i} R_{K_i})}{g_{K_i}}$ *exists and we have:*

$$\mathrm{BS}(\mathcal{K}) = 1 + \sum_q \phi_q \log \frac{q}{q-1} - \phi_{\mathbb{R}} \log 2 - \phi_{\mathbb{C}} \log 2\pi,$$

the sum beeing taken over all prime powers q.

We see that in the above theorem both the conditions (i) and (ii) of the classical Brauer–Siegel theorem are weakend. A natural supplement to the above theorem is the following result obtained by the author in [**36**]:

THEOREM 2.3 (Zykin). *Let $\mathcal{K} = \{K_i\}$ be an asymptotically bad family of almost normal number fields (i. e. a family for which $n_{K_i}/g_{K_i} \to 0$ as $i \to \infty$). Then we have $\mathrm{BS}(\mathcal{K}) = 1$.*

One may ask if the values of the Brauer–Siegel ratio $\mathrm{BS}(\mathcal{K})$ can really be different from one. The answer is "yes". However, due to our lack of understanding of the set of possible (ϕ_α) there are only partial results. Under GRH one can prove (see [**33**]) the following bounds on $\mathrm{BS}(\mathcal{K})$: $0.5165 \le \mathrm{BS}(\mathcal{K}) \le 1.0938$. The existence bounds are weaker. There is an example of a (class field) tower with $0.5649 \le \mathrm{BS}(\mathcal{K}) \le 0.5975$ and another one with $1.0602 \le \mathrm{BS}(\mathcal{K}) \le 1.0938$ (see [**33**] and [**36**]). Our inability to get the exact value of $\mathrm{BS}(\mathcal{K})$ lies in the inexplicitness of the construction: as it was said before, class field towers are hard to control. A natural question is whether all the values of $\mathrm{BS}(\mathcal{K})$ between the bounds in the examples are attained. This seems difficult to prove at the moment though one may hope that some density results (i. e. the density of the values of $\mathrm{BS}(\mathcal{K})$ in a certain interval) are within reach of the current techniques.

Let us formulate yet another version of the generalized Brauer–Siegel theorem proven by Lebacque in [**19**]. It assumes GRH but has the advantage of beeing explicit in a certain (unfortunately rather weak) sense:

THEOREM 2.4 (Lebacque). *Let $\mathcal{K} = \{K_i\}$ be an asymptotically exact family of number fields. Assume that* GRH *in true. Then the limit $\mathrm{BS}(\mathcal{K})$ exists, and we have:*

$$\sum_{q \le x} \phi_q \log \frac{q}{q-1} - \phi_{\mathbb{R}} \log 2 - \phi_{\mathbb{C}} \log 2\pi = \mathrm{BS}(\mathcal{K}) + \mathcal{O}\left(\frac{\log x}{\sqrt{x}}\right).$$

This theorem is an easy corollary of the generalised Mertens theorem proven in [**19**]. We should also note that Lebacque's apporoach leads to a unified proof of theorems 2.2 and 2.3 with or without the assumption of GRH.

3. Varieties over global fields

Once we are in the realm of higher dimensional varieties over global fields the question of finding a proper analogue of the Brauer–Siegel theorem becomes more complicated and the answers which are currently available are far from being complete. Here we have essentially three approaches: the one by the author (which leads to a fairly simple result), another one by Kunyavskii and Tsfasman and the last one by Hindry and Pacheko (which for the moment gives only plausible conjectures). We will present all of them one by one.

The proof of the cassical Brauer–Siegel theorem as well as those of its generalisations discussed in the previous section passes through the residue formula. Let $\zeta_K(s)$ be the Dedekind zeta function of a number field K and $\varkappa_K$ its residue at $s = 1$. By w_K we denote the number of roots of unity in K. Then we have the following classical residue formula:

$$\varkappa_K = \frac{2^{r_1}(2\pi)^{r_2} h_K R_K}{w_K \sqrt{D_K}}.$$

This formula immediately reduces the proof of the Brauer–Siegel theorem to an appropriate asymptotical estimate for $\varkappa_K$ as K varies in a family (by the way, this makes clear the connection with GRH which appears in the statement of the Brauer–Siegel theorem). So, in the higher dimensional situation we face two completely different problems:

(i) Study the asymptotic properties of a value of a certain ζ or L-function.
(ii) Find an (arithmetic or geometric) interpretation of this value.

One knows that just like in the case of global fields in the d-dimensional situation zeta function $\zeta_X(s)$ of a variety X has a pole of order one at $s = d$. Thus the first idea would be to take the residue of $\zeta_X(s)$ at $s = d$ and study its asymptotic behaviour. In this direction we can indeed obtain a result. Let us proceed more formally.

Let X be a complete non-singular absolutely irreducible projective variety of dimension d defined over a finite field $\mathbb{F}_q$ with q elements, where q is a power of p. Denote by $|X|$ the set of closed points of X. We put $X_n = X \otimes_{\mathbb{F}_q} \mathbb{F}_{q^n}$ and $\overline{X} = X \otimes_{\mathbb{F}_q} \overline{\mathbb{F}}_q$. Let Φ_{q^m} be the number of places of X having degree m, that is $\Phi_{q^m} = |\{\mathfrak{p} \in |X| \mid \deg(\mathfrak{p}) = m\}|$. Thus the number N_n of $\mathbb{F}_{q^n}$-points of the variety X_n is equal to

$$N_n = \sum_{m|n} m \Phi_{q^m}.$$

Let $b_s(X) = \dim_{\mathbb{Q}_l} H^s(\overline{X}, \mathbb{Q}_l)$ be the l-adic Betti numbers of X. We set $b(X) = \max_{i=1\ldots 2d} b_i(X)$. Recall that the zeta function of X is defined for $\operatorname{Re}(s) > d$ by the following Euler product:

$$\zeta_X(s) = \prod_{\mathfrak{p} \in |X|} \frac{1}{1 - N(\mathfrak{p})^{-s}} = \prod_{m=1}^{\infty} \left(\frac{1}{1 - q^{-sm}}\right)^{\Phi_{q^m}},$$

where $N(\mathfrak{p}) = q^{-\deg \mathfrak{p}}$. It is known that $\zeta_X(s)$ has an analytic continutation to a meromorphic function on the complex plane with a pole of order one at $s = d$. Furthermore, if we set $Z(X, q^{-s}) = \zeta_X(s)$ then the function $Z(X, t)$ is a rational function of $t = q^{-s}$.

Consider a family $\{X_j\}$ of complete non-singular absolutely irreducible d-dimensional projective varieties over $\mathbb{F}_q$. We assume that the families under consideration satisfy $b(X_j) \to \infty$ when $j \to \infty$. Recall (see [**18**]) that such a family is called asymptotically exact if the following limits exist:

$$\phi_{q^m}(\{X_j\}) = \lim_{j \to \infty} \frac{\Phi_{q^m}(X_j)}{b(X_j)}, \qquad m = 1, 2, \ldots$$

The invariants ϕ_{q^m} of a family $\{X_j\}$ are called the Tsfasman–Vlăduţ invariants of this family. One knows that any family of varieties contains an asymptotically exact subfamily.

DEFINITION 3.1. We define the Brauer–Siegel ratio for an asymptotically exact family as

$$\mathrm{BS}(\{X_j\}) = \lim_{j \to \infty} \frac{\log |\varkappa(X_j)|}{b(X_j)},$$

where $\varkappa(X_j)$ is the residue of $Z(X_j, t)$ at $t = q^{-d}$.

In §4 we prove the following generalization of the classical Brauer–Siegel theorem:

THEOREM 3.2. *For an asymptotically exact family $\{X_j\}$ the limit $\mathrm{BS}(\{X_j\})$ exists and the following formula holds:*

$$(3.1) \qquad \mathrm{BS}(\{X_j\}) = \sum_{m=1}^{\infty} \phi_{q^m} \log \frac{q^{md}}{q^{md} - 1}.$$

However, we come across a problem when we trying to carry out the second part of the strategy sketched above. There seems to be no easy geometric interpretaion of the invariant $\varkappa(X)$ (apart from the case $d = 1$ where we have a formula relating $\varkappa_X$ to the number of $\mathbb{F}_q$-points on the Jacobian of X). See however [23] for a certain cohomological interpretation of $\varkappa(X)$.

Let us now switch our attention to the two other approaches by Kunyavskii–Tsfasman and by Hindry–Pacheko. Both of them have for their starting points the famous Birch–Swinnerton-Dyer conjecture which expresses the value at $s = 1$ of the L-function of an abelian variety in terms of certain arithmetic invariants related to this variety. Thus, in this case we have (at least conjecturally) an interpretation of the special value of the L-function at $s = 1$. However, the situation with the asymptotic behaviour of this value is much less clear. Let us begin with the approach of Kunyavskii–Tsfasman. To simplify our notation we restrict ourselves to the case of elliptic curves and refer for the general case of abelian varieties to the original paper [17].

Let K be a global field that is either a number field or $K = \mathbb{F}_q(X)$ where X is a smooth, projective, geometrically irreducible curve over a finite field $\mathbb{F}_q$. Let E/K be an elliptic curve over K. Let $\mathrm{III} := |\mathrm{III}(E)|$ be the order of the Shafarevich–Tate group of E, and Δ the determinant of the Mordell–Weil lattice of E (see [30] for definitions). Note that in a certain sense III and Δ are the analogues of the class number and of the regulator respectively. The goal of Kunyavskii and Tsfasman in [17] is to study the asymptotic behaviour of the product $\mathrm{III} \cdot \Delta$ as $g \to \infty$. They are able to treat the so-called constant case:

THEOREM 3.3 (Kunyavskii–Tsfasman). *Let $E = E_0 \times_{\mathbb{F}_q} K$ where E_0 a fixed elliptic curve over $\mathbb{F}_q$. Let K vary in an asymptotically exact family $\{K_i\} = \{\mathbb{F}_q(X_i)\}$, and let $\phi_{q^m} = \phi_{q^m}(\{X_i\})$ be the corresponding Tsfasman–Vlăduţ invariants. Then*

$$\lim_{i \to \infty} \frac{\log_q(\mathrm{III}_i \cdot \Delta_i)}{g_i} = 1 - \sum_{m=1}^{\infty} \phi_{q^m} \log_q \frac{N_m(E_0)}{q^m},$$

where $N_m(E_0) = |E_0(\mathbb{F}_{q^m})|$.

Note that there is no real need to assume the above mentioned Birch and Swinnerton-Dyer conjecture as it was proven by Milne [24] in the constant case. The proof of the above theorem uses this result of Milne to get an explicit formula for $\mathrm{III} \cdot \Delta$ thus reducing the proof of the theorem to the study of asymptotic properties of curves over finite fields the latter ones being much better known.

Kunyavskii and Tsfasman also make a conjecture in a certain non constant case. To formulate it we have to introduce some more notation. Let E be again an arbitrary elliptic K-curve. Denote by $\mathcal{E}$ the corresponding elliptic surface (this means that there is a proper connected smooth morphism $f : \mathcal{E} \to X$ with the generic fibre E). Assume that f fits into an infinite Galois tower, i.e. into a commutative diagram of the following form:

$$
(3.2) \qquad
\begin{array}{ccccccccc}
\mathcal{E} = \mathcal{E}_0 & \longleftarrow & \mathcal{E}_1 & \longleftarrow & \cdots & \longleftarrow & \mathcal{E}_j & \longleftarrow & \cdots \\
\downarrow f & & \downarrow & & & & \downarrow & & \\
X = X_0 & \longleftarrow & X_1 & \longleftarrow & \cdots & \longleftarrow & X_j & \longleftarrow & \cdots,
\end{array}
$$

where each lower horizontal arrow is a Galois covering. For every $v \in X$ closed point in X, let $E_v = f^{-1}(v)$. Let $\Phi_{v,i}$ denote the number of points of X_i lying above v, $\phi_v = \lim_{i \to \infty} \Phi_{v,i}/g_i$ (we suppose the limits exist). Furthermore, denote by $f_{v,i}$ the residue degree of a point of X_i lying above v (the tower being Galois, this does not depend on the point), and let $f_v = \lim_{i \to \infty} f_{v,i}$. If $f_v = \infty$, we have $\phi_v = 0$. If f_v is finite, denote by $N(E_v, f_v)$ the number of $\mathbb{F}_{q^{f_v}}$-points of E_v. Finally, let τ denote the "fudge" factor in the Birch and Swinnerton-Dyer conjecture (see [30] for its precise definition). Under this setting Kunyavskii and Tsfasman formulate the following conjecture in [17]:

CONJECTURE 3.4 (Kunyavskii–Tsfasman). *Assuming the Birch and Swinnerton-Dyer conjecture for elliptic curves over function fields, we have*

$$\lim_{i \to \infty} \frac{\log_q(\mathrm{III}_i \cdot \Delta_i \cdot \tau_i)}{g_i} = 1 - \sum_{v \in X} \phi_v \log_q \frac{N(E_v, f_v)}{q^{f_v}}.$$

Let us finally turn our attention to the approach of Hindry and Pacheko. They treat the case in some sense "orthogonal" to that of Kunyavskii and Tsfasman. Here, contrary to the previous setting of this section, we consider the number field case as the more complete one. We refer to [14] for the function field case. As in the approach of Kunyavskii and Tsfasman we study elliptic curves over global fields. However, here the ground field K is fixed and we let vary the elliptic curve E. Denote by $h(E)$ the logarithmic height of an elliptic curve E (see [13] for the precise definition, asymptotically its properties are close to those of the conductor). Hindry in [13] formulates the following conjecture:

CONJECTURE 3.5 (Hindry–Pacheko). *Let E_i run through a family of pairwise non-isomorphic elliptic curves over a fixed number field K. Then*

$$\lim_{i \to \infty} \frac{\log(\mathrm{III}_i \cdot \Delta_i)}{h(E_i)} = 1.$$

To motivate this conjecture, Hidry reduces it to a conjecture on the asymptotics of the special value of L-functions of elliptic curves at $s = 1$ using the conjecture of Birch and Swinnerton-Dyer as well as that of Szpiro and Frey (the latter one is equivalent to the ABC conjecture when $K = \mathbb{Q}$).

Let us finally state some open questions that arise naturally from the above discussion.

- What is the number field analogue of theorem 3.2?

It seems not so difficult to prove the result corresponding to theorem 3.2 in the number field case assuming GRH. Without GRH the situation looks much more challenging. In particular, one has to be able to controll the so called Siegel zeroes of zeta functions of varieties (that is real zeroes close to $s = d$) which might turn out to be a difficult problem. The conjecture 3.4 can be easily written in the number field case. However, in this situation we have even less evidence for it since theorem 3.3 is a particular feature of the function field case.

- How can one unify the conjectures of Kunyavskii–Tsfasman and Hindry—Pacheko?

In particular it is unclear which invariant of elliptic curves should play the role of genus from the case of global fields. It would also be nice to be able to formulate some conjectures for a more general type of L-functions, such as automorphic L-functions.

- Is it possible to justify any of the above conjectures in certain particular cases? Can one prove some cases of these conjectures "on average" (in some appropriate sense)?

For now the only case at hand is the one given by theorem 3.3.

4. The proof of the Brauer–Siegel theorem for varieties over finite fields: case $s = d$

Recall that the trace formula of Lefschetz–Grothendieck gives the following expression for N_n — the number of $\mathbb{F}_{q^n}$ points on a variety X :

$$(4.1) \qquad N_n = \sum_{s=0}^{2d} (-1)^s q^{ns/2} \sum_{i=1}^{b_s} \alpha_{s,i}^n,$$

where $\{q^{s/2}\alpha_{s,i}\}$ is the set of of inverse eigenvalues of the Frobenius endomorphism acting on $H^s(\overline{X}, \mathbb{Q}_l)$. By Poincaré duality one has $b_{2d-s} = b_s$ and $\alpha_{s,i} = \alpha_{2d-s,i}$. The conjecture of Riemann–Weil proven by Deligne states that the absolute values of $\alpha_{s,i}$ are equal to 1. One also knows that $b_0 = 1$ and $\alpha_{0,1} = 1$.

One can easily see that for $Z(X, q^{-s}) = \zeta_X(s)$ we have the following power series expansion:

$$(4.2) \qquad \log Z(X,t) = \sum_{n=1}^{\infty} N_n \frac{t^n}{n}.$$

Combining (4.2) and (4.1) we obtain

$$(4.3) \qquad Z(X,t) = \prod_{s=0}^{2d} (-1)^{s-1} P_s(X,t),$$

where $P_s(X,t) = \prod_{i=1}^{b_i}(1 - q^{s/2}\alpha_{s,i})$. Furthermore we note that $P_0(X,t) = 1 - t$ and $P_{2d}(X,t) = 1 - q^d t$.

To prove theorem 3.2 we will need the following lemma.

LEMMA 4.1. *For $c \to \infty$ we have*

$$\frac{\log|\varkappa(X_j)|}{b(X_j)} = \sum_{l=1}^{c} \frac{N_l(X_j) - q^{dl}}{l} q^{-dl} + R_c(X_j),$$

with $R_c(X_j) \to 0$ uniformly in j.

PROOF OF THE LEMMA. Using (4.3) one has

$$\frac{\log|\varkappa(X_j)|}{b(X_j)} + d\frac{\log q}{b(X_j)} = \frac{1}{b(X_j)} \sum_{s=0}^{2d-1} (-1)^{s+1} \log|P_s(X_j, q^{-d})| =$$

$$= \frac{1}{b(X_j)} \sum_{s=0}^{2d-1} (-1)^{s+1} \sum_{k=1}^{b_s(X_j)} \log(1 - q^{(s-2d)/2}\alpha_{s,i}) =$$

$$= -\frac{1}{b(X_j)} \sum_{s=0}^{2d-1} (-1)^{s+1} \sum_{k=1}^{b_s(X_j)} \sum_{l=1}^{\infty} \frac{q^{(s-2d)l/2}\alpha_{s,i}^l}{l} =$$

$$= \frac{1}{b(X_j)} \sum_{l=1}^{c} \frac{q^{-dl}}{l} \left(\sum_{s=0}^{2d} (-1)^s q^{sl/2} \sum_{k=1}^{b_s(X_j)} \alpha_{s,i}^l - q^{dl} \right) +$$

$$+ \frac{1}{b(X_j)} \sum_{s=0}^{2d-1} (-1)^s \sum_{k=1}^{b_s(X_j)} \sum_{l=c+1}^{\infty} \frac{q^{(s-2d)l/2}\alpha_{s,i}^l}{l} =$$

$$= \sum_{l=1}^{c} \frac{N_l(X_j) - q^{dl}}{l} q^{-dl} + R_c(X_j).$$

An obvious estimate gives

$$|R_c(X_j)| \leq \frac{\sum_{s=0}^{2d} b_s(X_j)}{b(X_j)} \sum_{l=c+1}^{\infty} \frac{q^{-l/2}}{l} \to 0$$

for $c \to \infty$ uniformly in j. □

Now let us note that

$$\frac{1}{b(X_j)} \sum_{l=1}^{c} \frac{1}{l} \leq \frac{2}{b(X_j)} \log c \to 0$$

when $\log c/b(X_j) \to 0$. Thus to prove the main theorem we are left to deal with the following sum:

$$\frac{1}{b(X_j)} \sum_{l=1}^{c} \frac{q^{-ld}}{l} N_l(X_j) =$$

$$= \frac{1}{b(X_j)} \sum_{l=1}^{c} \frac{q^{-dl}}{l} \sum_{m|l} m\Phi_{q^m} = \frac{1}{b(X_j)} \sum_{m=1}^{c} \Phi_{q^m} \sum_{k=1}^{\lfloor c/m \rfloor} \frac{q^{-mkd}}{k} =$$

$$= \frac{1}{b(X_j)} \sum_{m=1}^{c} \Phi_{q^m} \log \frac{q^{md}}{q^{md}-1} - \frac{1}{b(X_j)} \sum_{m=1}^{c} \Phi_{q^m} \sum_{\lfloor c/m \rfloor+1}^{\infty} \frac{q^{-mkd}}{k}.$$

Let us estimate the last term:

$$\frac{1}{b(X_j)} \sum_{m=1}^{c} \Phi_{q^m} \sum_{k=\lfloor c/m \rfloor+1}^{\infty} \frac{q^{-mkd}}{k} \leq$$

$$\leq \frac{1}{b(X_j)} \sum_{m=1}^{c} \frac{N_m(X_j)q^{-md(\lfloor c/m \rfloor+1)}}{m(\lfloor c/m \rfloor + 1)(1-q^{-md})} \leq \frac{1}{b(X_j)} \sum_{m=1}^{c} \frac{N_m(X_j)q^{-cd}}{c(1-q^{-md})} \leq$$

$$\leq \frac{1}{b(X_j)} \sum_{m=1}^{c} \left(q^{md} + 1 + \sum_{s=1}^{2d-1} b_s q^{ms/2} \right) \frac{q^{-dc}}{c(1-q^{-md})} \leq$$

$$\leq \frac{1}{b(X_j)} \left(q^{cd} + 1 + \sum_{s=1}^{2d-1} b_s q^{cs/2} \right) \frac{q^{-dc}}{(1-q^{-1})} \to 0$$

as both $b(X_j) \to \infty$ and $c \to \infty$.

Now, to finish the proof we will need an analogue of the basic inequality from [31]. In the higher dimensional case there are several versions of it. However, here the simplest one will suffice. Let us define for $i = 0 \ldots 2d$ the following invariants:

$$\beta_i(\{X_j\}) = \limsup_j \frac{b_i(X_j)}{b(X_j)}.$$

THEOREM 4.2. *For an asymptotically exact family $\{X_j\}$ we have the inequality:*

$$\sum_{m=1}^{\infty} \frac{m\phi_{q^m}}{q^{(2d-1)m/2}-1} \leq (q^{(2d-1)/2}-1) \left(\sum_{i \equiv 1 \bmod 2} \frac{\beta_i}{q^{(i-1)/2}+1} + \sum_{i \equiv 0 \bmod 2} \frac{\beta_i}{q^{(i-1)/2}-1} \right).$$

PROOF. See [18], Remark 8.8.   □

Applying this theorem together with the fact that

$$\log \frac{q^{md}}{q^{md}-1} = O\left(\frac{1}{q^{dm}-1}\right) = O\left(\frac{m}{q^{(2d-1)m/2}-1}\right)$$

when $m \to \infty$, we conclude that the series on the right hand side of (3.1) converges. Thus the difference

$$\sum_{m=1}^{\infty} \phi_{q^m} \log \frac{q^{md}}{q^{md}-1} - \frac{1}{b(X_j)} \sum_{m=1}^{c} \Phi_{q^m} \log \frac{q^{md}}{q^{md}-1} =$$

$$= \sum_{m=1}^{c} \left(\phi_{q^m} - \frac{\Phi_{q^m}}{b(X_j)} \right) \log \frac{q^{md}}{q^{md}-1} - \sum_{m=c+1}^{\infty} \phi_{q^m} \log \frac{q^{md}}{q^{md}-1} \to 0$$

when $c \to \infty, j \to \infty$ and j is large enough compared to c. This concludes the proof of theorem 3.2.

Acknowledgements. I would like to thank my advisor Michael Tsfasman for very useful and fruitful discussions.

References

1. A. Baker. Linear Forms in the Logarithms of Algebraic Numbers. I, Mathematika **13** (1966), 204–216.
2. S. Bessassi. Bounds for the degrees of CM-fields of class number one, Acta Arith. **106** (2003), Num. 3, 213-245.
3. R. Brauer. On zeta-functions of algebraic number fields, Amer. J. Math. **69** (1947), Num. 2, 243–250.
4. R. Daileda. Non-abelian number fields with very large class numbers, Acta Arith. **125** (2006), Num. 3, 215-255.
5. W. Duke. Extreme values of Artin L-functions and class numbers, Compos. Math. **136** (2003), 103-115.
6. W. Duke. Number fields with large class groups, in: Number Theory (CNTA VII), CRM Proc. Lecture Notes **36**, Amer. Math. Soc., 2004, 117-126.
7. A. Garcia, H. Stichtenoth. A tower of Artin-Schreier extensions of function fields attaining the Drinfeld-Vlăduţ bound, Invent. Math. **121** (1995), Num. 1, 211-222.
8. A. Garcia, H. Stichtenoth. Explicit Towers of Function Fields over Finite Fields, in: Topics in Geometry, Coding Theory and Cryptography (eds. A. Garcia and H. Stichtenoth), Springer Verlag (2006), 1–58.
9. E. S. Golod, I. R. Shafarevich. On the class field tower, Izv. Akad. Nauk SSSSR **28** (1964), 261–272 (in Russian)
10. K. Heegner. Diophantische Analysis und Modulfunktionen, Math. Zeitschrift **56** (1952), 227–253.
11. H. A. Heilbronn. On the class number of imaginary quadratic fields, Quart. J. Math. **5** (1934), 150-160
12. H. A. Heilbronn, E. N. Linfoot. On the Imaginary Quadratic Corpora of Class-Number One, Quart. J. Math. **5** (1934), 293–301.
13. M. Hindry. Why is it difficult to compute the Mordell–Weil group, preprint.
14. M. Hindry, A. Pacheko. Un analogue du théorème de Brauer–Siegel pour les variétés abéliennes en charactéristique positive, preprint
15. J. Hoffstein. Some analytic bounds for zeta functions and class numbers, Invent. Math. **55** (1979), 37–47.
16. Y. Ihara. Some remarks on the number of rational points of algebraic curves over finite fields, J. Fac. Sci. Tokyo **28** (1981), 721-724.
17. B. E. Kunyavskii, M. A. Tsfasman. Brauer–Siegel theorem for elliptic surfaces, preprint, arXiv:0705.4257v2 [math.AG]
18. G. Lachaud, M. A. Tsfasman. Formules explicites pour le nombre de points des variétés sur un corps fini, J. reine angew. Math. **493** (1997), 1–60.
19. P. Lebacque. Generalised Mertens and Brauer–Siegel Theorems, preprint, arXiv:math/0703570v1 [math.NT]
20. P. Lebacque. On Tsfasman–Vlăduţ Invariants of Infinite Global Fields, preprint, arXiv:0801.0972v1 [math.NT]
21. S. R. Louboutin. Explicit upper bounds for residues of Dedekind zeta functions and values of L-functions at $s = 1$, and explicit lower bounds for relative class number of CM-fields, Canad. J. Math, Vol. **53**, Num. 6 (2001), 1194–1222.
22. S. R. Louboutin. Explicit lower bounds for residues at $s = 1$ of Dedekind zeta functions and relative class numbers of CM-fields, Trans. Amer. Math. Soc. **355** (2003), 3079–3098.
23. J. S. Milne. Values of zeta functions of varieties over finite fields, Amer. J. Math. **108** (1986), 297–360.
24. J.S. Milne. The Tate–Shafarevich group of a constant Abelian variety, Invent. Math. **6** (1968), 91–105.
25. A. M. Odlyzko. Some analytic estimates of class numbers and discriminants, Invent. Math. **29** (1975), 275–286.

26. J. P. Serre. Rational points on curves over finite fields, Lecture Notes, Harvard University, 1985.

27. C. L. Siegel. Über die Classenzahl quadratischer Zahlkörper, Acta Arith. **1** (1935), 83-86.

28. H. M. Stark. A Complete Determination of the Complex Quadratic Fields of Class Number One, Michigan Math. J. **14** (1967), 1–27.

29. H. M. Stark. Some effective cases of the Brauer–Siegel Theorem, Invent. Math. **23**(1974), 135–152.

30. J. Tate. On the conjectures of Birch and Swinnerton-Dyer and a geometric analog, Sém. Bourbaki, Vol. **9** (1995), Exp. 306, Soc. Math. France, Paris, 415–440.

31. M. A. Tsfasman. Some remarks on the asymptotic number of points, Coding Theory and Algebraic Geometry, Lecture Notes in Math. **1518**, 178–192, Springer—Verlag, Berlin 1992.

32. M. A. Tsfasman, S. G. Vlăduţ. Asymptotic properties of zeta-functions, J. Math. Sci. **84** (1997), Num. 5, 1445–1467.

33. M. A. Tsfasman, S. G. Vlăduţ. Asymptotic properties of global fields and generalized Brauer–Siegel Theorem, Moscow Mathematical Journal, Vol. **2**, Num. 2, 329–402.

34. M. A. Tsfasman, S. G. Vlăduţ, T. Zink. Modular curves, Shimura curves and Goppa codes better than the Varshamov–Gilbert bound, Math. Nachr. **109** (1982), 21-28.

35. S. Vlăduţ. Kronecker's Jugendtraum and modular functions, Studies in the Development of Modern Mathematics, 2. Gordon and Breach Science Publishers, New York, 1991.

36. A. Zykin. Brauer–Siegel and Tsfasman–Vlăduţ theorems for almost normal extensions of global fields, Moscow Mathematical Journal, Vol. **5** (2005), Num 4, 961–968.

ALEXEY ZYKIN
INSTITUT DE MATHÉMATIQUES DE LUMINY
MATHEMATICAL INSTITUTE OF THE RUSSIAN ACADEMY OF SCIENCES
LABORATOIRE PONCELET (UMI 2615)
INDEPENDENT UNIVERSITY OF MOSCOW

E-mail address: `zykin@iml.univ-mrs.fr`

Titles in This Series

TITLES IN THIS SERIES

7553 068

For a complete list of titles in this series, visit the
AMS Bookstore at **www.ams.org/bookstore/**.